Decentralized Identity Explained

Embrace decentralization for a more secure and empowering digital experience

Rohan Pinto

<packt>

Decentralized Identity Explained

The author acknowledges the use of cutting-edge AI, such as ChatGPT, with the sole aim of enhancing the language and clarity within the book, thereby ensuring a smooth reading experience for readers. It's important to note that the content itself has been crafted by the author and edited by a professional publishing team.

Group Product Manager: Dhruv J. Kataria
Publishing Product Manager: Prachi Sawant
Book Project Manager: Ashwini C.
Senior Editor: Adrija Mitra
Technical Editor: Rajat Sharma
Copy Editor: Safis Editing
Proofreader: Adrija Mitra
Indexer: Pratik Shirodkar
Production Designer: Alishon Mendonca
Senior DevRel Marketing Executive: Marylou De Mello
DevRel Marketing Coordinator: Shruthi Shetty

First published: July 2024

Production reference: 1140624

Published by Packt Publishing Ltd.
Grosvenor House
11 St Paul's Square
Birmingham
B3 1RB, UK

ISBN 978-1-80461-763-2

www.packtpub.com

This book on decentralized identity bears testimony to technology's ever-changing environment, a world that is more than simply a canvas of codes and algorithms, but a domain defining the future in which you both will proudly tread. As I write these technical insights on decentralized identity, they are more than simply a collection of words; they are a testament to the society I foresee for you – a world where privacy, security, and autonomy over one's digital identity are valued values.

I would like to first and foremost thank my loving and patient daughter and son for their continued support, patience, and encouragement throughout the long process of writing this book. Ciel has contributed immensely by providing all the illustrations and diagrams that form a crucial part of the book and adding additional visible flavor to the book's content.

To my son, Ryan, may these pages encourage you to negotiate the difficulties of the digital world with perseverance and curiosity as you travel through life. As you explore the world of decentralized identity, may you discover the means to protect not just your online presence but also the essence of who you are, allowing you to prosper in a future where technology is used for good.

To my daughter, Ciel, within these chapters, I hope you will find the spirit of independence and self-sovereignty that will guide you through the various pathways of the digital cosmos. May you perceive in these lines a mirror of your own power and the boundless possibilities that decentralized identity provides, allowing you to construct your own story.

Together, my sweethearts, you and I are the link between the past and the future. In this digital era, when every click leaves a trace, may the information contained within these pages serve as a guidepost, guiding you toward a future in which your identities, both physical and digital, are treated with the highest respect and protected with the best technical breakthroughs possible.

This book is dedicated to you, Ciel and Ryan, as a pledge – a commitment that the ideals inherent in the core of decentralized identity will become the basis upon which you create your digital experiences. May your travels be defined by knowledge, resilience, and the unshakeable certainty that your identities remain sovereign, safe, and distinctively yours in this enormous sea of data.

Love always,

Rohan Pinto

Contributors

About the author

Rohan Pinto, a cryptography geek with three decades of experience in security and identity management, has founded multiple businesses leveraging blockchain and identity management advancements. He specializes as a senior identity and access management architect, focusing on large-scale infrastructures for identity management, authentication, and authorization (RBAC, ABAC, RiskBAC, and TrustBAC). Rohan was the lead architect for the Government of Ontario's security infrastructure and British Columbia's **Health Information Access Layer** (**HIAL**), and he is currently developing the US Department of Defense's Security Access Layer using **Common Access Cards** (**CACs**). He mentors emerging talent through Techstars and Founder Institute and is a member of the Forbes Technology Council, Decentralized Identity Foundation, and FIDO Alliance. Rohan combines strategic vision with technical expertise to drive tech-based growth, enhancing security and compliance throughout project life cycles.

About the reviewers

Jeremy Swampillai been a technology consultant and entrepreneur for more than two decades, delivering solutions across financial services, healthcare, insurance, and telecom. His connection with the author has led to many conversations about the complexities, importance, and value of identity and access management. These conversations have sparked the pursuit of new technology solutions to integrate digital identity in an AI-enhanced landscape. Throughout his life's journey and career, he has cultivated a passion for seeking technological and offline solutions to modern challenges in a way that helps uplift and empower others.

I've been blessed to know Rohan for more than a decade, channeling a working relationship into a strong friendship. It's an honor to have worked with him in curating content for a topic that will form a renewed foundation of online identity. Yet I couldn't have participated without the love and support of a brilliant wife, a son whose curiosity makes me marvel, and a daughter who flexes her mental muscles with confidence and power. Blessed.

Nikki Mohan is a security strategist who navigates the ever-evolving cybersecurity landscape as a skilled security practitioner and a vocal advocate for women in the field. Her strategic mind, honed by an MBA from USC Marshall School of Business, allows her to translate complex business needs into innovative and secure access management solutions.

Table of Contents

5

Historical Source of Authority 87

6

The Relationship between Trust and Risk 101

7

Informed Consent and Why It Matters 119

8

IAM – the Security Perspective 143

Part 3 - Digital Identity Era: The Near Future

9

Self-Sovereign Identity 165

10

Privacy by Design in the SSI Space 203

11

Relationship between DIDs and SSI 219

12

Protocols and Standards – DID Standards 245

Part 4 - Digital Identity Era: A Probabilistic Future

Preface

Looking forward to mastering digital identity? This book will help you get to grips with complete frameworks, tools, and strategies for safeguarding personal data, securing online transactions, and ensuring trust in digital interactions in today's cybersecurity landscape. *Decentralized Identity Explained* delves into the evolution of digital identities, from their historical roots to the present landscape and future trajectories, exploring crucial concepts such as **Identity and Access Management** (**IAM**), the significance of trust anchors and sources of truth, and emerging trends such as **Self-Sovereign Identity** (**SSI**) and **Decentralized Identities** (**DIDs**). Additionally, you'll gain insights into the intricate relationships between trust and risk, the importance of informed consent, and the evolving role of biometrics in enhancing security within distributed identity management systems. Through detailed discussions on protocols, standards, and authentication mechanisms, this book equips you with the knowledge and tools needed to navigate the complexities of digital identity management in both current and future cybersecurity landscapes. By the end of this book, you'll have a detailed understanding of digital identity management and best practices to implement secure and efficient digital identity frameworks, enhancing both organizational security and user experiences in the digital realm.

Who this book is for

This book is designed for cybersecurity professionals and IAM engineers/architects seeking to understand how DID can enhance security and privacy. It provides insights into leveraging DID as a robust trust framework for effective identity management.

Overall, reading about distributed identity management can provide valuable insights into how these systems work, their potential benefits, and their potential drawbacks. It can also help individuals and organizations make informed decisions about whether to adopt these technologies and how best to implement them.

What this book covers

Chapter 1, The History of Digital Identity: The concept of digital identity has evolved over the past several decades as technology has advanced and the internet has become more ubiquitous. Overall, the history of digital identities is a story of how technology has enabled us to create and manage increasingly complex and sophisticated online identities, while also grappling with the challenges of security and privacy in the digital age.

Chapter 2, Identity Management Versus Access Management: IAM is a security framework that manages and controls access to an organization's systems, applications, and data. While both terms are related to security and access control, they have different meanings. Identity management refers to the process of identifying and authenticating users or entities who want to access a particular resource or service. It involves creating and managing user accounts, credentials, and permissions, as well as ensuring that the user's identity is verified before granting access. Access management, on the other hand, is concerned with managing the permissions and privileges of authenticated users and entities, and ensuring that they have access to the resources they need, while also preventing unauthorized access to sensitive data and applications. In simpler terms, identity management deals with the identification and authentication of users, while access management deals with the control and management of the access rights and permissions that those users have once they are authenticated.

Chapter 3, IAM Best Practices: Overall, implementing IAM best practices is a critical aspect of ensuring the security and efficiency of your organization's IT systems and data. By doing so, you can reduce the risk of data breaches, comply with regulatory requirements, improve efficiency, and save costs.

Chapter 4, Trust Anchors/Sources of Truth and Their Importance: Sources of truth refer to the authoritative sources of data or information that are considered reliable and accurate. In the context of identity verification and management, sources of truth are critical for establishing and maintaining trust in the identity of an individual. The importance of sources of truth in identity verification and management cannot be overstated. Establishing a reliable and accurate source of truth is essential to building trust in an individual's identity and preventing fraud and identity theft. Organizations that rely on inaccurate or unreliable sources of truth run the risk of exposing themselves to financial and reputational harm, as well as legal liability.

Chapter 5, Historical Source of Authority: Historically, there have been various sources of authority for verifying identities. These historical sources of authority have influenced the development of modern identity verification systems, which often rely on a combination of government-issued identification documents, institutional databases, and trusted third-party verification services. However, emerging technologies such as blockchain and SSI may change the way identities are verified and authenticated in the future.

Chapter 6, Relationships between Trust and Risk: Trust and risk are closely related concepts that have a significant impact on individual and organizational decision-making. Trust is a belief that an individual or organization will act in a reliable, responsible, and ethical manner. Trust is essential for building strong relationships between individuals and organizations, and for promoting cooperation and collaboration. Risk, on the other hand, refers to the potential for harm or loss associated with a particular decision or action. Risk can come from many sources, including financial, legal, reputational, and physical. The relationship between trust and risk is complex. On one hand, trust can help mitigate risk by providing a sense of security and confidence that an individual or organization will act in a responsible and ethical manner. For example, if an individual trusts a financial institution to manage their investments, they are more likely to take on higher levels of risk because they believe that the institution will act in their best interest.

Chapter 7, Informed Consent and Why It Matters: Informed consent is a process by which individuals are fully informed about the risks and benefits of a particular decision or action, and then make a voluntary decision based on that information. In healthcare, informed consent is required before any medical procedure or treatment is performed, but it is also important in many other areas, such as research, data privacy, and online services. Overall, informed consent is a critical principle that helps ensure that individuals have control over their own bodies and personal information, and that they are treated with respect and dignity. It is essential for promoting ethical and responsible behavior in healthcare, research, data privacy, and other areas, and for building trust between individuals and organizations.

Chapter 8, IAM – the Security Perspective: IAM is a security framework that focuses on managing user identities and controlling access to resources and applications within an organization. IAM systems provide a way to manage user authentication, authorization, and user account provisioning and deprovisioning. From a security perspective, IAM plays a critical role in protecting an organization's digital assets by controlling who has access to what information, applications, and resources. Overall, IAM plays a critical role in securing an organization's digital assets and must be designed and implemented with security considerations in mind.

Chapter 9, Self-Sovereign Identity: SSI is a new paradigm in digital identity that puts individuals in control of their own personal data. Instead of relying on centralized authorities or organizations to manage their identity, individuals create and manage their own digital identities, which are stored on a decentralized network. Overall, SSI brings a range of benefits to the table, including enhanced privacy, security, interoperability, trust, and flexibility. By giving individuals control over their own personal data, SSI has the potential to transform the way we manage digital identity and help build a more secure and trustworthy digital society.

Chapter 10, Privacy by Design in the SSI Space: This chapter underlines the critical role of **Privacy by Design** (**PbD**) in protecting digital identities and data in the digital era. It emphasizes the need for proactive privacy safeguards, user empowerment, and stringent security measures, as well as the relevance of PbD frameworks, user-centric privacy controls, and security best practices. The chapter covers the importance of data minimization strategies, permission management, selective dissemination, and end-to-end security in ensuring privacy and security. It also presents SSI as a method of reclaiming control over digital identity while maintaining privacy and security. The core message is that PbD is critical for organizations to reduce privacy risks, improve data protection, and build stakeholder trust, ensuring compliance with privacy regulations and maintaining the integrity and confidentiality of sensitive information in a rapidly changing digital landscape.

Chapter 11, Relationship between DIDs and SSI: The relationship between DIDs and SSI is that DIDs provide the foundation for SSI systems. DIDs allow individuals to create and manage their own digital identities, which can then be used in an SSI system to establish trust relationships and control access to personal information. By using DIDs, SSI systems can provide a secure and decentralized way for individuals to manage their digital identity, and give them complete control over their personal data.

Chapter 12, Protocols and Standards – DID Standards: Protocols and standards are essential for creating a digital society that is secure and efficient and respects individuals' privacy rights. Without them, digital systems would be more fragmented, less secure, and less interoperable, which would limit their potential to improve our lives and solve important societal challenges.

Chapter 13, DID Authentication: DID authentication is a method of authentication that relies on decentralized digital identities. DIDs are digital identities that are not controlled by any single organization or authority, but instead are created and managed by the individuals themselves. They are based on blockchain technology and use public-private key cryptography to secure and verify identity. Overall, DID authentication is an innovative and promising approach to identity authentication that has the potential to provide a high level of security, privacy, and control for users.

Chapter 14, Identity Verification: Identity verification is the process of confirming that a person's claimed identity matches their actual identity. It involves gathering and verifying information about an individual, such as their name, date of birth, social security number, and other personal identifying information. The goal of identity verification is to prevent identity theft, fraud, and other types of malicious activity by ensuring that the person accessing a system or service is who they claim to be. Identity verification is used in a variety of contexts, such as online account creation, financial transactions, and government services. By verifying a person's identity, organizations can help prevent identity theft and fraud, protect sensitive information, and ensure that their systems and services are used only by authorized individuals.

Chapter 15, Biometrics Security in Distributed Identity Management: Biometric security is an increasingly popular method of authentication in distributed identity management systems. Biometrics refers to physical or behavioral characteristics that can be used to identify an individual, such as fingerprints, facial recognition, iris scans, and voice recognition. In a distributed identity management system, biometric security can be used to provide a high level of security and convenience. Instead of relying on traditional passwords or tokens, users can authenticate their identity using their unique biometric characteristics. This can help prevent identity theft and fraud, as biometric traits are difficult to forge or replicate. Overall, biometric security has the potential to provide a highly secure and convenient authentication method in distributed identity management systems, but careful consideration must be given to security and privacy concerns and the practical limitations of different biometric technologies.

To get the most out of this book

Before diving into learning about distributed systems, it is recommended to have a good foundation in the following areas:

- **Computer networking**: A good understanding of computer networking concepts is essential to understand how distributed systems work. Concepts such as TCP/IP, routing, and protocols such as HTTP are fundamental to distributed systems.
- **Operating systems**: Understanding operating system concepts, such as process management, memory management, and filesystems, is important as distributed systems often involve multiple nodes with their own operating systems.

- **Data structures and algorithms**: Understanding basic data structures and algorithms, such as trees, graphs, hash tables, and search algorithms, is important, as they are used extensively in distributed systems for storing and retrieving data.
- **Programming**: Knowledge of a programming language such as Java, Python, or C++ is essential, as distributed systems are usually implemented using these languages.
- **Database systems**: Understanding basic concepts of database systems, such as data modeling, normalization, indexing, and transactions, is important since distributed systems need to store and access data across multiple nodes.
- **Cloud computing**: Knowledge of cloud computing concepts, such as virtualization, load balancing, and autoscaling, is important since many distributed systems are implemented on cloud infrastructure.

Having a good foundation in these areas will provide a solid basis for understanding distributed systems and their implementation. However, learning about distributed systems is an ongoing process that requires continuous learning and staying up to date with the latest technologies and best practices.

Conventions used

There are a number of text conventions used throughout this book.

`Code in text`: Indicates code words in text, database table names, folder names, filenames, file extensions, pathnames, dummy URLs, user input, and Twitter handles. Here is an example: "When all of this is put together, a DID can look like this: `did:method:identifier`."

A block of code is set as follows:

```
{
    "name": "John Doe",
    "age": 30,
    "organization": "XYZ Corporation"
}
```

Bold: Indicates a new term, an important word, or words that you see onscreen. For example, words in menus or dialog boxes appear in the text like this. Here is an example: "Furthermore, **Data Protection Impact Assessments (DPIAs)** are crucial in implementing informed consent and digital identity practices, particularly for high-risk processing operations."

> **Tips or important notes**
> Appear like this.

Get in touch

Feedback from our readers is always welcome.

General feedback: If you have questions about any aspect of this book, mention the book title in the subject of your message and email us at `customercare@packtpub.com`.

Errata: Although we have taken every care to ensure the accuracy of our content, mistakes do happen. If you have found a mistake in this book, we would be grateful if you would report this to us. Please visit `www.packtpub.com/support/errata`, selecting your book, clicking on the Errata Submission Form link, and entering the details.

Piracy: If you come across any illegal copies of our works in any form on the Internet, we would be grateful if you would provide us with the location address or website name. Please contact us at `copyright@packt.com` with a link to the material.

If you are interested in becoming an author: If there is a topic that you have expertise in and you are interested in either writing or contributing to a book, please visit `authors.packtpub.com`.

Reviews

Please leave a review. Once you have read and used this book, why not leave a review on the site that you purchased it from? Potential readers can then see and use your unbiased opinion to make purchase decisions, we at Packt can understand what you think about our products, and our authors can see your feedback on their book. Thank you!

For more information about Packt, please visit `packtpub.com`.

Share Your Thoughts

Once you've read *Decentralized Identity Explained*, we'd love to hear your thoughts! Scan the QR code below to go straight to the Amazon review page for this book and share your feedback.

`https://packt.link/r/1804617636`

Your review is important to us and the tech community and will help us make sure we're delivering excellent quality content.

Download a free PDF copy of this book

Thanks for purchasing this book!

Do you like to read on the go but are unable to carry your print books everywhere?

Is your eBook purchase not compatible with the device of your choice?

Don't worry, now with every Packt book you get a DRM-free PDF version of that book at no cost.

Read anywhere, any place, on any device. Search, copy, and paste code from your favorite technical books directly into your application.

The perks don't stop there, you can get exclusive access to discounts, newsletters, and great free content in your inbox daily

Follow these simple steps to get the benefits:

1. Scan the QR code or visit the link below

   `https://packt.link/free-ebook/9781804617632`

2. Submit your proof of purchase
3. That's it! We'll send your free PDF and other benefits to your email directly

Part 1 - Digital Identity Era: Then

In this part of the book, *Chapter 1* covers the evolution of digital identities, from institutional databases and access control lists to advancements in public key cryptography, the rise of the World Wide Web and social networks, biometric identity, IoT, and blockchain as a new identity model. *Chapter 2* delves into the distinctions between identity management and access management, clarifying the concepts and their implementation.

This part has the following chapters:

- *Chapter 1, The History of Digital Identity*
- *Chapter 2, Identity Management Versus Access Management*

1

The History of Digital Identity

The digital depiction of an individual, organization, or item in the internet world is referred to as a **digital identity**. On the internet, it is a collection of data and qualities that uniquely identify and differentiate a person or thing. This identification can contain a username, email address, biometric data, social media accounts, and other information. With the rise of the internet and the proliferation of online services, the notion of digital identity has evolved as the world progressively moved toward the digital era. In the early days of the internet, digital identification was frequently as easy as a username and password combination to access certain online services. As online behaviors became more complicated and prevalent, a more robust and secure system of digital identity management became necessary.

In this chapter, we will cover the following topics:

- The fundamentals and evolution of digital identity
- Institutional databases and access control lists
- Public key cryptography
- Introduction to blockchain-based identity management

What is digital identity?

You can consider your digital identity to be a unique online version of yourself. In the same way that you have a name, a face, and some information about yourself in real life, you have a name and information about yourself on the internet. This online version of you allows websites and applications to recognize your identity when you use them. It's similar to flashing your ID card when you want to enter a building. Your digital identity may include your email address, username, and maybe a photo of yourself. This allows websites to remember you and protect your information as you buy online, communicate with friends, or play games. Just as you are cautious about your physical identification, you should be cautious about your digital identity as well, so that only the proper individuals may access and use your information.

A digital identity, from an institutional standpoint, is a combination of electronic credentials and information that uniquely identifies an individual or entity in the online world. It's similar to a virtual ID card that organizations and systems use to verify and connect with online users. Individuals, organizations, and governments frequently develop digital identities, which comprise information such as a login, password, email address, and other personal characteristics. These identities are used to access online services, perform transactions, and participate in numerous internet activities.

When it comes to digital identities, institutional perspectives emphasize the necessity of security and privacy. They are concerned with putting safeguards in place to secure personal information, prevent identity theft, and guarantee that only authorized persons or organizations have access to particular resources and services. According to this viewpoint, a digital identity is an essential instrument for creating trust and responsibility in the digital environment, enabling secure online interactions and transactions while protecting sensitive information.

An institutional view of a person's digital identity journey refers to the perspective of a firm, government agency, or educational institution as it interacts with and administers individuals' digital identities through time. This journey includes the many stages and exchanges that occur between the individual and the institution during their partnership. The institutional perspective of a person's digital identification journey is shown in the following figure:

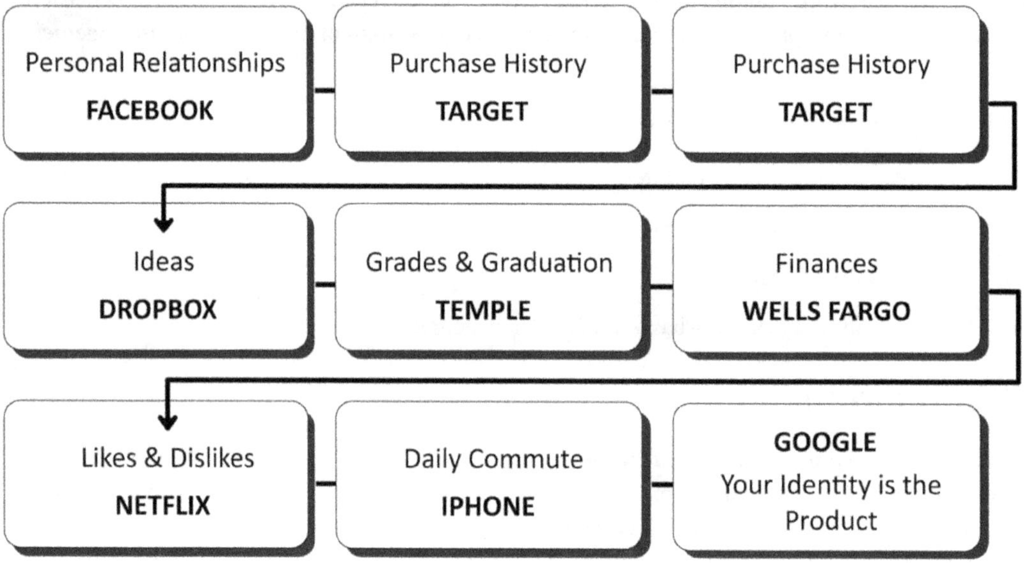

Figure 1.1 – An institutional view of digital identity

Trust, openness, and data privacy are key components for sustaining a healthy connection between the individual and the organization throughout the journey. Institutions may improve user pleasure, preserve user data, and promote their image by managing digital identities responsibly and offering a smooth and secure experience.

To address these difficulties, governments, organizations, and technology providers must work together to build safe, user-friendly, and privacy-aware digital identification solutions. Striking a balance between ease, security, and privacy will be critical in the future to develop a sustainable and inclusive digital identity ecosystem.

Now that we have covered the fundamentals of what a digital identity is, let's take a closer look at the evolution of digital identities.

The evolution of digital identities

The advancement of how individuals and entities establish and maintain their online presence, establish their validity, and control their personal information is referred to as the evolution of digital identities. This notion has developed greatly throughout time. Several things influenced this need:

- **Security**: Traditional username-password combinations were vulnerable to identity theft and hacking. As cybercrime became a major problem, more secure methods of identity verification and authentication were necessary.
- **Convenience**: Users required a smoother and simpler way to access many platforms without having to remember several usernames and passwords as online services and e-commerce proliferated.
- **Personalization**: By adapting information and services to individual tastes, service providers attempted to personalize user experiences. To do this, they needed a method to uniquely identify consumers across several platforms and services.
- **Trust and accountability**: To build trust and responsibility in online interactions, digital identity is required. It holds individuals and corporations accountable for their activities and supports legal and regulatory compliance in the digital domain.
- **Interoperability**: As the number of online services increased, a standardized method for verifying and authenticating digital identities across multiple platforms and apps became necessary.

Various digital identity systems have been created to meet these demands. Biometrics, **two-factor authentication** (**2FA**), digital certificates, **public key infrastructure** (**PKI**), and decentralized identity systems (for example, blockchain-based solutions) are among the technologies that are used in these solutions.

The subject of digital identification is evolving as new technologies emerge, such as artificial intelligence and machine learning, which are being used to improve identity verification and fraud detection procedures while protecting user privacy and security. Nonetheless, difficulties such as data privacy, user permission, and the balance between convenience and security in digital identity management persist.

The concept of digital identity has evolved over the past several decades as technology has advanced and the internet has become more ubiquitous. Here's a brief history of digital identities:

- **Digital identities**: In the early days of the internet, digital identities were often limited to usernames and passwords that users created to access online services.
- **Social networking**: With the rise of social networking platforms such as MySpace and Facebook in the mid-2000s, digital identities began to take on a more social dimension. Users could create profiles, share personal information, and connect with others in ways that were not previously possible.
- **Mobile devices**: The widespread adoption of smartphones and other mobile devices in the late 2000s and early 2010s further expanded the use of digital identities. Users could access their accounts from anywhere, and mobile apps made it easier than ever to create and manage digital identities.
- **Digital authentication**: As online services and transactions became more common, the need for secure digital authentication grew. 2FA, biometric authentication, and other advanced security measures became more widespread.
- **Blockchain technology**: In recent years, blockchain technology has emerged as a new way to manage digital identities. With blockchain, users can create a decentralized digital identity that isn't controlled by any single entity, which can provide greater privacy and security.

Overall, the history of digital identities is a story of how technology has enabled us to create and manage increasingly complex and sophisticated online identities, while also grappling with the challenges of security and privacy in the digital age.

Digital identity systems originated from institutional databases in the late 1960s and progressed with the invention of the internet and the surrounding ecosystem, including PKI, web identity federations, certificate authority reliance, and public identity providers (such as social networks). Today, digital identity is still evolving with biometrics, the **Internet of Things** (**IoT**), and the modern initiatives being taken toward self-sovereign models with the novel technology of blockchain.

To summarize, the evolution of digital identities shows a trend toward more secure, decentralized, and user-centric identification and verification mechanisms, while also taking privacy and convenience into account in the digital realm.

Now that we've covered the evolution of digital identities over time, let's dive deeper into how institutional databases play a role in the identity landscape.

Institutional databases

An institutional database is a systematic and centralized collection of digital information, records, and resources particular to an organization that allows for effective data administration and retrieval.

Before the arrival of the internet and its revolutionary influence in the mid-1990s, governments, corporations, and banks were the entities that owned and regulated digital identity databases to access and analyze the accumulated data on companies, employers, citizens, and customers. Think of how the consumer credit history in the mid-1960s used to shift to electronic storage by credit reporting agencies.

Traditional identity management systems rely heavily on institutional databases. In addition to serving as centralized repositories of personal information, these databases support identity-related processes and services offered by government agencies, financial institutions, and healthcare providers. The purpose of this chapter is to explore the characteristics, advantages, and challenges of institutional databases, which are commonly used for traditional identity management.

Characteristics of institutional databases

Traditional identity management institutions use institutional databases that possess several key characteristics:

- **Centralized storage**: The databases store a vast amount of personal data, including names, addresses, social security numbers, and other identification details. Data relating to identity can be accessed and managed easily through this centralization.
- **Scalability**: Data storage and processing can be handled by institutional databases on a large scale. As identity-related transactions and individuals increase, databases can scale up to accommodate the increase.
- **Security measures**: To secure the confidentiality, integrity, and availability of the stored data, strong security measures are put in place. Access restrictions, encryption, firewalls, intrusion detection systems, and regular security audits are examples of such safeguards.
- **Data integration**: Data from many sources and departments within an organization is frequently combined in institutional databases. This integration provides a full view of an individual's identity and makes identity verification processes more efficient.

Now, let's look at the merits and demerits of institutional databases.

Advantages of institutional databases

In the sphere of traditional identity management, institutional databases offer various advantages:

- **Streamlined processes**: Organizations can use centralized databases to simplify identification-related activities such as identity verification, document authentication, and identity credential issuance. As a result, service delivery is faster and more efficient.
- **Improved accuracy**: Organizations can limit the likelihood of duplicate or incorrect entries by keeping a centralized store of identification data. This increases the accuracy and dependability of identity-related data.

- **Enhanced fraud detection**: Organizations can use institutional databases to create sophisticated fraud detection systems. Organizations can discover suspected fraudulent activity and take the necessary action by analyzing trends and anomalies in recorded data.
- **Interoperability**: Interoperability across various systems and organizations can be facilitated through institutional databases. For example, government organizations may securely communicate identity information with other authorized organizations, supporting seamless service delivery across many sectors.

Disadvantages of institutional databases

While institutional databases provide several benefits, they also create issues that must be addressed:

- **Data privacy concerns**: Concerns regarding data privacy and the possibility of unauthorized access or exploitation arise when sensitive personal information is stored in centralized systems. To prevent these risks, organizations must develop comprehensive data protection procedures and comply with appropriate privacy rules.
- **Data breaches**: Because of their centralized character, institutional databases are appealing targets for hackers. Identity theft, financial fraud, and other criminal behaviors can result from data breaches. Organizations must invest in comprehensive cybersecurity measures to successfully prevent and respond to possible intrusions.
- **Data accuracy and quality**: It might be difficult to ensure the accuracy and quality of data that's kept in institutional databases. Incorrect or outdated information might cause problems during identity verification processes and impede service delivery. To solve these difficulties, regular data maintenance and quality control techniques are required.
- **System integration**: Integrating disparate systems and information inside and across organizations may be difficult. To allow smooth data sharing and interoperability, organizations must invest in comprehensive integration frameworks and standards.

Traditional identity management systems rely on institutional databases as a crucial infrastructure. They offer centralized storage, scalability, security measures, and data integration capabilities, all of which help to expedite identity-related procedures and improve service delivery. However, concerns relating to data privacy, breaches, accuracy, and system integration must be addressed to guarantee that these databases operate effectively and securely. As technology advances, new techniques to address some of these difficulties, such as decentralized identification systems and blockchain-based solutions, are being developed, providing alternatives to established institutional databases.

Up next, we will look at **access control lists** (**ACLs**).

ACLs

As technology advanced, computer systems that could manage databases based on identities and access were developed. ACLs have been used since the 1960s and 1970s, and they are still commonly utilized today. Despite recent updates to ACLs, operating systems continue to utilize them to determine which users have access privileges to a resource. Given this, how identity is conceptualized and executed is heavily affected. It is specifically in charge of encrypting passwords and usernames.

In conventional identity management systems, ACLs are routinely used to govern access to resources and sensitive information. ACLs are used to manage rights and enforce security restrictions based on user identities. This section investigates the use of ACLs in conventional identity management and evaluates their drawbacks.

Functions of ACLs in traditional identity management

In conventional identity management systems, ACLs are critical in the following respects:

- **Authorization**: Based on their identities, ACLs decide on the amount of access to be provided to people or organizations. Organizations can regulate who can access and change resources within their systems by allocating certain rights or privileges to individuals.
- **Resource protection**: ACLs guarantee that only those who are authorized can access sensitive information or conduct certain activities. Organizations can secure private data and prevent unauthorized use or disclosure by creating rules and limits based on user identities.
- **Compliance and auditability**: ACLs assist organizations in meeting regulatory obligations. Organizations may track and audit user activity by establishing identity-based access restrictions, guaranteeing accountability, and aiding compliance efforts.

Disadvantages of ACLs

While ACLs are frequently utilized in traditional identity management systems, they have significant drawbacks:

- **Complexity and maintenance**: ACL management may become increasingly difficult as organizations expand and adapt. The process of creating, setting, and maintaining access restrictions for many resources and identities necessitates considerable work and continual maintenance.
- **Inflexibility**: ACLs frequently have a static and inflexible structure. Changes to access rights or user roles may be time-consuming and difficult to implement, particularly in big organizations with complicated hierarchies. ACL rigidity can stymie adaptability and responses to changing business demands.

- **Role explosion**: To control access to diverse resources, organizations may wind up developing many roles to satisfy varying access needs. This can result in *role explosion*, a phenomenon in which the number of positions becomes unmanageable, resulting in role sprawl. Role explosion makes access control management more difficult and can present security problems.
- **Lack of contextual information**: Traditional ACLs are primarily concerned with user identities and permissions. They frequently lack contextual information, which allows for a more sophisticated assessment of user behavior and purpose. Organizations may fail to recognize and prevent insider threats or abnormal user behavior in the absence of contextual data.
- **Access creep and privilege abuse**: Access rights provided via ACLs can accrue over time, resulting in access creep. Access creep happens when individuals amass superfluous or excessive rights, either as a result of employment position changes or errors in access revocation. This raises the possibility of privilege misuse and insider threats.
- **Scalability and performance**: The speed and scalability of ACL-based systems might be difficult to maintain as the number of users and resources grows. Verifying access rights against complex ACLs can add delay and reduce system responsiveness, especially in high-demand scenarios.

Circumventing the drawbacks of ACLs

Organizations might consider applying the following techniques to alleviate the drawbacks of ACLs in conventional identity management:

- **Role-based access control (RBAC)**: RBAC offers a more organized and adaptable approach to access control. RBAC streamlines administration and decreases the danger of role explosion by defining roles and giving permissions based on job tasks or responsibilities.
- **Attribute-based access control (ABAC)**: ABAC makes access control choices based on factors other than user identification, such as time, location, and contextual data. ABAC allows organizations to build fine-grained policies based on numerous criteria, allowing for a more dynamic and contextual approach to access management.
- **Regular access reviews**: Periodic access evaluations can assist in identifying and removing superfluous access rights. Organizations may prevent access creep, decrease the risk of privilege abuse, and ensure that access restrictions fit with business objectives by assessing ACLs and user privileges regularly.
- **Automation and identity governance**: Identity governance systems can help to streamline access control management operations. To increase productivity and compliance, automation can help with granting and deprovisioning user access, enforcing the division of roles, and keeping audit trails.
- **Continuous monitoring and analytics**: Monitoring and analytics technologies can provide insights into user behavior and spot aberrant activity. Organizations can improve their capacity to detect and respond to security events by integrating ACLs with behavior-based monitoring and machine learning techniques.

ACLs have long been a key component of conventional identity management systems, allowing organizations to regulate resource access and secure critical data. They do, however, have drawbacks such as complexity, inflexibility, and access creep. To address these issues, organizations can use more complex access control models, such as RBAC and ABAC, as well as automation, identity governance, and continuous monitoring. Organizations may improve the efficiency, agility, and security of their identity management operations by using these solutions.

As we learn about managing large-scale data systems, we must not only grasp how information is stored and organized inside institutional databases but also how to guarantee that this information is accessed and altered safely and efficiently, hence why we covered ACLs.

Now that we've discussed the procedures, benefits, and drawbacks of ACLs for controlling and safeguarding data access, we'll shift our focus to another critical facet of information security: public key encryption. In the next section, we will look at how public key cryptography may be used to provide solid solutions for data encryption, authentication, and secure communications, in addition to the access control methods we've already discussed.

Public key cryptography – the origin of secure public networks

Public key cryptography is regarded as one of the most significant technological achievements of the 20th century. There would be no way to safeguard the public networks on which global communication and business rely without it. British government cryptographers discovered the approach in the mid-1970s, and US researchers Whitfield Diffie and Martin Hellman revealed it separately in 1970 and released a white paper.

The system operates based on a connected pair of keys, one private and one public. The public key can be widely distributed; however, the private key must be kept hidden to decode messages that have been encrypted by the public key. The crucial aspect is that it is computationally impossible to extract the private key from the public key; hence, while the public key may encrypt messages, only the private key holder can decode them.

Secure communication and data transfer are critical in today's digital world. Public networks are frequently used by organizations and people to transfer sensitive information such as financial transactions, personal data, and business interactions. The necessity for privacy and secrecy, on the other hand, has led to the creation of secure public networks that rely on PKI. In this section, we will look at the beginnings of these networks and how PKI became a critical tool for ensuring safe and trusted communication.

The evolution of public networks

The history of public networks, such as the internet, dates back to the 1960s. These networks were initially intended to enhance communication and data exchange between the government and academic organizations. As the internet grew in popularity and became more widely available, its potential as a global communication medium became clear. However, the open and decentralized structure of public networks faced serious security issues.

The need for secure communication

Concerns regarding data privacy and the interception of sensitive information arose as public networks proliferated. When it came to secure communication over public networks, traditional encryption approaches, such as symmetric encryption, had drawbacks. To build trust and confidentiality in these contexts, a breakthrough was required.

The emergence of PKI

PKI, the cornerstone for secure public networks, was first proposed in the 1970s. Whitfield Diffie and Martin Hellman, two British mathematicians, invented public key cryptography in 1976. Their revolutionary study transformed the science of cryptography by presenting a way for secure communication that did not require the use of a shared secret key.

Using each other's public keys, the Diffie-Hellman key exchange technique allows two parties to establish a shared secret key via an unsecured channel. This notion set the way for the creation of PKI, which was built on the idea of public key cryptography to establish a comprehensive framework for secure communication.

Components of PKI

PKI is made up of numerous components that work together to enable safe communication and trust between entities. These elements include the following:

- **Public/private key pair**: Each entity in the PKI ecosystem has a set of mathematically connected keys. The public key is freely accessible to others, but the private key is kept secret by its owner.
- **Digital certificates**: PKI relies heavily on digital certificates. These certificates are issued by a trusted third party known as a **certificate authority** (CA) and tie an entity's public key to its identity. Certificates provide information such as the entity's name, public key, and the digital signature of the CA.
- **Certificate authorities (CAs)**: CAs are in charge of certifying the legality and authenticity of companies obtaining digital certificates. They digitally sign these certificates, offering a mark of approval and establishing confidence in the entity's public key.
- **Certificate revocation**: PKI also includes procedures for revoking certificates. If a private key is hacked or an entity's identification is no longer legitimate, the related certificate can be revoked to prevent it from being abused.

Benefits and applications of PKI

PKI provides significant advantages for secure communication and has found uses in a variety of sectors. Among the notable advantages, we have the following:

- **Confidentiality**: PKI enables sensitive information to be encrypted, guaranteeing that only the intended receiver can decrypt and access it
- **Authentication**: PKI allows organizations to use digital certificates to verify one another's identities, lowering the danger of impersonation or unauthorized access
- **Integrity**: PKI ensures that data is not tampered with during transmission by allowing the use of digital signatures
- **Non-repudiation**: PKI enables digital transactions to be authenticated, prohibiting entities from denying their involvement

Drawbacks of PKI

While PKI is extensively used and regarded as a strong security system for storing digital certificates and enabling secure communication, it does have several shortcomings and challenges:

- **Complexity**: Implementing and administering a PKI may be complicated and time-consuming, particularly for organizations with big and remote systems. Infrastructure, CAs, and **certificate revocation lists** (**CRLs**) require technical competence and constant maintenance.
- **Cost**: PKI deployment can be costly in terms of hardware, software, and people training. Managing the lifespan of digital certificates also incurs continuous operational costs.
- **Single point of failure**: In classic PKI designs, the central CA is a single point of failure. If the CA's private key is hacked or becomes inaccessible, the entire PKI may be compromised or become unavailable.
- **Certificate revocation challenges**: It can be difficult to revoke certificates, especially in large-scale installations. To verify certificate validity, dependent parties must execute CRL or **Online Certificate Status Protocol** (**OCSP**) checks. CRLs, on the other hand, may become huge and inconvenient, and OCSP checks can add delay to the authentication process.
- **Scalability and performance**: The speed and scalability of the PKI infrastructure might become an issue as the number of users and devices grows. The time it takes to validate certificates during authentication might influence the user experience, especially in high-load settings.
- **Certificate management**: Certificate administration must be handled carefully by organizations so that they can avoid expired or invalid certificates, which can cause service interruptions and security issues.

- **Limited anonymity**: PKI commonly employs public and private key pairs, which intrinsically link a user's identity to their digital certificate. This lack of anonymity may be an issue for some apps that demand user privacy.
- **Trust in CAs**: The credibility of the CAs providing digital certificates is critical to the security of a PKI. If a CA is hacked or behaves maliciously, the entire PKI ecosystem is jeopardized.
- **Key management**: Private key protection and management are critical components of PKI security. Unauthorized access and potential data breaches can occur if private keys aren't safeguarded appropriately.
- **Interoperability**: Different systems and applications may support PKI and digital certificates to differing degrees, resulting in interoperability concerns.

Despite these disadvantages, PKI is still an important technique for secure digital communication and authentication. Many of the issues may be minimized by careful planning and following best practices in certificate administration, as well as the use of developing PKI technologies and standards.

Secure public networks and PKIs

Because of PKI being incorporated into public networks, secure communication protocols such as **HyperText Transfer Protocol Secure** (**HTTPS**) for web browsing, **Secure Shell** (**SSH**) for secure remote access, and **virtual private networks** (**VPNs**) for secure private connections over the internet have been developed. PKI is used in these protocols to create secure connections, encrypt data, and authenticate the identity of organizations.

The introduction of secure public networks based on PKI has transformed the way we interact and trade data. PKI lays the groundwork for establishing trust, maintaining secrecy, and facilitating secure transactions across public networks. Its continuing development and implementation will be critical in protecting sensitive information and preserving the integrity of our digital communication in the coming years.

With a thorough grasp of public key cryptography and its function in data security, we can now appreciate its practical applicability in real-world circumstances. The World Wide Web is one of the most widely used venues for public key cryptography. In the next section, we will look at how this core technology supports the security of online communications, e-commerce, and other aspects of the current digital world.

The World Wide Web

The World Wide Web, or simply the web, has transformed the way we interact, share information, and conduct business. Sir Tim Berners-Lee, a British computer scientist, envisioned a system that would let individuals explore and retrieve information stored on multiple computers effortlessly in the late 1980s. This concept resulted in the construction of the first web browser, **Hypertext Markup Language** (**HTML**), and the deployment of **Hypertext Transfer Protocol** (**HTTP**).

The first graphical web browser, launched publicly in August 1991, was supported by HTTP, HTML, and web servers. Then came PKI and CA technology, resulting in *secure HTTP* (HTTPS) websites with identities users could trust and with whom they could transmit information safely across different encrypted routes.

Identity management was not a significant concern in the early days of the web. Users could surf websites, get information, and connect with others while remaining anonymous. Websites were more concerned with delivering material and functioning than with confirming their users' identities. However, as the web grew in popularity and e-commerce grew in popularity, the necessity for dependable identity management became obvious.

The introduction of usernames and passwords was the first step toward web identity management. To access personalized features and secure their information, websites began requiring users to establish accounts and select a unique username and password combination. While this was an important step forward, it resulted in the proliferation of passwords and the difficulty of remembering them across numerous websites.

Single sign-on (**SSO**) solutions evolved to address the rising problem of password fatigue. SSO allows users to utilize a single set of credentials to log in to different websites and apps. OpenID and OAuth systems enabled users to identify themselves using a trusted third-party identity provider, such as Google or Facebook, minimizing the need to generate and remember multiple usernames and passwords.

The growth of social media platforms and online services altered the web's identity management environment even more. Platforms such as Facebook, Twitter, and LinkedIn have grown in popularity not only for social interactions but also as centralized identity suppliers. Many websites began to provide social login options, allowing users to check in using their social network profiles. This expedited the user registration process and gave websites access to verified user information.

While centralized identity providers provided convenience, they also posed privacy, security, and data management issues. Decentralized identification and **self-sovereign identity** (**SSI**) evolved as a result. Users may control their identification information with a decentralized identity, reducing the need for middlemen and central authority. Blockchain and **distributed ledger technology** (**DLT**) lay the groundwork for safe and verified identity management solutions.

The notion of verifiable credentials gained significance with the emergence of decentralized identities. Verifiable credentials are digital representations of identifying information supplied by reputable organizations. They can be cryptographically signed and tamper-proof, allowing people to exchange specified qualities or credentials with others while maintaining control over their data. Users may securely store and maintain their verified credentials with digital identity wallets supported by decentralized technology.

Privacy and user consent become increasingly important as identity management advances. Individuals should have control over the information they provide and the opportunity to adjust their data usage preferences. Zero-knowledge proofs and differential privacy, for example, can aid in striking a balance between identity verification and the protection of sensitive personal data.

With developing technologies, the future of identity management on the web looks bright. Biometric authentication systems, such as face recognition and fingerprint scanning, provide quick and safe identity verification. Artificial intelligence and machine learning algorithms can aid in the detection and prevention of identity fraud. Furthermore, developing standards such as the **Decentralized Identifier** (**DID**) specification from the Decentralized Identity Foundation seek to provide a uniform foundation for decentralized identity management.

Identity management has developed tremendously from the early days of the web to the present. What began as anonymous surfing evolved into a world of usernames, passwords, centralized identity providers, and, more recently, decentralized identity solutions. As technology advances, the future offers the promise of user-centric, privacy-preserving, and secure identity management systems on the web, allowing individuals to confidently and securely manage their digital identities.

Now that we've explored the evolution and influence of the World Wide Web, we'll turn our attention to how identities are managed and portrayed in this digital age. Identity 2.0 has gained traction in social networks, where user identity verification and control are critical. In the next section, we will look at Identity 2.0 and its implications for privacy, security, and user engagement on social networks.

Social networks – Identity 2.0

While most *identity* schemas of the 1980s and 1990s were supported in some way by the CA model, the next notable revolution in digital identity was accelerated by social networking sites. Social networking sites function by allowing individuals to publish their social graphs, institutions, hobbies, and beliefs, among other things, publicly. While billions of IDs can be accessed, social network identification has grown in popularity due to its ease of use for SSO and identity reuse. It is referred to as the Identity 2.0 generation.

As the internet evolved, a new wave of invention and connectedness arose, ushering in the internet's second generation. The emergence of social networks characterized this period, revolutionizing the way individuals interact, exchange information, and engage with one another online. Individuals might utilize social networks to build profiles, connect with friends and family, and exchange material in a more engaging and user-centric manner.

Social networking has its origins in the early 2000s when numerous pioneering platforms arose. Friendster, which debuted in 2002, is widely regarded as the first contemporary social networking service. Users could make profiles, interact with friends, and share updates. However, technological problems and scalability limitations hampered its long-term viability.

In 2004, a Harvard University student named Mark Zuckerberg founded Facebook, a social networking website. Initially confined to Harvard students, Facebook swiftly moved to other colleges before opening to the general public in 2006. Because of its simple design, emphasis on actual identities, and comprehensive privacy safeguards, Facebook has become enormously popular. It transformed the social networking scene and paved the way for the subsequent social media explosion.

Following Facebook's popularity, several social networking sites arose, each with its own set of features and target demographic. MySpace, which rose to prominence in the mid-2000s, enabled users to customize their accounts with music and themes. LinkedIn focuses on professional networking, connecting people based on their professional interests and achievements. Twitter, a microblogging site, pioneered the idea of sending out brief, real-time messages known as tweets.

People's online communication and interaction have been transformed by social networks. They provided a platform for people to keep in touch with friends and family even when they were separated by large distances. Real-time interactions and involvement were made possible by features such as private messaging, comments, and likes. Social networks also made it easier to find new connections and common interests, establishing online communities and virtual partnerships.

The capacity for users to generate and share content was a fundamental motivation behind the growth of social networks. Photos, videos, status updates, and blog articles have all become commonplace in online conversations. The advent of the sharing economy, in which individuals could monetize their talents, assets, and knowledge through platforms such as YouTube, Instagram, and TikTok, resulted from this shift toward user-generated content.

Concerns over security and privacy have become increasingly prominent as social networks have gained popularity. There were concerns about the handling and misuse of the large amounts of personal information being shared on these platforms. Social networks came under scrutiny due to issues such as privacy settings, data breaches, and targeted advertising. Regulations were increased as a result of these concerns and greater emphasis was placed on user privacy and consent.

In addition to news and information, social networks have become important sources of information for many users. Sharing news articles, opinions, and personal experiences shaped public discourse and spread information rapidly. In addition, this brought with it challenges such as the spread of misinformation, the creation of filter bubbles, and the manipulation of social networks as political tools. While maintaining an ecosystem of healthy information and upholding freedom of expression, platforms faced increasing pressure to address these issues.

The introduction of smartphones and mobile apps drove the expansion of social networks even further. Users can now access their social media profiles while they were on the road, share information, and communicate with others in real time. Mobile applications added elements such as location tracking, augmented reality filters, and live streaming to social networking experiences, making them more participatory and immersive.

The advent of social networks was a watershed moment in the growth of the internet. These platforms transformed the way people interacted, communicated, and exchanged material on the internet. Social networks provided a platform for self-expression, community building, and knowledge sharing, but they also included privacy, data security, and societal implications. As social networks expand, their impact on how we interact and engage in the digital world remains significant, determining the future of online connectedness and communication.

Building on our study of Identity 2.0 in social networks and its implications for digital identity management, let's move on to a more sophisticated and safe method of identity verification: biometric identification. In the next section, we'll look at how biometric technologies such as fingerprint and facial recognition are transforming identity identification and improving security across several platforms.

Biometric identity

The practice of identifying people based on observable physical characteristics, such as fingerprints, dates back to the 18th century. To identify people, modern biometrics rely on digital abstractions of physiological and behavioral features. The use of unique physical or behavioral traits to verify and validate an individual's identification is referred to as biometric identity management. Biometric technologies make use of these characteristics' uniqueness to improve security, minimize fraud, and provide a convenient method of identification verification. The different biometric modalities and their applications in identity management will be discussed in this section.

Fingerprint recognition is one of the most established and commonly used biometric modalities. Each person's fingers have a distinct pattern of ridges and furrows that may be collected and saved as a biometric template for authentication purposes. Fingerprint recognition is utilized in a variety of applications, including unlocking cellphones, gaining access to protected facilities, and validating identities at border crossings.

Face recognition analyses and matches people's unique facial traits to determine their identity. Because of the availability of high-resolution cameras and powerful algorithms, facial recognition technology has advanced significantly in recent years. It is used in access control systems, surveillance systems, and digital identity verification processes.

To validate an individual's identification, voice recognition analyses the acoustic properties of their voice, such as pitch, tone, and pronunciation. In voice assistants, phone banking, and contact center authentication systems, speech recognition is employed. It provides a simple and non-intrusive technique for verifying identification.

For identity verification, behavioral biometrics collect distinctive behavioral characteristics such as typing rhythm, stride, or hand gestures. These characteristics are more difficult to imitate or fabricate, offering an extra degree of protection. Continuous authentication systems and fraud detection use behavioral biometrics.

Compared to traditional approaches, such as passwords or PINs, biometric identity management provides a better level of security. Because biometric features are unique to each individual, impostors find it difficult to imitate someone else. Furthermore, biometric data is difficult to copy or fabricate, providing an additional degree of security against identity theft.

Biometric authentication is both convenient and user-friendly. Users may just exhibit their biometric feature to authenticate themselves, removing the need to memorize and manage several passwords or PINs. This shortened procedure saves time and lowers user annoyance.

Biometric data is extremely private and sensitive. Its collection and storage raise concerns about privacy, data security, and potential abuse. To preserve biometric information and guarantee that it is only used for authorized reasons, certain protections must be in place. In biometric identity management systems, transparent consent methods and respect for privacy standards are critical.

Biometric identity management is used by government agencies and law enforcement organizations to improve border security, identify offenders, and prevent identity fraud. For identification reasons, biometric data such as fingerprints and face photographs are saved in databases.

Biometric authentication is used by banks and financial organizations to improve the security of financial transactions. Biometrics provides an extra layer of security against unauthorized account access, lowering the risk of fraud and identity theft.

Biometric identity management in healthcare guarantees precise patient identification, safeguards access to electronic health information, and avoids medical identity theft. Biometric systems can be used to verify the identity of healthcare personnel, patients, and anyone seeking access to regulated medications.

In business settings, biometric authentication is used to manage access to protected locations, computer systems, and sensitive information. It improves physical and logical security by allowing only authorized individuals to enter.

The subject of biometric identity management is evolving as a result of technological improvements and the demand for increased security. Let's look at some future breakthroughs and trends:

- **Multi-modal biometrics**: Combining various biometric modalities, such as face and iris recognition, improves identity verification accuracy and resilience. Multi-modal systems provide better performance and resistance against spoofing attacks.
- **Mobile biometrics**: Biometric sensors integrated into smartphones and wearable gadgets provide on-the-go identification verification. Mobile biometrics provides safe authentication while making mobile payments, using digital wallets, or accessing applications or services.
- **Biometric encryption**: Biometric encryption approaches encrypting biometric templates, ensuring that stored biometric data is safe even if a data breach occurs. Biometric encryption adds an extra degree of security to meet privacy issues.
- **Continuous authentication**: To enable continual identity verification, continuous authentication systems monitor biometric features in real time. This method provides enhanced security by constantly confirming the user's identity throughout an active session.

Biometric identity management has completely transformed how identities are validated and confirmed. Biometric qualities enable greater security, ease, and a smooth user experience because of their distinct properties. While privacy and ethical concerns remain, continued improvements and the adoption of robust security measures continue to make biometric identity management a critical component of modern security systems across numerous industries.

Now that we've explored the improvements and uses of biometric identification for safe authentication, let's turn our attention to the developing idea of the "identity of things." In the next section, we'll look at how the IoT assigns unique identities to things and devices, allowing for seamless interaction and communication in an increasingly linked world.

IoT and the identity of things

The flourishing gadgetry has generated a whole ecosystem of *smart* things that are wirelessly connected to the internet for data exchange. Household devices such as Amazon Alexa and a variety of smart appliances are now the most desired goods in the 21st century. This is commonly referred to as IoT. Industrial data recording and analysis devices are also prevalent in agriculture, robots (industrial IoT), and supply chains.

IoT is a network of interconnected physical devices, automobiles, appliances, and other items that are embedded with sensors, software, and connections to gather and share data. This section delves into the history and evolution of IoT, from its inception to the present.

The concept of linking gadgets and machines to improve communication and automation dates back to the early 1980s. In 1982, one of the earliest internet-connected appliances was a customized Coca-Cola vending machine at Carnegie Mellon University. However, Kevin Ashton, a British entrepreneur and technological pioneer, created the phrase *Internet of Things* considerably later, in 1999.

Advances in **radio-frequency identification** (**RFID**) technology opened the path for IoT in the late 1990s and early 2000s. RFID tags, which use radio waves to identify and track objects, enabled automatic product and asset identification and tracking. This technique laid the groundwork for the creation of sensor networks capable of collecting and transmitting data via the Internet.

Smart home technologies emerged in the early 2000s to help connect household appliances, lighting systems, and security devices for enhanced convenience and energy efficiency. Simultaneously, industrial IoT applications gained steam, with the introduction of **machine-to-machine** (**M2M**) communication and remote monitoring systems in industries such as manufacturing, logistics, and healthcare.

In the 2000s and early 2010s, the spread of wireless technologies such as Wi-Fi and Bluetooth spurred the rise of IoT. These technologies enabled devices to communicate with one another and with the internet without the limitations imposed by physical cables. Because of widespread connection, IoT has grown into sectors such as wearable gadgets, smart cities, and linked cars.

The introduction of cloud computing and data analytics was critical in the emergence of IoT. The infrastructure and storage capacity required to manage the large volumes of data created by IoT devices were given by cloud platforms. Data analytics allows organizations to gain relevant insights from acquired data, resulting in better decision-making and efficiency across sectors.

The necessity for standardization and interoperability became obvious as the IoT ecosystem evolved. To ensure interoperability and easy communication between IoT devices and systems, many industry consortiums and standards organizations, such as the **Industrial Internet Consortium** (**IIC**) and the **International Electrotechnical Commission** (**IEC**), created frameworks and protocols.

The growth of IoT has raised serious security and privacy issues. The networked nature of IoT devices enhanced fraudsters' attack surface, increasing data breaches and unauthorized access. Assuring the security and privacy of IoT devices and data became a crucial emphasis, which resulted in the creation of rigorous security mechanisms and best practices.

IoT is evolving as a result of technological breakthroughs such as 5G, edge computing, artificial intelligence, and blockchain. These advancements are projected to open up new opportunities for IoT, such as real-time analytics, autonomous systems, and smart infrastructure. Furthermore, the rise of edge computing is projected to result in quicker reaction times and less reliance on cloud connection.

Since its birth, IoT has transformed the way we interact with our surroundings, allowing a more connected and smarter world. From smart homes to industrial automation and beyond, IoT has the potential to revolutionize industries, increase efficiency, and improve people's quality of life. As technology advances, IoT's future promises limitless potential, with advancements that will transform the way we live, work, and interact with the world around us.

After exploring the identity of things and its function in the linked world of IoT, let's turn our attention to a revolutionary solution to identity management: blockchain technology. In the next section, we will look at how blockchain is developing as a new paradigm for identification, providing decentralized, secure, and transparent methods for confirming and maintaining identities in the digital era.

Blockchain – a new model for identity

Disrupting the present model of identity, well-matched with customary username and password, biometrics, SSO, and IoT, new identity models are arising, based on a new way of executing federated identities: using blockchain. They aim to bring back users' privacy, data transparency, and control.

Here are some best practices for managing traditional digital identities that involve safeguarding credentials such as user IDs and passwords:

- **Use strong passwords**: Encourage users to choose strong passwords that contain a combination of uppercase and lowercase letters, digits, and special characters. Passwords should be 8-12 characters long and should not contain personal information or popular terms.
- **Enforce password policies**: Implement policies that require users to change their passwords regularly and prohibit the reuse of old passwords. Consider using a password manager to generate and store strong, unique passwords for each user.
- **Educate users**: Provide training and resources to educate users about the importance of password security and best practices for creating and managing passwords.

- **Implement multi-factor authentication** (**MFA**): Consider adding MFA to user logins to offer an extra layer of protection. In addition to the user's password, MFA requires the user to submit a second form of verification, such as a fingerprint or a one-time code that's sent to their mobile device.
- **Monitor user activity**: Keep track of user activity and look for signs of suspicious behavior, such as multiple failed login attempts or unusual access patterns.
- **Regularly review and update security policies**: Maintain awareness of the most recent security risks and evaluate and update security policies and procedures regularly to guarantee their effectiveness.

By following these best practices, organizations can help ensure that their traditional user IDs and passwords are secure and protected from unauthorized access.

Traditional digital identities, such as usernames and passwords, are widely used in a variety of contexts. Here are some common use cases:

- **Online account management**: Usernames and passwords are used to set up and manage online accounts such as email, social networking, and online banking
- **E-commerce**: Users provide their digital identities to make purchases and complete transactions on e-commerce sites
- **Employee access**: Access to internal corporate systems and resources, such as email, file sharing, and project management tools, is controlled by usernames and passwords
- **Healthcare**: Patients use digital identities to access their personal healthcare information, such as medical records and test results
- **Education**: Students and faculty members use digital identities to access online course materials and manage their academic records
- **Government services**: Digital identities are used to access a range of government services, including tax filings, social security benefits, and voter registration

Overall, traditional digital identities are a fundamental aspect of online security and privacy. They enable users to access a wide range of online services while also providing a layer of protection against unauthorized access and identity theft.

Summary

This chapter provided a detailed overview of the history of digital identity, with a focus on the transition to more secure, decentralized, and user-centric identification and verification systems. It examined the problems and opportunities given by social networks, mobile technologies, and biometric identity in developing the digital identity environment. The significance of institutional databases in traditional identity management was emphasized, as was the introduction of innovative methodologies such as decentralized identification systems and blockchain-based applications. Additionally, this chapter discussed the need for security, privacy, and user-centric methods in managing digital identities responsibly, as well as the influence of IoT and the need for standardization and interoperability. Overall, it underlined the necessity for a mix of simplicity, security, and privacy in developing a sustainable and inclusive digital identity ecosystem.

In the next chapter, we'll dive into identity and access management and the differences between the two.

2

Identity Management Versus Access Management

Identity and Access Management (IAM) is a security framework that manages and regulates access to systems, applications, and data inside an organization. Both phrases, while linked to security and access control, have distinct meanings.

The process of identifying and authenticating persons or entities who wish to access a certain resource or service is referred to as identity management. It entails creating and managing user accounts, passwords, and permissions, as well as verifying the user's identity before giving access.

Access management, on the other hand, is concerned with controlling the permissions and privileges of authenticated individuals and entities, as well as ensuring that they have access to the resources they require while preventing unauthorized access to sensitive data and applications.

Identity management is concerned with identifying and authenticating users, whereas access management is concerned with controlling and administering the access rights and permissions that those users have once validated.

IAM is a comprehensive strategy for managing user identities and regulating access to resources safely and efficiently that incorporates both identity management and access management.

In this chapter, we will cover the following topics:

- What is identity management and access management?
- The difference between access and identity management
- The pitfalls of traditional IAM systems

What is identity management?

Everyone has a sense of self. In this digital age, our identities take the shape of database records and qualities. The tendency for online businesses is to collect these qualities to better serve us or to create a unique user experience based on the data that's obtained about our dynamic and static attributes.

A unique feature distinguishes us from other types of web users. An email address, social security number, or phone number may be used as this property. We obtain these characteristics from our employers in the form of jobs or titles, the business division to which we belong, the responsibilities that we do in projects, or the tier to which we belong in the organizational structure. Attributes associated with our personal and professional lives fluctuate and alter throughout time when we go through life changes such as changing jobs, getting married, and so on.

The following diagram depicts the critical components of an identity management system:

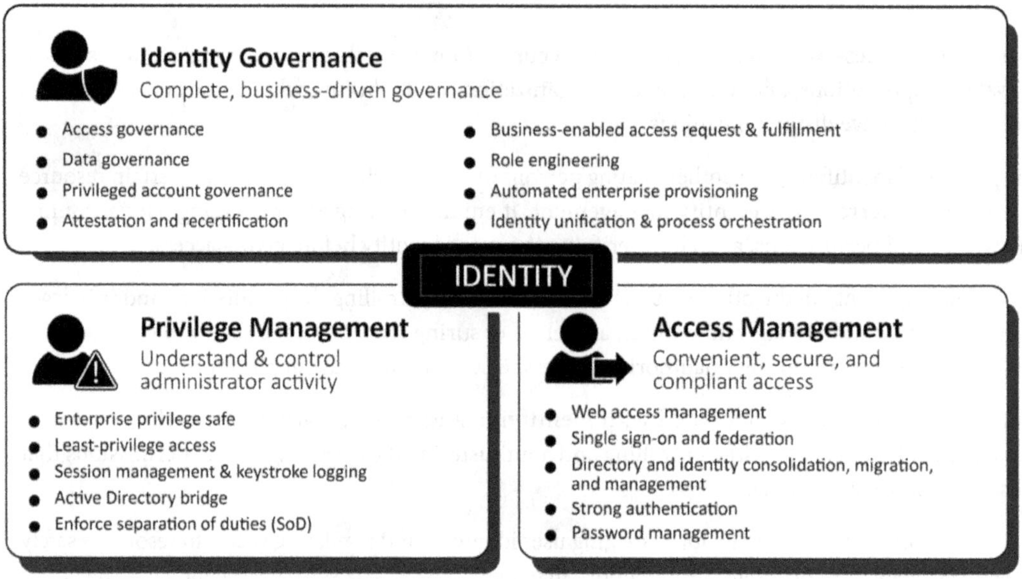

Identity Governance
Complete, business-driven governance
Access governance
Data governance
Privileged account governance
Attestation and recertification
Business-enabled access request & fulfillment
Role engineering
Automated enterprise provisioning
Identity unification & process orchestration
IDENTITY
Privilege Management
Understand & control administrator activity
Enterprise privilege safe
Least-privilege access
Session management & keystroke logging
Active Directory bridge
Enforce separation of duties (SoD)
Access Management
Convenient, secure, and compliant access
Web access management
Single sign-on and federation
Directory and identity consolidation, migration, and management
Strong authentication
Password management

Figure 2.1 – The pillars of identity management

When you register, you create your online persona. Some attributes are collected and kept in the database during the registration process. The registration process varies depending on the sort of digital identification you will receive. The government's electronic identity is supplied through a fairly comprehensive process; however, you may register for social networking sites with completely false (and hence unconfirmed) identification data.

Identity management is about managing qualities. You, your employer, your organization's HR representative, the e-commerce site service personnel, and the IT administrator are just a few of the people who can be held responsible for creating, changing, or even removing qualities that are personally connected with you.

Identity management is becoming increasingly vital in today's linked and digital society. Traditional identity management systems have been in existence for a long time, relying on centralized authorities and processes to establish and validate people's identities. Let's examine the characteristics, problems, and limits of conventional identity management in a centralized environment:

- **Centralized identity authorities**: A central authority or organization serves as the custodian of identity information in centralized identity management systems. Identity papers such as passports, driver's licenses, and national identification cards are often issued and verified by these authorities. They are critical in building trust and giving confirmation of identification for a variety of purposes, such as service access, travel, and financial transactions.
- **Identity verification processes**: For identity verification, traditional identity management primarily depends on human operations. This frequently entails handing over tangible papers, such as identification cards or passports, to authorized workers who physically match the information supplied with data stored by the centralized authority. This procedure can take a long time and is prone to human mistakes.
- **Limited portability**: The lack of mobility is one of the key constraints of traditional identity management in the centralized world. Identity documents issued by a certain authority are typically only valid inside the jurisdiction or nation in which they were issued. Individuals who need to establish their identities in cross-border transactions or while accessing services in a foreign jurisdiction may face difficulties as a result.
- **Security and privacy concerns**: Identity theft and fraud are attracted to centralized identity management systems. A breach or penetration of the systems of the centralized authority might have far-reaching effects, possibly revealing the sensitive personal information of millions of people. Furthermore, centralized systems might raise privacy concerns since they frequently gather and retain massive quantities of personal data, raising questions about who has access to this information and how it is utilized.
- **Lack of individual control**: Individuals have minimal control over their own identities in centralized identity management. The centralized authority has a strong grip over the issuing and verification of identification documents. Because of this lack of control, individuals may become reliant on other parties to validate their identities, which may be cumbersome and may not coincide with their personal preferences.
- **Inefficiencies and costs**: Traditional identity management can be inefficient and costly due to its dependence on physical papers and manual verification methods. Maintaining centralized identification systems takes tremendous administrative labor, infrastructure, and resources. The requirement for physical presence, such as visiting government offices or authorized agencies, adds to the process's overall complexity and discomfort.

- **Potential for Improvement**: While conventional identity management in the centralized world has limits, technological innovations provide the potential for change. Blockchain and decentralized identification systems, for example, offer the potential to overcome some of the issues associated with centralized identity management. These technologies attempt to offer people more control over their identities, improve security and privacy, and allow for interoperability between systems and governments.

Traditional identity management has long been the main strategy for establishing and validating identities in the centralized world. It is not without its flaws, however, such as restricted mobility, security problems, a lack of individual control, and inefficiencies. As technology advances, there is an increasing need to investigate alternative models that solve these limits and enable individuals to have greater control over their identities while ensuring security and privacy. The next section will look into the notion of decentralized identity and its potential to change the identity management environment.

What is meant by access?

Yes/no choices are used to describe access decisions. When access control is established, it is associated with making a yes/no decision when an online user seeks to enter or utilize the resource. The online service can and usually does have several access control points. At the highest level, we may locate an access control point that's attempting to determine if the user is authorized to enter the site. The access control point, on the other hand, is tied to the individual files on the hard drive at the lowest level. Some access control points are visible to the end user, requiring them to take action. The most obvious example would be authentication.

Access control in general

Access control is critical in guaranteeing the security and integrity of digital identification systems in today's digital environment. Traditional access control strategies have been widely utilized in these systems to secure sensitive information and control user access to diverse resources. Let's consider the core ideas and components of traditional access control for digital identification systems:

- **Access control models**: A conceptual foundation for designing and implementing access control policies is provided by access control models. The **discretionary access control** (**DAC**) model and the **mandatory access control** (**MAC**) model are two prevalent approaches that are used in digital identification systems:
    - **DAC**: DAC allows the owner of an object to control access privileges and decide who has access to it. DAC can be implemented in digital identification systems using **access control lists** (**ACLs**) or **access control matrices** (**ACMs**). ACLs list people or groups together with their access privileges, whereas ACMs represent access control permissions in a matrix format.

    - **MAC**: The MAC access control paradigm is more rigorous, enforcing access decisions based on preset security criteria. The system administrator determines resource access in MAC, and users have limited authority over access permissions. MAC versions such as the Bell-LaPadula, which focuses on enforcing confidentiality and ensuring that users can only access information at their clearance level or lower, and Biba models, which focus on enforcing integrity and ensuring that information is not modified by unauthorized users or processes, are often employed in high-security areas where data secrecy and integrity are crucial.

- **Authentication and authorization**: In digital identity systems, authentication and authorization are critical components of access control:

    - **Authentication**: Authentication confirms the identification of people seeking to gain access to the system. Username and password combinations, biometric authentication (such as via fingerprints and facial recognition), and smart-card-based authentication are all examples of traditional authentication techniques. **Multi-factor authentication** (**MFA**) is increasingly being used to improve security by integrating several authentication factors, such as something the user knows (password), something the user owns (smart card), and something the user is (biometric data).

    - **Authorization**: After a person has been authenticated, authorization defines their level of access to the digital identity system. This is usually determined by their position, group membership, or access control restrictions. To handle authorization in digital identification systems, access control technologies such as **role-based access control** (**RBAC**) and **attribute-based access control** (**ABAC**) are often used.

- **Access control enforcement**: Access control enforcement tools guarantee that access control policies are executed consistently and accurately. These mechanisms can be implemented using a variety of methods, including the following:

    - **ACLs**: ACLs link access permissions to specific resources and determine who may do what with those resources. To regulate access at a granular level, ACLs can be connected to files, directories, or other system objects.

    - **RBAC**: RBAC gives roles to users and grants those roles permissions. Users inherit access rights depending on the roles they have been allocated. By combining users with comparable access needs and offering a centralized role assignment mechanism, RBAC simplifies access management.

    - **Access control policies**: The rules and conditions under which access is allowed or denied are defined by access control policies. These restrictions might be based on user traits, access time, or the sensitivity of the resource being accessed.

Traditional access control measures are at the heart of safe digital identity systems. Organizations may guarantee that only authorized users have access to critical information and resources by implementing access control models, authentication, authorization, and access control enforcement methods. However, as technology advances, new access control methods and technologies arise, necessitating the constant adaptation and enhancement of access control practices.

Traditional access control for web applications

Web applications are an important element of our digital life since they provide a variety of services and functionality. Traditional access control measures are critical for safeguarding online applications from unauthorized access, data breaches, and other security concerns. This section delves into the essential components and methods of conventional web access control:

- **RBAC**: RBAC is a popular web application access control paradigm. RBAC assigns access rights to roles, and users are allocated to one or more roles. By combining users with comparable access needs and offering a centralized role assignment mechanism, RBAC simplifies access management. RBAC capabilities are frequently implemented in web application frameworks, allowing developers to establish roles, permissions, and user-role mappings.
- **ACLs**: ACLs are extensively used in web applications to impose access control. Access rights are associated with individual resources (such as files, database records, and URLs) and determine who may do specified activities on those resources. Depending on the needs of the application, web application developers can implement ACLs at various levels, such as file-level ACLs, object-level ACLs, or URL-level ACLs.
- **Authentication and session management**: Authentication is a critical component of web application access control. Username/password combinations, social login (for example, OAuth), and MFA are all common authentication techniques. After a user has been authorized, a session is created to record their activities within the online application. Session management systems guarantee that users are authenticated during their session and that unauthorized access to critical resources is prevented.
- **Input validation and security controls**: SQL injection, **cross-site scripting** (**XSS**), and **cross-site request forgery** (**CSRF**) are all threats to web applications. To reduce these risks, proper input validation and security procedures are required. User inputs should be validated and sanitized, secure coding practices should be enforced, and security measures such as input/output filtering, parameterized queries, and output encoding should be implemented.

- **Audit logging and monitoring**: Audit logging and monitoring are key components of online application security. Organizations can discover possible security breaches and take immediate action by documenting access attempts, authorization changes, and crucial system events. Monitoring tools and intrusion detection systems aid in identifying and responding to suspicious actions, adding an extra layer of security.
- **Session and access expiration**: Web applications frequently use session and access expiration restrictions to improve security. These policies specify the length of a user's session and when access rights expire. Web applications can reduce the danger of unauthorized access by automatically logging out inactive users and canceling access credentials when they are no longer needed.

Traditional access control measures are required for web application security and sensitive data protection. Organizations may reduce the risk of unauthorized access and protect the confidentiality, integrity, and availability of their online applications by adopting RBAC, ACLs, effective authentication and session management, input validation and security measures, audit logging, and monitoring. To meet growing security threats and emerging technologies, it is critical to regularly analyze and upgrade access control systems.

Now that we've reviewed the many aspects of identity management, let's shift our focus to access control. In the next section, we'll look at how access management systems handle and monitor user rights to ensure secure and efficient access to resources inside an organization.

Access management

Can the user access the service after establishing their identity? Nope, they can't. After authentication, an access control option must be selected. The decision about whether or not the user can access a particular service is based on the information available about the user. Personality qualities come into play at this point. However, if the authentication technique is adequate for supplying the required set of characteristics to the access control decision point, the system assesses the attributes to make a yes/no decision.

A formalized decision point is made by an authorization policy. The authorization policy in the IAM domain can be applied centrally, locally, or even in both places. The identity provider's duty is to aggregate the available identity attributes and make high-level access decisions from the online service side. Making an authorization policy framework that is service-level functional is a bad concept since it adds complexity and maintenance costs, is difficult to update rapidly, and is prone to errors.

So, the key differences to note between identity management and access management are as follows:

- Identity management is related to managing the attributes associated with the user.
- Access management involves assessing the attributes based on rules, policies, and yes/no decisions.

Access management versus access controls in traditional centralized digital identity systems

Access management and access controls are two key components of conventional centralized digital identification systems' security and integrity. While they are linked, they serve diverse functions and have unique properties. In this section, we'll discuss the contrasts between access management and access restrictions in typical centralized digital identification systems. Let's take a look:

- **Access controls**: The techniques and rules that are used to manage and restrict user access to resources inside a digital identification system are referred to as access controls. Access controls are designed to enforce established rules and permissions to guarantee that only authorized users have access to certain resources. This often includes systems for authentication, authorization, and access control enforcement.
- **Authentication**: Authentication confirms the identities of people seeking to gain access to the system. It validates users' credentials, such as usernames and passwords, or other authentication elements, such as biometrics or smart cards, to confirm that they are who they say they are. Authentication is an important part of access restrictions since it sets the initial degree of trust and determines whether or not a person should be permitted access.
- **Authorization**: The level of access that an authenticated user has within the digital identity system is determined by authorization. It entails setting access rights and permissions for users depending on their position, group membership, or special access control regulations. To manage and enforce access rights, authorization systems such as RBAC or ABAC are often used.
- **Access control enforcement**: Access control enforcement tools guarantee that access control policies are executed consistently and accurately. ACLs, RBAC, and other policy enforcement methods are used. These procedures regulate access to specified resources and guarantee that users only do actions that are authorized.
- **Access management**: Access management refers to a larger range of actions that try to manage the complete lifespan of user access inside a digital identity system. It extends beyond access restrictions to include a broader approach to user provisioning, authentication, authorization, and continuous access review.
- **User provisioning**: The process of generating and managing user accounts inside a digital identification system is known as user provisioning. User registration, account creation, and granting appropriate access privileges depending on the user's position or duties are all part of this. User provisioning guarantees that only those who are authorized have access to the system and its resources.
- **Identity life cycle management**: Identity life cycle management focuses on managing user identities inside the digital identity system during their existence. When a person leaves the organization or no longer requires access, it comprises processes such as user onboarding, account changes, access revocation, and account de-provisioning. Access privileges are linked to the user's current status and responsibilities when identity life cycle management is used effectively.

- **Access request and approval processes**: Establishing methods for users to seek extra access or modifications to their existing access permissions is part of access management. These requests are usually subject to approval by certain people or functions inside the organization. The access request and approval processes guarantee that changes to access are reviewed, authorized, and as per the organization's security policies and compliance needs.

Access management and access restrictions are required for traditional centralized digital identification systems to preserve their security and integrity. While access controls focus on enforcing resource access rules and permissions, access management involves a larger range of operations such as user provisioning, identity life cycle management, and access request and approval procedures. Organizations may establish strong and comprehensive plans to safeguard their digital identity systems by recognizing the distinctions between access management and access controls.

Access management versus access controls in web applications

Web applications are an essential component of our digital ecosystem, and maintaining safe access is critical. Access management and access controls perform diverse but interrelated functions to prevent unauthorized access to online applications and preserve data integrity. In the context of online applications, let's examine the distinctions between access management and access restrictions:

- **Access controls**: The techniques and policies that are used to govern user access to specified resources or capabilities inside the program are referred to as access controls in web applications. The purpose of access controls is to enforce specified rules and permissions to guarantee that only authorized users may access and interact with the application's functionality.
    - **RBAC**: RBAC is a widely used access control technology in online applications. It entails giving roles to users and granting those roles rights. Users inherit access rights depending on the roles they have been allocated. By combining users with comparable access needs and offering a centralized role assignment mechanism, RBAC simplifies access management.
    - **ACLs**: ACLs are another access control mechanism employed in web applications. They associate access permissions with individual resources, such as files, database records, or URLs. ACLs define who can perform specific actions on those resources, allowing granular control over access.
- **Access management**: Access management in web applications refers to a larger range of operations that aim to control the whole user access life cycle and ensure the application's overall security and usability.
- **User registration and authentication**: User registration and authentication are critical components of web application access management. User registration entails creating user accounts, collecting relevant user information, and establishing login credentials. Authentication confirms user identities and gives access to authorized users depending on the credentials they supply, such as usernames and passwords or other authentication criteria.

- **Identity and access provisioning**: Identity and access provisioning in web applications entails handling user identities and access privileges throughout the lifespan of the program. Tasks such as user onboarding, role assignment, access revocation, and de-provisioning when users leave the system fall under this category. Effective provisioning ensures that users have access to what is suitable for their jobs and responsibilities.
- **Single sign-on (SSO) and federation**: Implementing SSO and federation techniques are frequently used in web application access management. SSO allows users to authenticate once and access various apps without needing to log in again. Users can utilize their identity and access credentials from one trusted system to access resources in another system or domain, increasing ease and security.
- **Access policy enforcement and auditing**: To maintain compliance and security, access management entails enforcing access regulations and monitoring user activities. It includes methods such as access policy enforcement, which guarantees that access rules are consistently enforced, as well as auditing capabilities, which track and report user access attempts, permission changes, and crucial system events.

Access management and access restrictions are critical components of online application security. While access controls focus on enforcing resource access rules and permissions, access management involves a larger range of operations, such as user registration and authentication, identity and access provisioning, SSO and federation, and access policy enforcement. Organizations may maintain a safe and user-friendly web application environment while protecting sensitive data and resources by properly implementing both access management and access controls.

Now that we have an adequate understanding of IAM, we must address its limits. In the next section, we will look at the flaws of traditional IAM systems, highlighting the issues and inefficiencies that result in more modern solutions needing to be created.

The pitfalls

Traditional identity management solutions have various flaws. For instance, they frequently lack a centralized identity store, which results in inconsistent and asynchronous user data. Manual provisioning and de-provisioning operations take time and are prone to mistakes, resulting in delays in providing or withdrawing access permissions. Traditional systems, which rely on rudimentary RBAC models, also fail to manage complicated access policies. Inadequate identity verification and authentication mechanisms, as well as restricted scalability and flexibility, contribute to traditional identity management systems' flaws.

Traditional access management methods have problems. In creating access controls, they frequently lack granularity and flexibility, relying on RBAC models that may not give fine-grained permissions. Delays and mistakes are caused by complex and inefficient access provisioning processes. Monitoring and tracking user access activities is difficult due to limited visibility and auditability. Administrative difficulties and potential security threats are created by ineffective access revocation processes and the absence of user-friendly self-service options. Finally, ineffective access review and recertification processes can lead to unused access rights and compliance difficulties. Recognizing these problems can assist organizations in identifying the limitations of existing methods and in seeking more modern solutions to these difficulties.

The pitfalls of traditional identity management systems

Traditional identity management systems have been widely utilized in many organizations to handle user identities, access limits, and authentication. These systems, however, are not without their flaws and hazards. Let's look at some of the typical drawbacks of traditional identity management systems:

- **Lack of a centralized identity repository**: The lack of a centralized identity repository is one of the major flaws of traditional identity management systems. User identities and related information are frequently dispersed across several apps and databases in such systems. This decentralization makes maintaining consistency and synchronization of user data difficult, resulting in inconsistencies and mistakes. It also makes efficient identity provisioning and maintenance more difficult.
- **Manual identity provisioning and de-provisioning**: For user provisioning and de-provisioning, traditional identity management systems frequently rely on manual processes. This method is time-consuming, prone to errors, and can result in delays in issuing or canceling access permissions. Manual processes can make tracking and managing user access quickly and accurately more difficult, increasing the risk of unauthorized access or account misuse.
- **Limited support for complex access policies**: When it comes to managing complicated access controls, many traditional identity management solutions have drawbacks. These systems frequently rely on simple RBAC models, which may not meet fine-grained access control requirements or dynamic access restrictions based on contextual variables properly. As a result, organizations may find it difficult to implement exact access restrictions that meet their unique security and compliance requirements.
- **Inadequate identity verification and authentication**: Traditional identity management systems frequently depend on ineffective or obsolete authentication mechanisms. Static passwords, which are vulnerable to password-related assaults such as brute-force or dictionary attacks, are a common mistake. The lack of MFA alternatives aggravates the situation by making user accounts vulnerable to hacking. Unauthorized access and identity theft can occur as a result of inadequate identity verification and authentication methods.
- **Limited scalability and flexibility**: Many conventional identity management solutions struggle with scalability and flexibility. These systems may struggle to support an increasing number of users, particularly in big organizations or those undergoing fast growth. Furthermore, integrating new apps or adjusting access controls and user roles can be time-consuming and complex, impeding agility and adaptation.

In conclusion, investigating access controls exposes a complex environment of issues that go beyond conventional information management. While these constraints may appear to provide security and order, they frequently have unexpected effects that impede growth, innovation, and equal development. As we traverse the digital world, it is critical to find a balance between protecting sensitive data and encouraging open communication. By tackling these difficulties head-on and adopting a comprehensive strategy for access control, we can pave the way for a more inclusive and dynamic future in which the advantages of unfettered information flow may be fully realized while maintaining privacy and security concerns.

The pitfalls of traditional access management systems

Traditional access management solutions are intended to enforce organizational access rules, permits, and policies. They are not, however, immune to traps and restrictions. Let's look at some of the typical shortcomings of traditional access management solutions:

- **Lack of granularity and flexibility**: The lack of granularity and flexibility in creating access controls is one of the flaws of traditional access management systems. These systems frequently rely on RBAC, which may lack the granularity required to set fine-grained access permissions. Traditional systems may fail to handle increasingly sophisticated access rules due to user traits, context, or resource sensitivity.
- **Complex and inefficient access provisioning**: Traditional access management solutions can entail complex and time-consuming access-providing processes. This might cause delays in giving access permissions to new users, reducing their productivity and the efficiency of the organization. Complexity in provisioning can potentially lead to mistakes or inconsistencies, resulting in improper access permissions or violations of security regulations.
- **Limited visibility and auditability**: Many traditional accesses control solutions are deficient in terms of visibility and auditability. This presents difficulties in monitoring and tracking user access activities. Organizations may struggle to detect and investigate suspicious or unauthorized access attempts in the absence of adequate audit records and monitoring tools. Limited visibility can stymie compliance efforts and jeopardize the capacity to detect and respond to security events.
- **Ineffective access revocation**: Access revocation is an important part of access management. Traditional systems, on the other hand, sometimes have difficulties in efficiently revoking access privileges when individuals change positions, leave the organization, or no longer require particular capabilities. Unauthorized access, data breaches, and other security issues can occur if access is not revoked swiftly.
- **Lack of user-friendly self-service features**: Traditional access management systems may lack user-friendly self-service capabilities, requiring users to rely on IT or administrative professionals for access-related requests and modifications. This reliance might result in bottlenecks and delays in access provisioning or changes, lowering user productivity and raising administrative expenses.

- **Inefficient access review and recertification**: Periodic access reviews and recertification are frequently required by access management systems to ensure that access permissions are still acceptable and aligned with business requirements. Traditional systems, on the other hand, may lack effective methods and instruments for completing these evaluations. As a result, access permissions may not be examined thoroughly, resulting in superfluous access privileges, increased security risks, and noncompliance with regulatory standards.

Traditional IAM solutions are not without their flaws and limitations. Some of the common pitfalls organizations face are a lack of centralized identity repositories, manual provisioning processes, limited support for complex access policies, inadequate authentication methods, scalability challenges, a lack of granularity, complex access provisioning, limited visibility and auditability, ineffective access revocation, a lack of user-friendly self-service features, and inefficient access review and recertification. Recognizing these problems can assist organizations in evaluating their current systems and investigating more sophisticated and resilient solutions to these difficulties.

Summary

This chapter delved into the intricacies and limits of traditional IAM systems, highlighting the necessity for a complete approach to access control in the digital era. It emphasized the shortcomings of existing systems, such as the lack of centralized identity repositories, manual provisioning processes, limited support for complicated access controls, and insufficient identity verification and authentication techniques. This chapter also discussed the importance of traditional access control measures such as RBAC, ACLs, authentication, session management, input validation, and security controls, as well as the challenges of limited visibility, auditability, and complex access provisioning. Furthermore, it investigated the significance of striking a balance between safeguarding sensitive information and fostering free communication in the digital age, paving the path for a more inclusive and dynamic future while maintaining privacy and security concerns.

In the next chapter, we'll look at various best practices for implementing a strong IAM system.

Part 2 - Digital Identity Era: Now

This part explores current practices and challenges in identity management. *Chapter 3* discusses IAM best practices, including **Single Sign-On (SSO)**, **Privileged Access Management** (**PAM**), and **Identity Governance and Administration** (**IGA**). *Chapter 4* examines trust anchors and sources of truth, highlighting the importance of government-issued IDs, biometric data, and online activity. *Chapter 5* looks at historical sources of authority, such as census records and birth certificates. *Chapter 6* analyzes the relationship between trust and risk, addressing issues such as compromised identity and social engineering. *Chapter 7* emphasizes the importance of informed consent for privacy, transparency, and trust. *Chapter 8* reviews IAM from a security perspective, focusing on IAM principles and role-based access.

This part has the following chapters:

- *Chapter 3, IAM Best Practices*
- *Chapter 4, Trust Anchors/Sources of Truth and Their Importance*
- *Chapter 5, Historical Source of Authority*
- *Chapter 6, Relationships between Trust and Risk*
- *Chapter 7, Informed Consent and Why It Matters*
- *Chapter 8, IAM – the Security Perspective*

3

IAM Best Practices

Identity and access management (**IAM**) best practices are important for several reasons:

- **Security**: Implementing IAM best practices helps to ensure the security of your organization's data and systems. By properly managing user access and permissions, you can prevent unauthorized access to sensitive information and reduce the risk of data breaches.
- **Compliance**: Many industries and organizations are subject to regulatory requirements regarding the protection of sensitive data. Implementing IAM best practices can help you comply with these regulations and avoid costly fines and penalties.
- **Efficiency**: Properly managing user access and permissions can also help improve efficiency within your organization. By providing employees with the appropriate access to the systems and information they need, you can help them work more efficiently and effectively.
- **Cost savings**: Implementing IAM best practices can also help reduce costs associated with managing user access and permissions. By automating access management processes and reducing the number of manual tasks required, you can save time and money.

Overall, putting IAM best practices into action is a vital part of guaranteeing the security and efficiency of your organization's IT systems and data. You may lower the risk of data breaches, comply with legal standards, enhance efficiency, and save money by doing so.

In this chapter, we will cover the following topics:

- An overview of the service components of an IAM system
- Building a comprehensive IAM strategy
- User lifecycle management and secure data-sharing practices
- Incident response and recovery and processes for regular evaluation

An overview of the service components of an IAM system

A typical IAM system is made up of various service components that collaborate to provide safe access and administration of user identities and resources. Authentication, authorization, user management, and directory services are examples of these components:

- **Authentication**: This is a critical component of IAM, responsible for confirming user identities. It entails checking user credentials, such as usernames and passwords, to ensure that only authorized users have access to the system. For increased security, authentication techniques may include single-factor authentication (for example, username and password) or multi-factor authentication (for example, combining passwords with biometrics or tokens).
- **Authorization**: This is yet another key component that controls what activities or resources a user is permitted to access. It upholds the idea of least privilege by allowing users the rights they require depending on their positions and responsibilities inside the organization. Authorization mechanisms set and enforce access regulations, ensuring that users have the necessary permissions to complete their duties while prohibiting unauthorized access.
- **User management**: The methods and tools used to establish, change, and remove user accounts inside the IAM system are referred to as **user management**. It entails user provisioning, which automates the establishment and assignment of user accounts and access privileges. User deprovisioning, which eliminates user accounts and revokes access when users change positions, leave the organization, or no longer require certain capabilities, is also part of user management.
- **Directory services**: This provides a centralized location for storing and managing user IDs and related data. They provide a single source of truth for user data and serve as a basis for IAM systems. Directory services enable rapid and accurate user identification and authentication by allowing for efficient user lookup and retrieval. They also provide features such as password management, user self-service, and system synchronization.

These service components provide the backbone of a standard IAM system, allowing organizations to implement secure access restrictions, maintain user identities, and streamline authentication and authorization operations. Organizations can maintain a comprehensive IAM architecture that secures resources, protects sensitive data, and provides adequate governance over user access by efficiently using authentication, authorization, user management, and directory services:

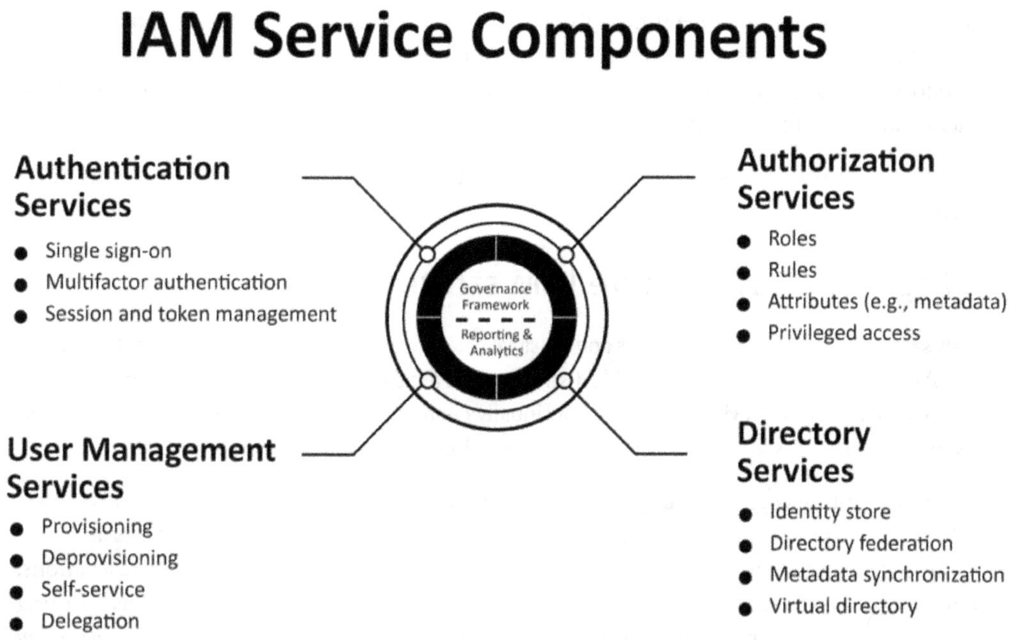

Figure 3.1 – Service components of an IAM system

In accordance with best practices, IAM enforces multi-factor authentication for all privileged IAM uses.

IAM also offers security token-based authentication, in which a six-digit number is generated based on a password-generation process. This value must be entered by the user on the second web page displayed when signing in. A security token might be a physical or virtual device that is assigned to an IAM user. A virtual device is often a software program that runs on a mobile device. However, security built into a physical device outperforms software installed on a mobile device. Each of these devices is unique, and the numerical numbers generated by them are unlikely to be shared among users.

Another IAM practice is SMS authentication. The main method is the same as described in the *Security token-based authentication* section, except instead of a physical or virtual device, a **one-time password (OTP)** with a six-digit numeric value is transmitted to the user's mobile device.

Another good IAM practice is not to use or share the (for example, an AWS account) root user account credentials. Your AWS account root user has complete access to all AWS services and resources, including billing information. To produce programmatic queries to AWS, you can utilize an access key that includes the access key ID and secret access key. You cannot reduce the permissions associated with your AWS account's root ID. As a result, protecting your root user access key is critical.

To boost security, you must employ third-party technologies. For greater results, an administrator must frequently go outside of the service itself. For example, logging solutions, such as AWS CloudTrail, make it easier to trace API queries. Other technologies, such as Chalice, help to automate IAM policy development, saving administrative time. Although these technologies lower hazards and speed up management chores, they lack the necessary flexibility and must be examined on a regular basis by administrators. They should be routinely updated, along with normal AWS upgrades, to ensure they are efficient and do not become unproductive.

In each of the upcoming sections, we will cover the various essential service components of an IAM system.

Building a comprehensive IAM strategy

To properly safeguard their precious resources and sensitive data, organizations must prioritize the deployment of a solid IAM strategy. A comprehensive IAM strategy is critical in today's complex and linked digital ecosystem, where threats and cyberattacks are continually changing. This section goes into the suggested practices that organizations should think about when developing a comprehensive IAM strategy.

Firstly, businesses must recognize the importance of having a well-defined IAM strategy that corresponds with their overall business objectives and security requirements. A solid IAM strategy acts as a road map, laying out the organization's goals and objectives for identity management, access restrictions, and authentication procedures. Organizations may successfully direct their efforts towards boosting security, improving user experience, maintaining compliance, and preserving important assets by clearly outlining the purpose and objectives of the IAM strategy.

Conducting a thorough risk assessment is a critical step in building an IAM strategy. Potential vulnerabilities and threats to an organization's resources and data must be analyzed and evaluated. This evaluation should take into account various elements, including the nature of the organization's activities, the sensitivity of the data being handled, and the regulatory obligations that must be met. Organizations may better understand their particular IAM requirements by identifying risks and vulnerabilities and prioritizing their activities appropriately, concentrating on areas that offer the greatest dangers and require quick attention.

Strong authentication procedures are an essential component of any IAM approach. Traditional username and password combinations are no longer enough to defend against sophisticated assaults. MFA solutions that require users to give several kinds of verification, such as something they know (password), something they have (smart card or token), or something they are (biometrics), should be used by organizations.

For privileged users, IAM provides **multi-factor authentication** (**MFA**). Strong passwords are required to safeguard company data and networks, yet simply having strong passwords is sometimes insufficient. The majority of breaches occur as a result of compromised authentication. As a result, security experts recommend multi-factor authentication. It requires the following elements before granting access to any system:

- Something you already know (for example, a password)
- Something you already possess (for example, a hardware token or an OTP on a mobile device)
- Something you naturally have (for example, a retina scan or fingerprint)

Implementing MFA greatly improves an organization's security posture by providing an extra layer of protection and lowering the chance of unauthorized access owing to compromised credentials.

Another important practice in IAM is enforcing the concept of least privilege. The idea of least privilege requires users to be granted just the access rights and privileges required to fulfill their tasks and obligations. Organizations can reduce the potential effect of insider threats, human mistakes, or malevolent behaviors by adhering to this guideline. It is critical to assess and update user privileges on a regular basis based on changing business demands and responsibilities to ensure that access permissions are aligned with the principle of least privilege.

Identity governance mechanisms must be robust in order for IAM to be effective. Identity governance refers to the methods and rules that organizations employ to manage user identities throughout their lifespan. Processes such as user provisioning, deprovisioning, and periodic access reviews fall under this category. Organizations may streamline the administration of user identities by establishing identity governance practices, ensuring that access privileges are given properly and quickly. Regular access reviews aid in the identification of any inconsistencies or unauthorized access, maintaining compliance with security standards and reducing the risk of unauthorized access.

Another critical part of the IAM approach is user lifecycle management. To expedite user onboarding, organizations should focus on building automated user provisioning procedures. Not only does automated provisioning save administrative costs, but it also lowers mistakes and delays in giving access to new employees. It is also critical to ensure that access privileges are withdrawn immediately when people change positions, leave the organization, or no longer require certain permissions. Failure to revoke access in a timely way may result in unauthorized access and potential data breaches, jeopardizing the security of the organization.

Maintaining a centralized user directory is critical for ensuring user information consistency and synchronization across systems and applications. A centralized user directory serves as a single source of truth for user IDs, reducing the conflicts and inaccuracies that might occur in decentralized identity repositories. The regular synchronization of user data across systems helps to keep user information correct and up to date, ensuring that access provisioning and management operations are efficient and dependable.

Self-service capabilities for users are a successful practice that improves user experience while reducing administrative responsibilities. Allowing users to maintain their own accounts, change passwords, and seek resource access empowers them and decreases their reliance on IT or administrative people. Organizations may enhance customer happiness, speed up access request procedures, and minimize the stress on IT and administrative employees by integrating self-service capabilities.

Secure authentication practices are critical in IAM. To safeguard sensitive data during transit and storage, organizations should prioritize the use of encryption techniques. Strong encryption methods and effective key management practices keep data private and safe. When exchanging files with others, organizations should use secure file transfer protocols (for example, SFTP or FTPS) that include encryption and integrity safeguards to reduce the risk of data breaches. Organizations should also put in place **data loss prevention** (**DLP**) procedures to monitor and prevent unauthorized data exchange or leakage. Organizations can safeguard sensitive information from falling into the wrong hands by establishing rigorous security measures during data exchange.

An effective IAM approach must include continuous monitoring and auditing. Organizations should put in place log management systems to gather and analyze security logs, allowing them to detect suspicious activity and respond quickly to security issues. To identify and respond to unauthorized access attempts, malware, or other security risks, **intrusion detection and prevention systems** (**IDPSs**) should be installed. Regular security evaluations, including penetration testing and vulnerability scanning, should be performed to discover and correct gaps in IAM systems. Organizations may detect and mitigate security problems by regularly monitoring and auditing IAM systems, assuring the continuous integrity and confidentiality of their resources and data.

An effective IAM approach must include user awareness and training. Organizations should create complete security awareness programs to teach users the value of strong passwords, safe authentication practices, and identifying social engineering threats. Training sessions should be held on a regular basis to keep users up to date on new dangers, best practices, and data management processes. Organizations may empower their workforce to be active participants in ensuring a safe IAM environment by training users and establishing a security-conscious culture.

In today's data-driven world, compliance with regulatory standards is critical. Organizations must keep current on the applicable rules affecting their operations and verify that their IAM practices comply with these standards. Compliance should be audited on a regular basis, and compliance reports should be created for internal and external stakeholders as needed. Organizations demonstrate their commitment to securing sensitive information and adhering to legal and industry-specific standards by including compliance concerns in their IAM strategy.

The reaction to incidents and recovery planning are essential components of the IAM strategy. Organizations should create a thorough incident response strategy that specifies the steps for recognizing, containing, investigating, and correcting IAM-related security problems. To ensure the availability of IAM systems and data in the case of a security incident or system failure, robust backup and recovery methods should be created. Organizations may reduce the impact of security events, ensure business continuity, and quickly resume normal operations by proactively preparing for incident response and recovery.

Regular reviews and improvements are required to keep an IAM program running well. Periodic security evaluations should be performed by organizations to discover flaws and vulnerabilities in IAM systems. Organizations may improve the overall security posture of their IAM environment by resolving detected concerns as soon as possible. Staying up to date on industry advancements is critical, as growing risks and new technology transform the IAM environment. To strengthen their IAM program and react to emerging security issues, organizations should regularly assess and incorporate best practices and innovations.

Organizations must have a strong IAM strategy in place to safeguard their resources and data. The following are the recommended practices for creating an all-encompassing IAM strategy:

- **Identify IAM objectives**: Define your IAM strategy's aims and objectives clearly. Determine your goals, such as boosting security, improving user experience, or guaranteeing compliance.
- **Conduct a thorough risk assessment**: A thorough risk assessment is required to establish whether and how your organization is vulnerable to possible vulnerabilities and threats. The findings of this evaluation will allow you to establish what specific IAM needs you have and then prioritize your efforts on a regular basis.
- **Implement strong authentication mechanisms**: Use robust authentication technologies, such as MFA and biometrics, to improve user verification. As a result, the risk of unauthorized access due to compromised credentials is reduced, lowering the possibility that unauthorized persons can gain access to critical systems or information. This proactive approach improves overall security by protecting against possible attacks caused by compromised login credentials.
- **Enforce the least privilege principle**: Ensure that you adhere to the principle of least privilege by allowing users only the access rights they need in order to perform their roles and responsibilities. Maintain a regular review and update of user privileges based on any changes in the requirements of the organization.
- **Establish identity governance processes**: The implementation of robust identity governance processes is crucial for managing user identities efficiently and effectively. As part of this process, provisioning, deprovisioning, and periodic access reviews should be carried out to ensure that appropriate access rights have been granted.

To summarize, organizations must have a solid IAM policy in place to successfully defend their resources and data. Organizations may build a complete IAM strategy that matches their business objectives, mitigates risks, assures compliance, and promotes a safe and efficient access management environment by following the suggested practices mentioned in this section.

User lifecycle management and secure data-sharing practices

IAM is inextricably tied to user lifecycle management. Effective user lifecycle management is critical for organizations to provide secure access to their resources and data integrity. This section goes into the best practices for managing users' lifecycles inside an IAM framework.

User lifecycle management includes a variety of tasks, such as user provisioning, authentication, authorization, and access revocation. It all starts with the onboarding process, which grants new users access to the organization's systems and resources. Organizations should focus on creating efficient and automated user provisioning procedures at this level. Organizations may expedite the onboarding process, decrease administrative costs, and reduce the risk of errors or delays in issuing access credentials to new workers by automating user provisioning.

As users become part of the organization, effective identity verification and authentication processes must be established. Strong authentication mechanisms, such as MFA, should be used by organizations to improve security and guarantee that only authorized persons have access to the organization's systems. MFA requires users to give multiple means of authentication, such as passwords, fingerprints, or hardware tokens, lowering the risk of unauthorized access due to compromised credentials dramatically.

Authorization is critical in user lifecycle management. Organizations must follow the concept of least privilege, allowing users to have just the access privileges that are required for their jobs and responsibilities inside the organization. Organizations that follow this approach reduce the danger of unauthorized access or the inadvertent exploitation of sensitive resources. Regular access evaluations are critical to ensuring that access permissions stay matched with the organization's developing demands. Organizations may maintain a secure access environment and avoid the needless accumulation of access permissions by assessing and changing user privileges on a regular basis.

When users change positions, leave the organization, or no longer require certain access credentials, effective user lifecycle management includes prompt access revocation. To quickly remove user permissions and reduce the danger of unauthorized access, organizations must develop well-defined policies and procedures for access revocation. Failure to remove access in a timely way can lead to security flaws and data breaches. When users no longer require access permissions, organizations should consider establishing automated deprovisioning processes to guarantee rapid and precise access revocation.

For successful user lifecycle management, a centralized user directory is essential. It acts as a centralized repository for user IDs and related data. Organizations may assure consistency and accuracy of user data across systems and apps by keeping a centralized user directory. This not only streamlines user management operations, but also reduces disparities and the mistakes that may occur when utilizing decentralized identity stores.

Organizations should empower users by allowing them to manage their own accounts, passwords, and access requests through self-service capabilities. Users may take control of their identities and access permissions using self-service capabilities, decreasing their reliance on IT or administrative professionals. Organizations may improve user experience, efficiency, and administrative overhead by allowing users to maintain their accounts, reset their passwords, and request access to resources.

Regular user lifecycle management training and awareness programs are essential. Users should be educated on best practices, security rules, and their roles in access management and data protection. Organizations may build a security culture and encourage users to actively engage in safeguarding the integrity and confidentiality of resources by raising user awareness.

User lifetime management requires regular monitoring and auditing. Organizations should constantly monitor user behaviors, access patterns, and security records for any suspicious or unauthorized activity. To track user access and changes in access rights, strong logging and auditing tools must be implemented. Organizations can quickly detect and handle possible security breaches or policy violations by performing frequent audits and examining records.

User lifecycle management practices should incorporate compliance considerations. Organizations must ensure that their processes and procedures are in accordance with applicable regulatory requirements, industry standards, and internal rules. Organizations demonstrate their commitment to data security, privacy, and regulatory compliance by implementing compliance standards into user lifecycle management.

IAM relies heavily on managing user lifecycles. The recommended practices for successful user lifetime management are as follows:

- **Automated user provisioning**: In order to streamline new employee onboarding, minimize errors, and expedite access provisioning for new employees, it is necessary to implement automated user provisioning processes.
- **Timely access revocation**: Ensure that people who change positions, leave the organization, or no longer require particular rights have their access revoked in a timely manner. This reduces the possibility of unauthorized access and data breaches.
- **Centralized user directory**: A centralized user directory should be maintained in order to consolidate user information and ensure consistency across systems as much as possible. Regularly synchronize the data of users in order to prevent a discrepancy from occurring.
- **Self-service user management**: Provide users with self-service capabilities so they can manage the profile themselves, reset their passwords, and request access to the services they require. By doing so, administrative burdens can be reduced and user experience can be improved.

To summarize, effective IAM requires successful user lifecycle management. Organizations may expedite user provisioning, authentication, authorization, and access revocation procedures by adopting the suggested practices mentioned in this section. Automated user provisioning, robust authentication systems, and effective access revocation practices improve security and reduce risks. Users are empowered with improved user experience via centralized user directories, self-service capabilities, and user awareness programs. Monitoring, auditing, and compliance considerations are carried out on a regular basis to guarantee continuous security and regulatory compliance. Organizations may accomplish effective user lifecycle management and a strong IAM framework by using these practices.

Secure authentication practices

The authentication process is a vital and essential component of IAM. It acts as the entrance point for authorized personnel to obtain access to organizational resources. This section delves into a variety of recommended practices for secure authentication inside an IAM framework, emphasizing the necessity of safeguarding user identities and preventing unauthorized access.

To start with, organizations should use robust authentication mechanisms to improve the security of their IAM systems. Traditional username and password combinations are no longer enough to deal with increasingly complex cyberattacks.

Adaptive authentication systems should also be prioritized by organizations. Adaptive authentication considers contextual information and dynamically analyzes the risk of each authentication attempt. Adaptive authentication modifies the amount of security necessary by assessing parameters such as the user's location, the device utilized, and behavioral habits. If a user tries to access sensitive data from an unknown location or device, the system may request further verification procedures. Adaptive authentication assists organizations with striking a balance between user convenience and security, ensuring that the right degree of authentication is used in each case based on the perceived risk.

Another suggested practice for safe authentication under IAM is the use of **single sign-on (SSO)** systems. SSO allows users to authenticate once and receive access to numerous apps and systems without having to enter credentials several times. SSO enhances user experience and productivity by reducing the number of login prompts while lowering the danger of weak passwords or password reuse. Furthermore, SSO enables organizations to control user access centrally, making it easier to enforce security standards and streamline authentication procedures across several apps.

Implementing safe password practices is another critical part of secure authentication. Organizations should educate users on the necessity of generating strong, unique passwords and changing them on a regular basis. Passwords should be complicated, including a mix of uppercase and lowercase letters, numbers, and special characters. It is critical to discourage the use of passwords that are readily guessable or regularly used. Organizations should also enforce password expiration rules and provide ways for users to update their passwords on a regular basis. Password management technologies, such as password managers or self-service password reset systems, can also improve password security and user comfort.

Organizations should consider employing biometric authentication technologies in addition to secure passwords. Biometrics, which rely on unique physical traits, such as fingerprints, facial recognition, or iris scans, give a high level of security. Biometric authentication provides an additional level of assurance, guaranteeing that only authorized personnel have access to critical resources. Organizations, on the other hand, must take strong steps to protect biometric data, such as encrypting and securely storing biometric templates.

To avoid unauthorized interception or alteration, authentication data must be transmitted securely. To encrypt authentication data during transit, organizations should use secure communication protocols such as HTTPS or SSL/TLS. This keeps critical information, such as usernames, passwords, and authentication tokens, private and secure from eavesdropping or man-in-the-middle attacks.

Finally, organizations should assess and update their authentication practices on a regular basis to ensure that they are in line with developing security requirements and emerging technology. Organizations should proactively improve their authentication processes to address new and emerging risks by remaining updated about industry advancements and best practices. Regular security audits, vulnerability scanning, and penetration testing are required to discover and resolve any flaws or vulnerabilities in the authentication process.

The authentication process is one of the most critical components of IAM. There are a number of best practices for secure authentication that are discussed as follows:

- **Multi-factor authentication** (**MFA**): Implement MFA to provide an extra layer of protection by asking users to submit several kinds of verification, such as a password, biometrics, or a hardware token.
- **Adaptive authentication**: Use adaptive authentication techniques that evaluate the risk of each authentication attempt and alter the level of security accordingly.
- **Single sign-on** (**SSO**): SSO solutions may be used to streamline the authentication process for users across a wide range of apps. As a consequence, the user experience is improved, and several sets of credentials are no longer required.

To summarize, safe authentication is an important part of IAM, and organizations must follow best practices to protect user identities and prevent unauthorized access. Several suggested practices have been addressed in this section, including the usage of robust authentication mechanisms such as MFA, adaptive authentication, and SSO. Encouraging secure password practices, investigating biometric authentication, ensuring secure authentication data transfer, and remaining current with industry standards are all critical aspects of developing a strong and secure authentication process. Organizations may enhance their IAM systems and lower the risk of unauthorized access and possible data breaches by applying these recommended practices.

Security token-based authentication

Security token-based authentication is a strong technique inside IAM frameworks and plays an important role in improving the security posture of digital environments. At its foundation, this authentication mechanism is based on the creation and use of security tokens, which are unique, time-sensitive codes or cryptographic keys. The usage of security tokens enhances the standard username-password authentication architecture, reducing the dangers associated with compromised credentials.

In the context of IAM security, security token-based authentication presents a dynamic and multi-faceted method of identity verification. Unlike static passwords, security tokens are transitory, updating every few minutes. This time-sensitive nature dramatically reduces the ability of malevolent actors to exploit stolen credentials because the acquired token rapidly becomes obsolete. This element is aligned with the notion of continuous authentication, a core pillar of IAM best practices that ensures user identification is continually confirmed throughout the life of the session.

Time-based one-time passwords (TOTPs), a widely recognized standard, represent one well-known example of security token-based authentication. TOTPs create time-bound tokens, which are generally valid for 30 seconds and serve as a dynamic second factor of authentication. The creation of these tokens requires a shared secret between the server and the client device, which improves security via cryptographic principles.

Furthermore, security token-based authentication is very successful at resolving the typical weaknesses seen in traditional authentication techniques. The dynamic nature of security tokens helps to prevent phishing attempts, in which malevolent actors try to deceive users into exposing their credentials. Even if a user unintentionally reveals their login and password, the time-sensitive token renders the collected data unusable for a lengthy duration.

From a best practices standpoint, the implementation of security token-based authentication should be accompanied by secure token storage methods. The usage of hardware tokens or secure mobile applications as token generators improves overall security. Furthermore, combining MFA with security tokens raises the authentication process by forcing users to provide various kinds of verification to obtain access.

Another aspect of security token-based authentication is the use of **public key infrastructure** (**PKI**) for digital signatures. In this case, security tokens serve as cryptographic keys, improving the authentication process by utilizing asymmetric encryption. This method is especially useful in contexts where robust cryptographic authentication is required, such as when safeguarding sensitive transactions or accessing vital systems.

As enterprises negotiate the changing threat landscape, the use of security token-based authentication is consistent with industry trends preferring stronger and more adaptable security solutions. This strategy not only strengthens IAM frameworks but also corresponds with legislative requirements that highlight the significance of strong authentication procedures. In the context of zero trust security models, where continuous verification is a key component, security token-based authentication appears to be an appropriate approach, giving a sophisticated and robust layer of defense against unwanted access.

To summarize, security token-based authentication is an effective ally in the world of IAM security, providing a dynamic and time-sensitive layer of protection. Its effectiveness at reducing the risks associated with compromised credentials, together with its flexibility to varied deployment settings, solidifies it as a cornerstone of modern cybersecurity methods. As companies continue to pursue strong IAM policies, the inclusion of security token-based authentication acts as a proactive strategy to strengthen digital identities and protect critical data in an ever-changing threat scenario.

Access control and authorization

It is critical to have adequate access control and authorization mechanisms in place to protect data confidentiality and prevent unauthorized access to sensitive information. This section focuses on recommended practices for access control inside an IAM architecture.

Access control is critical in preventing unauthorized disclosure, modification, or the destruction of organizational data. Organizations may guarantee that only authorized personnel can access certain resources based on their roles and responsibilities by adopting effective access control methods. Implementing **role-based access control** (**RBAC**) is one of the essential best practices for access control. RBAC assigns access privileges based on preset organizational roles. Users are given access privileges that correspond to their respective responsibilities, ensuring that they can only access the resources required to accomplish their tasks. Regular job-assignment reviews and modifications are required to ensure that access privileges stay matched with organizational demands and changes in personnel responsibilities.

Organizations should consider using **attribute-based access control** (**ABAC**) techniques in addition to RBAC. Access control choices can be based on user traits, contextual information, or resource sensitivity using ABAC. This fine-grained approach to access management allows for greater flexibility and accuracy in granting or refusing access based on certain traits or criteria. ABAC allows organizations to design complicated access policies that take into account elements such as access time, location, device type, and user traits. Organizations may achieve more granular and context-aware access control by using ABAC, guaranteeing that resources are accessible only by authorized personnel under acceptable conditions.

The notion of least privilege is another key best practice. When giving access permissions, organizations should follow the concept of least privilege. This concept states that users should be given just the rights required to fulfill their job tasks efficiently. Organizations can reduce the risk of unauthorized access, inadvertent data breaches, or malicious activity by restricting access to just what is necessary. Access reviews and audits should be performed on a regular basis to ensure that access privileges are consistent with the concept of least privilege. To prevent possible security threats, any unneeded access rights should be swiftly canceled.

Another suggested practice in access control is the use of **access control lists** (**ACLs**) or permissions. To restrict access based on user identities or groups, organizations should build and enforce appropriate access control lists at the resource level. This helps to guarantee that critical resources are only accessed by authorized individuals or groups. To prevent unauthorized access and data loss, ACLs should be defined and managed with care.

Strong authentication mechanisms must be used for effective access control. To validate user identities, organizations should adopt robust authentication systems such as MFA. MFA requires users to submit more than one type of confirmation, such as something they know (a password), something they own (a token or smart card), or something about themselves (biometrics). Using MFA greatly improves the security of access control for enterprises by providing an extra layer of precaution against unauthorized access in the event that credentials are compromised.

Regular access evaluations and recertification processes are required to keep access control effective. Organizations should assess and validate user access privileges on a regular basis to ensure that they are still suitable and aligned with business needs. Examining user roles, group memberships, and individual access rights is part of this process. Managers or data owners who can attest to the appropriateness of access privileges should be included in access recertification processes. Organizations can reduce the risk of unauthorized access and maintain continuing compliance with security policies and laws by performing frequent access evaluations and recertifications.

Finally, access control is an important part of IAM since it helps maintain data security and prevent unauthorized access to sensitive information. Several best practices have been emphasized in this section, including the use of RBAC and ABAC, adherence to the concept of least privilege, the use of access control lists, the use of robust authentication mechanisms, and frequent access reviews and recertification. Organizations may develop comprehensive access control and authorization systems by applying these best practices, successfully preserving their data and lowering the danger of unauthorized access.

It is critical to have appropriate access control and authorization systems in place to ensure data confidentiality and prevent unauthorized access to data. This section discusses access control best practices:

- **Role-based access control** (**RBAC**): RBAC should be used to assign access privileges based on user roles and responsibilities. Review and update jobs on a regular basis to ensure they are still relevant to the demands of the organization.
- **Attribute-based access control** (**ABAC**): Leverage the advantages of ABAC to establish access controls based on user traits, contextual data, and resource sensitivity. This allows for more detailed, context-aware access restrictions.
- **Regular access reviews**: Conduct frequent access evaluations to verify that access permissions are current and in line with business needs. Remove superfluous access privileges as soon as possible to reduce security concerns.

To summarize, effective access control and authorization procedures are critical in protecting digital assets and maintaining the integrity of sensitive information, ensuring that only authorized entities have the necessary permissions to access resources in a secure and regulated environment.

Secure data-sharing practices

Organizations must use safe data-sharing practices to protect sensitive information. This section discusses best practices for secure data transmission in the context of IAM.

Because sensitive information may be exposed to unauthorized persons or harmful entities, data sharing has inherent dangers. Organizations should prioritize the development of effective data encryption technologies to prevent these dangers. Data is encrypted when it is converted into an unreadable format, rendering it worthless to unauthorized parties. Organizations may guarantee that sensitive data stay secret and secure during transit and storage by utilizing robust encryption algorithms and good key management practices.

Furthermore, when exchanging data with others, organizations should use secure file transfer protocols. Protocols such as **secure file transfer protocol** (**SFTP**) and **FTP secure** (**FTPS**) provide encryption and integrity features to protect data while they are being transmitted. It is critical to avoid employing unprotected protocols, such as standard FTP, that might jeopardize the security and integrity of shared data. Organizations reduce the danger of interception or unauthorized access to shared information by using secure file transfer protocols.

Another essential practice in safe data sharing is the use of **data loss prevention** (**DLP**) procedures. DLP systems monitor and prevent unauthorized data exchange or leakage. These systems can detect the patterns and policies specified by the organization in order to detect and prevent efforts to transfer sensitive information across multiple channels. Organizations may considerably minimize the risk of unintentional or purposeful data breaches by using DLP procedures, ensuring that sensitive information is secured.

When exchanging data, organizations should also consider setting access restrictions and permissions. Organizations may guarantee that only authorized individuals or groups have access to shared data by designing and implementing suitable access control procedures. User or group permissions, file-level permissions, or folder-level permissions are all examples of access restrictions. Granular access controls give an extra layer of protection and limit the danger of unauthorized access or the inadvertent disclosure of sensitive data.

Organizations should also include audit trails and monitoring methods to track data-sharing actions. Organizations can discover and investigate any security issues or policy violations by keeping extensive logs and recordings of data exchange occurrences. Regularly analyzing these logs and undertaking proactive monitoring aids in the detection of any unauthorized or irregular data-sharing actions, allowing for rapid reaction and risk minimization.

User education and training programs are critical in supporting secure data-sharing practices. It is critical to educate users on the necessity of data protection, the hazards associated with inappropriate data sharing, and the recommended practices for safe data interchange. Organizations may promote a culture of security-conscious users who understand their responsibilities in securing sensitive information during data-sharing activities by increasing awareness and offering training.

Finally, secure data exchange practices are critical for preventing unauthorized access or exposure to sensitive information. Several recommended practices within IAM have been highlighted in this section, including data encryption, the use of secure file transfer protocols, the implementation of DLP measures, the enforcement of access controls and permissions, the implementation of audit trails and monitoring, and user awareness and training. Organizations may develop a safe data-sharing framework by applying these practices, reducing the risks associated with unauthorized data access or disclosure and maintaining the confidentiality and integrity of shared information.

To protect sensitive information, secure data-sharing practices are essential. Within IAM, this section explores recommended practices for safe data exchange:

- **Implement data encryption**: Encrypt sensitive data during transport and storage to prevent unauthorized access. Make use of robust encryption methods and key management procedures.
- **Use secure file transfer protocols**: When exchanging files with others, use secure file transfer protocols (for example, SFTP and FTPS). Using unsecured protocols can jeopardize data integrity and confidentiality.
- **Implement DLP measures**: DLP systems may be used to monitor and prevent unauthorized data exchange or leakage. Implement procedures to detect and prevent the improper sharing of sensitive data.

In conclusion, implementing secure data-sharing methods not only strengthens data integrity but also develops a trustworthy and compliant workplace, finding a balance between co-operation and protecting sensitive information in today's linked digital ecosystem.

Continuous monitoring and auditing

Continuous monitoring and audits are critical for discovering and resolving any security concerns in IAM systems. This section focuses on the best practices for monitoring and auditing IAM systems to guarantee a proactive security strategy.

A solid log management and analysis system is required for successful monitoring and auditing. To obtain insights into user activity, access attempts, and system events, organizations should collect and analyze security logs from various IAM components and connected systems. Organizations can discover suspicious actions, possible security breaches, and policy violations by monitoring logs. Log analysis assists with identifying trends or abnormalities that may signal security issues, allowing for quick investigation and response.

Intrusion detection and prevention systems (**IDPSs**) are critical components of IAM system monitoring. IDPSs aid in the detection of and response to unauthorized access, malware infections, and other security risks. Organizations may improve their capacity to detect and mitigate possible vulnerabilities by employing IDPS capabilities, assuring the continuing security of their IAM environment. To respond to emerging threats and retain maximum performance, the IDPS requires regular upgrades and adjustments.

Regular security evaluations are essential for monitoring and auditing IAM systems. Periodic assessments, including vulnerability scanning and penetration testing, should be performed by organizations to discover any gaps or vulnerabilities in the IAM infrastructure. These evaluations aid in identifying vulnerabilities that attackers may exploit, allowing organizations to resolve them proactively. Organizations can keep one step ahead of possible attacks and maintain the continuous integrity and confidentiality of their resources and data by routinely reviewing the security of IAM systems.

Furthermore, organizations should include incident response methods in their monitoring and auditing practices. The measures to be taken in the case of a security issue or breach inside the IAM system are defined in incident response plans. Organizations with well-defined policies and procedures may respond quickly to security issues, minimizing the effect and avoiding subsequent risks. The incident response plan should be tested and updated on a regular basis to verify its efficacy and conformity with current threats and organizational demands.

Audits are essential for determining the overall efficacy of IAM systems. These audits determine if security policies, industry requirements, and internal controls are being followed. Organizations can detect gaps, flaws, or non-compliance concerns by performing audits, allowing them to take remedial steps and enhance their IAM environment. Audit reports offer useful information to management and stakeholders, allowing for transparency and accountability in security practices.

Finally, ongoing monitoring and audits are required to ensure the security and integrity of IAM systems. Organizations may proactively detect and respond to possible security concerns by adopting comprehensive log management and analysis, utilizing IDPS, conducting frequent security assessments, developing incident response protocols, and undertaking audits. These recommended practices assist organizations in ensuring that their IAM systems are successfully monitored in compliance with legislation and that they are capable of detecting and responding to security problems as soon as possible.

Continuous monitoring and audits are required to identify and respond to possible security problems. The recommended practices for monitoring and auditing IAM systems are as follows:

- **Log management and analysis**: To gather and analyze security logs, use log management solutions. Monitor records for unusual activity and examine any potential security problems as soon as possible.
- **Intrusion detection and prevention systems (IDPSs)**: Use IDPSs to identify and respond to unauthorized access, malware, and other security risks. Update and fine-tune the IDPS on a regular basis to keep up with emerging threats.
- **Regular security assessments**: Conduct frequent security evaluations, including penetration testing and vulnerability scanning, to detect and remediate gaps in IAM systems.

To summarize, continuous monitoring and auditing methods serve as vigilant guardians, offering real-time insights into system activity and guaranteeing compliance. Organizations that adopt a proactive strategy can spot abnormalities quickly, strengthening cybersecurity defenses and ensuring the resilience of their digital infrastructure.

User awareness and training

User education and training are critical components of implementing safe IAM practices. This section digs into the best practices for user education, emphasizing the necessity of providing users with the information and skills they need to actively engage in the maintenance of a secure IAM environment.

Raising user knowledge about the importance of IAM and its influence on security is one of the main best practices. Users should be aware of the dangers of weak passwords, phishing attempts, social engineering, and other frequent hazards. Organizations may educate users on recommended practices, such as setting strong passwords, recognizing and reporting suspicious activity, and following security rules by delivering comprehensive security awareness programs.

Regular training sessions are required to bring users up to date on emerging risks, new security practices, and IAM system changes. Secure authentication mechanisms, data security, access control, and incident reporting should all be included in the training. Organizations may encourage good security behaviors, establish a culture of security awareness, and empower users to be proactive in maintaining a safe IAM environment by continuously educating users.

Organizations should also consider making user-friendly materials and tools available to support safe IAM practices. This includes methods for self-service password resets, standards for safe password development, and simple documentation on access request procedures. User-friendly resources empower users to actively participate in IAM procedures, minimizing the pressure on IT or administrative employees and instilling a sense of security in their own hands.

Furthermore, organizations should convey security updates and reminders to users on a regular basis. Email notifications, newsletters, and internal communication channels can all be used to do this. Organizations may ensure that users remain attentive and comply with security measures by keeping them informed about current risks, security policy changes, or system updates.

Finally, user education and training are critical components of promoting secure IAM practices. Organizations may empower users to actively contribute to maintaining a safe IAM environment by raising user awareness, giving frequent training sessions, providing user-friendly materials, and communicating security updates. User education contributes to the development of a culture of security awareness and accountability, lowering the risk of security incidents and improving the overall efficacy of IAM systems.

User education and training are critical to supporting secure IAM practices. This section focuses on best practices in user education:

- **Security awareness programs**: Create security awareness programs to teach users the value of strong passwords, safe authentication, and identifying social engineering threats
- **Regular training sessions**: Conduct frequent training sessions to bring users up to date on emerging threats, security best practices, and sensitive data management

In conclusion, user awareness and training are critical components of IAM best practices, providing users with the information they need to navigate the digital realm safely. Organizations that cultivate a culture of cybersecurity awareness strengthen their entire defense against emerging threats, laying the groundwork for successful IAM.

Compliance and regulatory considerations

Compliance with regulatory requirements is critical for firms seeking to operate legally and ethically. This section discusses the recommended practices for incorporating compliance into IAM practices, emphasizing the need to align IAM operations with regulatory requirements.

Conducting a comprehensive examination of the applicable legislation and standards that affect the organization's operations is one of the important best practices. This covers industry-specific restrictions, data protection laws, privacy legislation, and any other legal obligations that may apply. Organizations can determine the precise compliance duties that must be met within their IAM architecture by knowing the regulatory landscape.

Organizations should develop and adopt clear rules and procedures that reflect compliance requirements in their IAM practices. This involves setting access control policies, user provisioning and deprovisioning processes, and regulatory-compliant authentication needs. Organizations may guarantee that IAM practices comply with regulatory rules and ease continuing compliance by recording and sharing these policies.

Audits and evaluations should be performed on a regular basis to examine the efficacy of IAM practices in satisfying compliance standards. Internal reviews, external assessments, or third-party audits should all be included in these audits to give an unbiased assessment of the organization's conformity to regulatory requirements. Organizations can maintain compliance by detecting gaps or inadequacies and implementing necessary adjustments.

Compliance considerations in IAM systems necessitate continual monitoring and documentation. To show compliance with regulatory obligations, organizations should keep detailed records of IAM actions, access logs, and user permissions. Regular IAM system monitoring ensures that access rules are constantly implemented, user rights are matched with compliance duties, and any deviations or anomalies are discovered and remedied as soon as possible.

Finally, organizations should keep current on legislative developments and alter their IAM practices accordingly. Compliance standards change over time, and it is critical to assess and update IAM processes on a regular basis to be in line with evolving rules. Organizations may ensure that their IAM practices continue to satisfy regulatory needs and successfully secure sensitive information by being educated and proactive.

Compliance with regulatory standards is critical for businesses. The best practices for incorporating compliance into IAM practices are as follows:

- **Stay updated on regulations**: Maintain awareness of applicable rules and verify that IAM practices are in line with industry-specific regulations (for example, GDPR and HIPAA).
- **Regular audits and reporting**: Conduct frequent audits to verify that all applicable requirements are being followed. As needed, generate compliance reports for internal and external stakeholders.

To summarize, including compliance within IAM practices is critical for organizations to achieve regulatory standards while also operating legally and ethically. Organizations may effectively align their IAM practices with compliance requirements by performing assessments, setting clear rules, conducting frequent audits, monitoring IAM operations, and staying current on regulatory changes. This section gave insights into best practices for integrating compliance concerns into IAM frameworks and demonstrating regulatory compliance commitment.

Incident response and recovery and processes for regular evaluation

It is critical to have an effective incident response and recovery plan in place to mitigate the impact of security vulnerabilities. This section delves into the best practices for incident response and recovery in IAM.

By recognizing that security events are unavoidable, organizations should prioritize the establishment of a well-defined incident response strategy. This strategy lays out the steps for discovering, containing, investigating, and resolving security problems inside the IAM system. Organizations may minimize the inconvenience caused by security events and guarantee a prompt and efficient response by implementing an organized and documented strategy.

Establishing defined roles and duties for incident management is an essential aspect of incident response. Delegating responsibility for incident detection, analysis, response co-ordination, and communication to people or teams is critical for effective incident management. Well-defined responsibilities aid in ensuring a coordinated and streamlined response, reducing confusion and allowing for quick decision-making in crucial situations.

The need to perform frequent incident response drills and exercises is also emphasized throughout the section. Organizations may uncover weaknesses, enhance co-ordination, and adjust procedures by simulating security events and assessing the effectiveness of any response plans. These exercises aid in the development of incident response skills, improve preparation, and give useful insights into updating and improving incident response plans.

In addition to crisis response, organizations should prioritize the development of effective recovery procedures. Implementing suitable backup and restoration processes to maintain the availability and integrity of IAM systems and data is part of this. Regular backups, both on-site and off-site, enable the restoration of systems and data in the case of a security breach or system failure. It is also vital to test the restoration procedure to ensure the efficacy of backups and the capacity to recover key systems and data.

The importance of communication and reporting in incident response and recovery cannot be overstated. Organizations should establish clear communication channels and mechanisms for reporting and escalating issues. This allows incident response teams, stakeholders, and management to communicate more quickly and effectively. Following each security issue, detailed incident reports should be prepared, recording the nature of the occurrence, the response activities performed, the lessons learned, and any recommendations for enhancing future incident response.

Finally, a well-structured incident response and recovery strategy is critical for limiting the consequences of security concerns inside an IAM architecture. Organizations can effectively respond to security incidents and minimize their impact by implementing best practices such as developing an incident response plan, defining roles and responsibilities, conducting regular drills and exercises, ensuring robust recovery procedures, and establishing effective communication channels. This section delves into IAM best practices for incident response and recovery, allowing organizations to improve their incident management skills and overall security posture.

It is vital to have an effective incident response and recovery plan in place to mitigate the effects of security issues. This section delves into IAM best practices for incident response and recovery:

- **Incident response plan**: Create a thorough incident response strategy that includes methods for recognizing, containing, investigating, and resolving IAM-related security problems
- **Backup and recovery processes**: Implement comprehensive backup and recovery mechanisms to ensure IAM system operation and data availability in the event of a security incident or system failure

To summarize, a strong incident response and recovery plan is a critical component of IAM best practices, providing an organized way to quickly address and mitigate security events. This proactive approach guarantees that businesses may effectively negotiate problems, mitigate possible harm, and recover quickly while retaining the integrity of their identity and access management frameworks.

Regular evaluation and improvement

To keep an IAM program successful, it must be evaluated and improved on a regular basis. This section digs into the methodologies proposed for continuously reviewing and enhancing IAM practices.

Organizations should prioritize frequent evaluations to evaluate the success of an IAM program to guarantee its seamless functioning. These evaluations include assessing many components of IAM, such as user provisioning, authentication systems, access controls, and compliance adherence. Organizations can discover any gaps or opportunities for development in their IAM practices by performing detailed evaluations.

Organizations should adopt suitable actions to improve their IAM program after identifying areas for improvement. This might include revising rules and procedures, implementing new technology, or improving existing processes. Organizations may adapt to evolving issues, rectify any detected weaknesses, and improve the overall efficacy and security of their IAM program by always aiming for improvement.

Organizations should also develop **key performance indicators** (**KPIs**) and metrics to assess the effectiveness of their IAM program. User onboarding time, access request turnaround time, incident response time, and compliance levels are examples of metrics. Monitoring these indicators on a regular basis helps organizations track progress, spot trends, and take proactive steps to rectify any deviations or areas of concern.

Organizations can also solicit input from stakeholders such as users, IT employees, and management. This feedback can give useful insights into the IAM program's strengths and flaws and assist with driving ongoing improvement initiatives. By establishing a feedback loop and developing an open communication culture, the IAM program may remain aligned with organizational priorities and meet any growing requirements.

Finally, ongoing evaluation and improvement are required to keep an IAM program running well. This section outlines the suggested practices for regularly assessing and improving IAM practices. Organizations may ensure that their IAM program stays effective, efficient, and capable of handling the increasing problems of identity and access management by performing evaluations, applying improvement methods, creating KPIs, and incorporating stakeholder input.

Continuous review and improvement are required to keep an IAM program running well. This section discusses the recommended practices for assessing and improving IAM practices on a regular basis:

- **Regular security assessments**: Conduct frequent security reviews of IAM systems to discover any flaws and vulnerabilities. To improve overall security, address any discovered flaws as soon as possible.
- **Stay current with industry developments**: Keep up to date on IAM industry trends, emerging threats, and new technology. To develop your IAM program, always assess and implement best practices and innovations.
- **Solicit user feedback**: Collect input from users to discover areas for improvement in user experience and usability. Incorporate user input to improve IAM procedures and systems.

Implementing best practices for IAM may greatly improve an organization's security posture and expedite access management. Organizations can strengthen their digital fortresses by implementing these recommendations for a comprehensive IAM strategy that includes user lifecycle management, secure authentication practices, access control and authorization, secure data exchange, continuous monitoring and auditing, user awareness and training, and compliance and regulatory adherence. These IAM best practices work together to establish an integrated and proactive strategy that reduces risks, ensures regulatory compliance, and fosters a security culture, therefore protecting the organization's digital assets and retaining trust in today's evolving cybersecurity world.

Summary

This chapter highlighted critical best practices for IAM that safeguard the security and integrity of corporate resources and data. It highlighted the value of user lifecycle management, secure data-sharing procedures, access control and authorization, secure authentication systems, continuous monitoring and auditing, user awareness and training, compliance, incident response, and continual review and improvement. These practices attempt to reduce risks, maintain regulatory compliance, and develop a security-conscious culture, eventually boosting an organization's security posture in today's changing cybersecurity environment.

In the next chapter, we cover sources of truth and trust anchors and their importance in securing identity management frameworks.

4

Trust Anchors/Sources of Truth and Their Importance

In the digital era, correct identification data verification is critical for building confidence, combating fraud, and assuring safe access to systems and resources. Sources of truth are critical in this procedure because they serve as authoritative references for authenticating and confirming identification information. This chapter examines the role of sources of truth in identity data verification, emphasizing their relevance in preserving data integrity, enhancing identity assurance, and allowing effective identity management.

In this chapter, we will cover the following topics:

- Identifying and defining trustworthy sources of truth
- Use of sources of truth in identity management and its challenges
- Deep dive into the web of trust model
- Real-world use cases of leveraging the WoT model

Sources of truth

Sources of truth are authoritative sources of facts or information that are trusted and truthful. Sources of truth are crucial for establishing and sustaining confidence in an individual's identity in the context of identity verification and management. Some examples of truth sources are as follows:

- **Government-issued identification documents**, such as passports, driver's licenses, and national identity cards, are regarded as the primary source of truth for determining an individual's identity. These documents are issued by a reputable government body and are often recognized as identification evidence by other organizations.

- **Institutional databases**, such as those kept by banks, educational institutions, and employers, can also be used to verify an individual's identification. These databases frequently contain verified personal information, such as a person's name, address, and date of birth, which may be used to confirm an individual's identification.
- **Biometric data**, such as fingerprints, face recognition, and iris scans, can be used to authenticate an individual's identification. Biometric data is unique to each person and may be used to validate their identification with high accuracy.
- **Trusted third-party verification services**, such as credit reporting agencies and background check providers, can also be used to verify an individual's identification. These services gather and verify personal information from many sources in order to develop a full profile of an individual's identity.

The value of truth sources in identity verification and management cannot be underestimated. Building trust in an individual's identity and combating fraud and identity theft require establishing a trusted and accurate source of truth. Organizations that rely on erroneous or untrustworthy sources of truth risk financial and reputational loss, as well as legal consequences.

Defining sources of truth

Establishing *sources of truth* is a core idea in information management that ensures data consistency, correctness, and dependability across several applications and systems. These sources, which are frequently regarded as authoritative repositories, serve an important role in presenting a single, unified version of reality, reducing differences, and instilling trust in decision-making procedures:

- **Concept and role**: Sources of truth are stores or systems providing specific identity information that are regarded as authoritative. They give trustworthy, accurate, and current data that may be used to verify and confirm an individual's identification. Government-issued identity documents, official databases, biometric databases, and trustworthy third-party systems are common examples of truth sources.
- **Reliability and integrity**: The dependability and integrity of truth sources are crucial in identity verification. These sources are built and maintained with stringent controls to assure data correctness, consistency, and anti-tampering security. To protect the integrity and security of the identification information contained in various sources, strict data management practices, encryption, and access restrictions are used.

Ensuring data accuracy and consistency

Ensuring data quality and consistency is critical in the field of information management, and this work is founded on the notion of sources of truth. In this section, we look at the crucial role that authoritative repositories play in ensuring data trustworthiness, eliminating inconsistencies, and laying the groundwork for informed decision-making across a wide range of applications and systems:

- **Eliminating data discrepancies**: By providing a standardized and accepted reference, sources of truth help to reduce inconsistencies in identification data. Cross-referencing several sources, such as different government-issued identity cards, can assist in detecting and resolving any conflicts or contradictions in the data. Organizations may verify that the identification information they rely on is correct and consistent by relying on authoritative sources.
- **Data synchronization**: Integrating and synchronizing identification data across systems with trustworthy sources of truth is critical for consistency. Organizations may guarantee that identification information is up to date and consistent across systems and apps by using data synchronization mechanisms. This synchronization aids in the prevention of mistakes, duplication, and out-of-date data, which can jeopardize the accuracy and reliability of identity verification operations.

After identifying the many sources of truth in identity management, we can now work on increasing the dependability and security of these identities. In the next section, we will look at options for increasing identity assurance levels and providing more robust and reliable verification processes.

Enhancing identity assurance

In today's fast-changing digital environment, identity verification has become a critical component of safe online interactions. This section digs into the multidimensional domain of improving identity assurance and examining innovative approaches and technology aimed to strengthen verification procedures and assure users' authenticity in different digital transactions. From biometric improvements to cutting-edge cryptographic approaches, the quest for increased identity assurance is not just a technological necessity but also a necessary response to the growing issues of identity fraud and cyber threats. In the field of identity management, the use of reliable sources is critical not only in creating trust but also in strengthening defenses against fraudulent actions. Organizations that depend on authoritative sources for identity validation may build confidence, expedite verification procedures, and strengthen their defenses against identity theft and fraud.

The two most important factors for establishing trust are as follows:

- **Authoritative sources**: Using authoritative sources to validate identification data fosters confidence between organizations and individuals. When individuals provide identity information from recognized and authoritative sources, organizations gain trust in the data's accuracy and validity. This trust promotes more efficient and easier identity verification procedures, minimizing the need for manual intervention or further examination.
- **Strengthening anti-fraud measures**: Identity theft is a major threat to a variety of businesses. Organizations can improve their anti-fraud procedures by depending on reliable sources of information. Comparing supplied identification data to authoritative sources aids in the detection of fake or altered data, lowering the danger of identity theft, impersonation, or unauthorized access.

Enabling effective identity management

In the complicated network of modern identity management, the notion of sources of authority is critical in laying the groundwork for trust and accuracy. This section explores the multifaceted terrain of allowing successful identity management by examining the sources of identity information. From government-issued credentials to trusted organizational databases, the investigation encompasses a wide range of authoritative repositories that serve as the foundation for verifying and maintaining personal identities. This section describes some of the complexities of these sources of authority, uncovering their vital role in building dependable, secure, and efficient identity management methods:

- **Reliable user onboarding**: Sources of truth help organizations to correctly confirm people's identities throughout user onboarding procedures. Organizations may validate the legality of the identity given by cross-referencing it against trustworthy sources, ensuring that only authorized persons have access to systems and resources.
- **Streamlining identity verification**: The use of sources of truth simplifies the identification verification procedure. Organizations can use automated methods to access and validate data from reputable sources rather than relying on manual checks or subjective judgments. During identification verification, this automation increases efficiency, eliminates mistakes, and improves the entire user experience.
- **Compliance with regulatory requirements**: Many sectors are governed by regulations governing identity verification, data privacy, and data protection. Organizations can show conformity with these laws by depending on recognized sources of truth. Using official databases or government-issued identity papers as sources of truth helps to guarantee that legal and regulatory duties are met.

Finally, this section has highlighted the vital role of sources of authority in contemporary identity management, emphasizing their relevance in providing reliable user onboarding, expediting identity verification procedures, and regulatory compliance. Organizations may create safe and efficient identity management systems that promote confidence and accuracy in today's complicated network of identity verification by using government-issued credentials and trusted corporate databases.

Challenges and considerations

In the complicated tapestry of identity management, using sources of truth presents both possibilities and difficulties that must be carefully considered. This section delves further into the issues and concerns that come with capturing correct and trustworthy identification information. From negotiating difficulties of data quality and interoperability to dealing with emerging privacy concerns, this discussion will shed light on the complexities that companies confront when depending on sources of truth to manage identities. The following are some of the obstacles that one may encounter when seeking to investigate solutions for resolving concerns in the dynamic area of identity information management:

- **Data privacy and security**: While sources of truth are important for confirming identification, they may contain sensitive personal information. To secure the data acquired from various sources, organizations must establish stringent privacy and security safeguards. Encryption, access restrictions, data reduction, and compliance with data protection standards are all part of this.
- **Data sharing and interoperability**: Accessing and integrating data from several sources of truth might be difficult in some situations owing to technological or legal restrictions. To successfully exploit various sources of truth, organizations must develop safe data-sharing agreements, adhere to data protection requirements, and enable interoperability across diverse systems and sources.

Finally, sources of truth are critical for properly and reliably confirming identification data. Organizations may increase data integrity, improve identity assurance, and speed up identity verification procedures by relying on authoritative and reliable references. Data accuracy, consistency, and trust between organizations and people are critical components of good identity management. To effectively reap the benefits of sources of truth, organizations must handle issues such as privacy, security, data sharing, and interoperability. Organizations may successfully use sources of truth and boost their identity verification procedures in the digital age by following best practices and complying with legal regulations.

Building on our previous discussion of increasing identity assurance levels, we now look at a notion strongly anchored in decentralized trust: the web of trust. In the next section, we will look at how the web of trust model develops trust connections between users while providing a decentralized method for identity verification and authentication.

Web of trust

Before we can grasp the web of trust and the trust anchors, we must first comprehend the concept of self-sovereign identity. Self-sovereign identification employs a number of somewhat sophisticated levels of cryptographic ceremonies to return complete data ownership to the user while ensuring online interoperability (trust). In other words, you may control the personal information on your local devices, such as your online activity, medical information, financial data, and so on, but share it with third-party organizations while retaining validity and trust through the use of cryptographic signatures. Self-sovereign identification denotes complete control and authority over one's digital identity. Users may have full control over their private information using decentralized systems and carefully orchestrated cryptographic ceremonies, maintaining data authenticity and integrity as it is safely exchanged across a network.

The **web of trust** (**WoT**) is a revolutionary concept that arose with the advent of blockchain technology. A **trust anchor** is defined by NIST as follows:

> *A public or symmetric key that is trusted because it is directly built into hardware or software, or securely provisioned via out-of-band means, rather than because it is vouched for by another trusted entity (e.g. in a public key certificate).*

It is a critical challenge: who to completely believe in the absence of evidence or references to support the machine's statements?

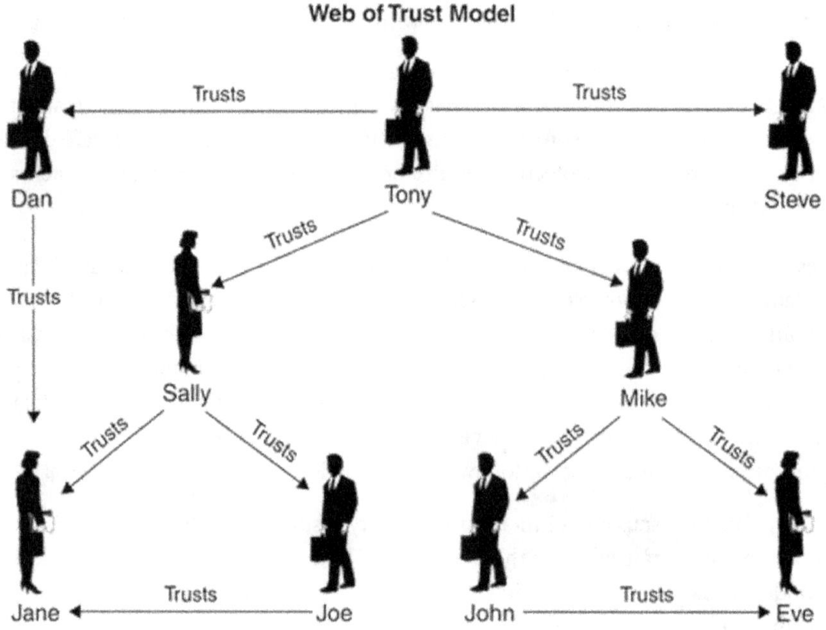

Figure 4.1 – The WoT ecosystem

To determine which machine identities to trust, current PKI procedures may be used to confirm the identity. In the case when the updated certificate must be provisioned, you may be reasonably certain that the new certificate is sent to the device that is qualified to accept it.

Establishing trust is critical for secure transactions and interactions in the field of identity data verification. The WoT paradigm, which is commonly used in **public key infrastructures** (**PKIs**) and identity management systems, is important in establishing trust connections. The WoT paradigm and its link with trust anchors in identity data verification are discussed in this chapter. We will look at the notion of trust anchors, their function in creating trust, and the benefits and drawbacks of this paradigm.

Understanding the WoT model

In the interconnected world of the WoT, the WoT model serves as a critical foundation for determining how devices interact and collaborate. This section begins an investigation into the complexities of the WoT paradigm, diving into its fundamental concepts and the orchestration of devices in a harmonic and compatible environment. Understanding the WoT model is critical for navigating the **Internet of Things** (**IoT**) environment and maximizing the potential of linked devices, which interact, share data, and perform activities using a standardized and globally recognized paradigm.

As we explore through the heart of the WoT model, we uncover the architecture that enables the seamless integration of various devices and services. The WoT paradigm establishes a standardized way to define and interact with IoT devices, facilitating cross-domain compatibility. We analyze the levels of abstraction and processes that help developers and systems understand and interact with the broad WoT world, ranging from particular description needs to the scripting capabilities of the WoT Scripting API. This section seeks to clarify the complexities of the WoT paradigm, allowing practitioners and enthusiasts to confidently and efficiently traverse the changing environment of connected devices:

- **Definition and principles**: The WoT paradigm is a distributed way to build trust in digital contexts. The WoT paradigm emphasizes peer-to-peer trust interactions, as opposed to hierarchical trust models that rely on centralized authority. Individuals in the network vouch for one another's identities, resulting in the formation of a *web* of interconnected trust connections.
- **Trust metrics and reputation**: Individuals are assigned trust metrics or reputation ratings under the WoT model based on their track record of trustworthiness. In most cases, these indicators are decided by criteria such as the amount and quality of endorsements obtained, previous transactions, and community comments. Higher trust metrics suggest a better reputation and, as a result, a greater degree of trust.

Trust anchors in identity data verification

In the context of identity data verification, the notion of trust anchors is critical for creating confidence and authenticity. From centralized trust anchors, which are frequently rooted in authoritative organizations such as government agencies, to decentralized trust anchors scattered over a network, the investigation explores the landscape of trustworthiness in validating individual identities. Trust anchors include a variety of organizations, ranging from official IDs to trusted persons, and are adapted to the unique requirements of identity verification procedures. Trust anchors, which are key to the design of the WoT, serve as the foundation for building trust and confirming identity data integrity, explained as follows:

- **Definition and role**: Trust anchors, also known as key anchors or identity anchors, are entities that are widely recognized and trusted within the WoT concept. They are trustworthy sources for confirming the validity and integrity of identification data. Trust anchors lay the groundwork for further trust connections on which identity claims may be built.

- **Types of trust anchors**: Depending on the context of identity data verification, trust anchors might take numerous shapes. Government-issued identification cards, notaries, well-known organizations, trustworthy certifying authorities, or people with high trust metrics within the WoT model can all be considered. The trust anchors used are determined by the unique demands and requirements of the identity verification procedure.

Advantages of the WoT model and trust anchors

Exploring the intricate domain of the WoT, let's look at the numerous benefits provided by the WoT model, particularly with regard to the idea of trust anchors. The WoT model provides a standardized framework for smooth device interaction and interoperability, as well as a uniform language for devices to communicate inside the IoT ecosystem. When combined with the strategic deployment of trust anchors, this combination improves the dependability and security of data exchanges, guaranteeing that information passing via networked devices is verified and trustworthy. In this section, we investigate the interdependent relationship between the WoT model and trust anchors, revealing how this collaboration helps to construct a strong and secure basis for the future of networked devices:

- **Decentralized trust establishment**: The WoT approach allows for decentralized trust formation, giving individuals more discretion over whom they trust. This technique is especially useful in situations when centralized authorities are not readily available or trusted. Individuals can verify identities based on trustworthy sources rather than relying entirely on centralized institutions by relying on trust anchors.
- **Flexibility and scalability**: In terms of identity data verification, the WoT paradigm provides flexibility and scalability. Trust anchors can be added or deleted in response to changing trust dynamics and requirements. Because of this versatility, the model may handle a wide range of trust connections, broadening the breadth and reach of trust in identity verification procedures.
- **Empowerment and privacy**: Individuals are empowered by the WoT approach because they can actively engage in confirming identities and developing trust. Individuals have the ability to pick which trust anchors to rely on based on their own criteria and judgment. Furthermore, the WoT architecture allows for privacy-conscious individuals who choose peer-to-peer trust connections over providing personal information to centralized authority.

Challenges and considerations

The WoT paradigm was intended to address the inherent obstacles and critical considerations connected with this complex network of interactions. As the WoT model aims to generate trust in a decentralized fashion, it faces complex difficulties such as maintaining the dependability of trust assertions and reducing the possibility of trust manipulation. Interactions within the WoT include subjective assessments, which complicate subjective trust evaluation and highlight the ever-changing nature of trust connections. This discussion digs into the delicate balance necessary to maintain a durable and trustworthy WoT model while managing the hurdles given by differing opinions, trust erosion, and the dynamic nature of online interactions:

- **Trustworthiness of trust anchors**: The reliability of trust anchors is an important aspect of the WoT paradigm. The trust placed in identification data validated by these anchors is directly influenced by their reputation and dependability. To preserve the integrity of the entire trust network, it is critical to assess the trustworthiness of trust anchors using rigorous evaluations, reputation systems, and community input.
- **Vulnerability to Sybil attacks**: Sybil attacks, in which malevolent actors assume several identities to manipulate trust connections, make the WoT concept susceptible. This hazard jeopardizes the model's integrity and dependability. Implementing measures to identify and counteract Sybil assaults may include integrating reputation systems, implementing multi-factor verification, and using community validation. Such proactive measures are critical to improving the security and trustworthiness of the WoT paradigm.
- **Adoption and standardization**: The WoT paradigm has issues due to widespread adoption and standardization. Building trust across groups, domains, and situations necessitates a common understanding and acceptance of trust anchors. For smooth and trustworthy identity data verification, standardization initiatives, collaboration, and interoperability across different WoT implementations are required.

Future trends and innovations

As we traverse the present environment of the WoT model, we must look to the future and investigate the trends and innovations that will influence this dense network of interactions. From improvements in decentralized technology such as blockchains to the incorporation of artificial intelligence for more nuanced trust ratings, the future of the WoT model promises to be dynamic and transformational. This section looks into the crystal ball of innovation, revealing future developments that will transform trust in the digital age and catapult the WoT model to new heights of efficacy and resilience:

- **Blockchain technology and distributed ledger systems**: Blockchain technologies and distributed ledger systems have the potential to improve the WoT concept. These technologies enable the recording of trust relationships, endorsements, and identity claims in a tamper-resistant and transparent manner. They provide enhanced security, auditability, and attack resistance, further enhancing the trust infrastructure inside the WoT architecture.
- **Machine learning and artificial intelligence**: Machine learning and artificial intelligence approaches have the potential to improve the accuracy and reliability of trust indicators in the WoT paradigm. These systems can identify trustworthy trust anchors and detect suspicious activity by analyzing patterns, behaviors, and reputation data. Incorporating machine learning techniques can help to make trust formation procedures more resilient and flexible.

The WoT concept, with trust anchors at its heart, transforms identity data verification by moving away from centralized trust authority and towards a decentralized peer-to-peer trust network. Individuals may build trust connections based on recognized and trustworthy sources thanks to trust anchors, which provide a reliable framework for identification verification. While there are obstacles, such as assuring trustworthiness and dealing with Sybil attacks, the WoT concept has advantages in terms of flexibility, privacy, and scalability. As blockchain technology and machine intelligence advance, they have the potential to improve the WoT paradigm, empower individuals, and build trust in identity data verification procedures.

Blockchains are changing the way individuals and organizations manage their personal data. By design, it promotes more responsibility and enforces a higher level of privacy. The data flow is governed by distributed ledger technology and is not centralized. Building a data marketplace creates new revenue opportunities by enforcing strict access controls using blockchains and a trustworthy data flow. The customer will be the largest winner, since they will get more control over, and even return for, their data, fostering confidence among all parties.

Bitcoin and Ethereum are described as public, permissionless blockchains with three characteristics that serve as the foundation of trust. To begin, anybody can participate by giving computing resources—no prior affiliation to any other node in the system is required. Second, adding a new block to the blockchain is prohibitively expensive computationally because the consensus method is designed to take a fixed amount of wall-clock time to complete regardless of network size. Finally, it is impossible to predict who will be the first to add the following block. The primary benefit of blockchains is the ability to construct a history of transactions that is extremely difficult to modify.

After examining the ideas and functionality of the WoT architecture, we now turn our attention to its augmentation using blockchain technology. In the next section, we will look at how blockchain technology improves the WoT paradigm by providing greater security, transparency, and decentralization in identity verification and authentication procedures.

Enhancing the WoT model through blockchain infrastructure

The WoT paradigm has evolved as a novel technique for establishing trust and security in distributed systems. It is based on a network of trusted entities vouching for others' authenticity and integrity. The WoT concept, on the other hand, poses issues in terms of scale, tamper resistance, and transparency. This section investigates how using blockchain infrastructure might overcome these issues and improve the WoT paradigm. We investigate how blockchain technology provides value to the WoT architecture and highlight its major properties. Furthermore, we present real-world use cases that show how blockchains may be used to improve trust and security within the WoT.

In this section, we will look at how blockchain technology can be integrated into the WoT architecture. We will begin by outlining the WoT paradigm, its principles, and the issues it confronts. Then, we'll go through the foundations of blockchain technology, such as its properties, network kinds, consensus methods, and smart contracts.

Following that, we'll look at how blockchain technology may improve the WoT paradigm by using its major characteristics. We'll talk about how decentralization and distributed consensus enable a trustless network in which members may validate each other's transactions. Blockchain data storage's immutability and tamper resistance will be investigated, demonstrating how it protects the integrity and security of trust-related information. In addition, we will investigate blockchain's transparency and auditing features, which give a verifiable record of trust interactions.

We will provide real-world use cases that highlight how blockchain improves trust and security within the WoT concept to give practical insights. Applications such as supply chain management, identity verification, voting systems, intellectual property rights, and decentralized social networks will be discussed.

While blockchain technology has tremendous advantages, we will also discuss the difficulties and issues involved in integrating it with the WoT concept. Scalability and throughput difficulties, privacy and confidentiality concerns, governance and consensus methods, interoperability obstacles, and energy consumption are among them.

Finally, we will summarize the important findings and suggest potential future routes for incorporating blockchains into the WoT architecture. We will emphasize the importance of blockchains in improving trust, security, and dependability in decentralized systems.

Building trust and security inside decentralized systems has always been a difficulty. The WoT concept arose as an alternate method to addressing these difficulties, depending on a network of trustworthy institutions to vouch for others' legitimacy and integrity. This concept has gained popularity because of its ability to construct a decentralized and trust-based ecosystem. The WoT concept, on the other hand, has limits in terms of scalability, tamper resistance, and transparency. The integration of blockchain infrastructure has been offered as a possible method to solve these issues. Blockchain technology, recognized for its decentralized and immutable nature, has the ability to improve the WoT paradigm and offer a strong foundation for trust and security in decentralized systems.

The major goal of this chapter is to investigate how blockchain infrastructure integration provides value to the WoT architecture. We hope to explain how blockchain technology overcomes the shortcomings of the old WoT paradigm by analyzing its fundamental traits and advantages. In addition, we will look at real-world examples that demonstrate how blockchains may be used to improve trust and security within the WoT.

The WoT model in the decentralized space

This section describes the WoT paradigm, including its concepts and components. It covers the old WoT model's difficulties, emphasizing the need for improvements in scalability, tamper resistance, and transparency:

- **Principles and components**: There are important concepts and components that govern the system in the context of a WoT model combined with blockchain technology:
    - The first concept is decentralization, which guarantees that confidence is dispersed among participants rather than centralized in a central authority. Decentralization is facilitated by blockchain technology, which maintains a distributed ledger across several nodes, with each participant having a copy of the whole transaction history. Because of its decentralized structure, trust may be formed without relying on a centralized institution.
    - Another important component for developing trust within the WoT paradigm is transparency. Blockchains are transparent platforms that record all transactions on a public ledger. This openness enables participants to verify and audit trust-related interactions, improving system accountability and integrity. It enables players to judge the reputation and trustworthiness of others based on their transactions' transparent history.
    - Immutability is a core feature of blockchains that ensures the integrity of trust-related data. Once a transaction has been recorded on the blockchain, it is almost impossible to change or tamper with it. This immutability assures the stability and validity of trust-related data, safeguarding it from malicious modification and serving as a dependable source of truth for trust evaluations.
    - Smart contracts, a component of blockchain technology, are important components of the WoT paradigm. Smart contracts are self-executing agreements in which the terms of the agreement are encoded directly into code. They enable participants to engage in an automated and trustless manner. Smart contracts may be used inside the WoT architecture to specify and enforce the rules and circumstances for creating trust. They enable safe and transparent interactions, removing the need for an intermediary and improving the efficiency and dependability of trust-building procedures.

  By adhering to these principles and utilizing the components of blockchains, the WoT model integrated with blockchain infrastructure can establish a decentralized, transparent, and secure ecosystem. These principles and components collectively contribute to building trust and ensuring the integrity of trust-related interactions within the system.

- **Limitations and challenges**: While blockchain technology provides major benefits to the WoT paradigm, it is not without limitations and obstacles. Understanding these limits is critical for incorporating blockchains into the WoT paradigm efficiently. Here are some of the most significant restrictions and challenges:

  - **Scalability**: Scalability difficulties plague blockchain networks, particularly in public and permissionless networks. As the number of participants and transactions grows, so do the consensus methods and data storage needs. This can result in slower transaction processing times and higher expenses. To overcome this issue, sharding and layer-2 protocols are being investigated as scalability options.

  - **Privacy and confidentiality**: While blockchains offer openness, it also faces privacy and confidentiality concerns. Because of a blockchain's transparency, all transactions are accessible to participants. This amount of openness may not be desired or compatible with privacy needs in some instances. To address these problems, solutions such as zero-knowledge proofs and privacy-focused blockchains are being developed.

  - **Governance and consensus processes**: In order to make collective choices and protect the system's security and integrity, blockchain networks require robust governance structures and consensus processes. Because diverse stakeholders may have competing interests, designing and implementing fair and efficient governance models may be difficult. Furthermore, finding a suitable consensus method that matches the WoT model's aims and criteria is critical.

  - **Interoperability**: This refers to the capacity of multiple blockchain networks to communicate and share information in real time. Interoperability is difficult to achieve since separate blockchains may use different protocols, consensus procedures, and data formats. To allow for the smooth integration of blockchain-based WoT systems, interoperability solutions such as cross-chain communication protocols and standardization activities are required.

  - **Energy consumption**: Blockchain networks require a lot of energy, especially those that use proof-of-work consensus processes. This energy use is a source of worry from both an environmental and a cost-efficiency standpoint. This difficulty can be mitigated by developing energy-efficient consensus processes or moving to alternate consensus mechanisms such as proof of stake.

  - **User experience and adoption**: Blockchain technology remains highly complicated for normal consumers, necessitating technical expertise and comprehension. Improving the user experience and making it easier to engage with blockchain-based WoT technologies is critical for wider adoption. User-friendly interfaces, intuitive design, and educational initiatives can all help to overcome usability hurdles.

Addressing these limitations and challenges is crucial to ensuring the successful integration of blockchains into the WoT model. Ongoing research and development efforts are focused on overcoming these obstacles and refining blockchain technology for broader applicability in trust-based decentralized systems.

As we conclude our initial look into the WoT, we find ourselves at the intersection of innovation, resilience, and the dynamic growth of digital interactions. The voyage through this chapter emphasizes the necessity of building trust in a wide landscape of interrelated identities, leaving us with a deep respect for the WoT model's role in defining the future of safe and trustworthy digital interactions.

Blockchain technology

Establishing trust in online settings is critical in a society that is more reliant on digital connections. The WoT paradigm, which relies on decentralized trust networks built by individuals vouching for the authenticity and integrity of others, offers a novel method to address this difficulty. However, when the WoT model grows in size, it runs into difficulties in terms of trust scalability, privacy, and security. This is where blockchain technology comes into play as a critical component of the WoT business.

As an immutable and distributed ledger, the blockchain is a great alternative for improving the WoT's integrity and stability. The WoT architecture can safely store and authenticate trust endorsements by exploiting a blockchain's cryptographic principles and decentralized nature. Each trust endorsement may be cryptographically connected to the identity of the party recommending it, guaranteeing network openness and accountability.

Furthermore, the consensus processes of blockchains ensure that trust data remains consistent across the network, reducing the possibility of manipulation or criminal activity. The blockchain's decentralized structure also eliminates the need for central authority, allowing people to take control of their identities and reputations.

Furthermore, blockchain technology has the potential to overcome privacy problems in the WoT paradigm. Personal information can be retained off-chain while trust endorsement data is kept on the blockchain. Users may keep control over their sensitive data in this manner, lowering the danger of identity theft or privacy breaches.

Finally, integrating blockchains into the WoT architecture improves trust and scalability, data integrity, privacy, and security. The WoT model's symbiotic relationship with blockchain technology has the potential to revolutionize how we develop and retain trust in our digital interactions, enabling a more secure and dependable online environment.

Characteristics of blockchains

A blockchain serves as a basic technology in the WoT architecture, enhancing trust, transparency, and security within the network. The following are the properties of a blockchain under the WoT model:

- **Decentralization**: In the WoT paradigm, a blockchain functions as a decentralized and distributed ledger. Rather than depending on a centralized authority or middleman, trust data is kept and maintained throughout a network of nodes, guaranteeing that no single party has complete control over the system. Because of this decentralized character, the network is more resilient and less prone to manipulation or censorship.

- **Immutable records**: The immutability of a blockchain is one of its most important qualities. Once data is stored on the blockchain, it is very hard to change or erase it. This implies that trust endorsements and identification information are safely and permanently kept on the blockchain in the context of the WoT architecture. As a result, members may depend on the validity and integrity of the trust data with confidence, increasing the network's overall reputation.
- **Cryptographic security**: To safeguard the data held in blockchains, complex cryptographic techniques are used. Each trust endorsement and transaction on the blockchain is cryptographically linked to the identity of the endorsing party, guaranteeing that the network is only accessible to authorized persons. Furthermore, consensus techniques such as **proof of work** (**PoW**) or **proof of stake** (**PoS**) authenticate and secure new blocks, protecting the blockchain against fraudulent or malicious activity. This cryptographic security strengthens the WoT concept, fostering a trustworthy and secure environment in which members may engage and exchange information.

In conclusion, we now have a thorough knowledge of a blockchain's decentralization, immutability, transparency, and security. The distinct combination of these characteristics not only defines the core of blockchain technology but also drives it into a transformational force, set to restructure sectors and reinvent the landscape of trust and transparency in the digital age.

Types of blockchain networks

Certain types of blockchain networks are more suitable for ensuring the appropriate amount of trust, security, and scalability in a WoT architecture. The following are examples of blockchain networks that complement the WoT model:

- **Public blockchains**: Bitcoin and Ethereum are examples of public blockchains that provide a high level of decentralization and openness. They are available for participation by anybody, allowing anyone to join the WoT network without requiring permission. The solid consensus processes and broad user bases of public blockchains improve security and make it difficult for hostile actors to falsify trust data. Furthermore, the openness and transparency of public blockchains encourage faith in trust endorsements, supporting accountability and integrity within the WoT paradigm.
- **Permissioned blockchains**: Permissioned blockchains, also known as private or consortium blockchains, limit participation to a small number of verified and recognized companies. Permissioned blockchains can be useful in the context of a WoT paradigm since they offer regulated access and allow organizations or communities with predetermined trust connections to communicate safely. This configuration is especially effective for applications requiring a small number of participants, such as enterprise-level identity management systems or government-related trust networks.
- **Hybrid blockchains**: Hybrid blockchains incorporate the best characteristics of both public and permissioned blockchains. They provide flexibility and customization by allowing for different degrees of data visibility and engagement. A hybrid blockchain under a WoT paradigm might allow a core network of trustworthy people or institutions to keep sensitive trust data secret while distributing verifiable endorsements on a public blockchain for broader validation and reputation development.

- **Layer-2 solutions**: Layer-2 solutions developed on top of current blockchains can help improve scalability and lower transaction costs within a WoT paradigm. State channels and sidechains, for example, process transactions off-chain or more efficiently before anchoring the final result back to the main blockchain. The WoT concept may accept a greater number of trust endorsements and interactions by utilizing layer-2 solutions without congesting the underlying blockchain.

Finally, the blockchain network type for a WoT model is determined by the unique use case, the intended amount of decentralization, and the members' scalability and privacy needs.

Consensus mechanisms

Consensus methods are critical in blockchain networks built for a WoT infrastructure because they ensure that participants agree on the state of the blockchain as well as the authenticity of transactions and trust endorsements. Here are some common consensus processes seen in such blockchains:

- **Proof of trust (PoT)**: The PoT consensus process can be used in a WoT paradigm when trust endorsements are critical. Participants in PoT vouch for the authenticity and credibility of others, hence certifying their trustworthiness within the network. The degree of influence each person has in the consensus process is dependent on the trust they have earned through time. This consensus method is well-suited for WoT blockchains since it is consistent with the model's primary notion of utilizing trust to validate transactions and maintain the blockchain's integrity.
- **Proof of authority (PoA)**: PoA is another viable consensus technique for WoT blockchains, especially in private or consortium situations. Validators in PoA are well-known and reliable entities, generally pre-approved by network administrators. These validators produce blocks in turn, and their authority is determined by their reputation and identity inside the WoT network. PoA is more efficient and less energy-intensive than PoW, making it a good choice for WoT blockchains where trust ties between participants are already formed.
- **Delegated proof of stake (DPoS)**: DPoS is a consensus process in which WoT network stakeholders elect a group of delegates in charge of block production and validation. These delegates are chosen based on their network reputation and trustworthiness. DPoS enables quicker block confirmation speeds and scalability, making it ideal for WoT blockchains with a high volume of trust endorsements and interactions.
- **Proof of reputation (PoR)**: PoR is a consensus technique that uses participants' reputations to assess their impact on block validation and decision-making processes. Participants' reputations are formed on their activities and contributions to the network, such as trust endorsements and prior transaction accuracy. PoR can be a useful strategy on a WoT blockchain, where trust is essential, to guarantee that credible and respectable players have larger voices in the consensus process. The consensus process used for a blockchain created for WoT infrastructure architecture is determined by criteria such as the required amount of decentralization, the number of participants, scalability requirements, and the desired level of trust validation within the network.

Smart contracts

Smart contracts are used in a WoT model network to automate and enforce trust-related agreements and interactions between users. Smart contracts are self-executing programs that operate on a blockchain and execute predetermined actions automatically when specific criteria are satisfied. They have a substantial impact on the functionality and security of a WoT network. In a WoT paradigm, smart contracts are employed as follows:

- **Trust endorsements**: Smart contracts make the process of trust endorsements in the WoT network easier. A trust endorsement is recorded as a transaction on the blockchain when one person vouches for the trustworthiness of another. Smart contracts may be programmed to check the legitimacy of these endorsements and guarantee that the party making the endorsement has the power to do so. This automation aids in the preservation of trust data and the prevention of fake endorsements.
- **Reputation management**: Smart contracts are used to compute and maintain the reputations of WoT network participants. Each trust recommendation adds to a participant's reputation score, which represents their trustworthiness. Smart contracts may be programmed to automatically update and compute these reputation ratings depending on the frequency and accuracy of trust endorsements. Participants with greater reputation ratings are more likely to be trusted by others, boosting the WoT model's overall efficacy.
- **Dispute resolution**: Smart contracts may be built to manage dispute resolution in the case of a disagreement between parties or contradictory trust endorsements. They can use preset rules and processes to settle differences and restore network confidence. This enables a fair and open method for addressing disputes, preserving the WoT model's trust.
- **Trust-based transactions**: Smart contracts enable safe and trust-based transactions between WoT network participants. When a participant interacts with another party, the smart contract can validate both parties' trustworthiness and only execute the transaction if the needed degree of trust is fulfilled. This lowers the likelihood of fraudulent or harmful activity, creating a more dependable and secure environment for performing transactions.

A WoT model network can use smart contracts to automate trust-related activities, keep correct reputation records, and enforce trust agreements in a transparent and efficient manner. This automation improves the network's trustworthiness and dependability, providing participants with a safe and accountable environment for their interactions.

Integrating blockchain infrastructure into the WoT model

By integrating blockchain infrastructure into the WoT concept, we reach a world where decentralized and secure technologies combine to revolutionize digital interactions. Blockchain's intrinsic properties of immutability, transparency, and decentralized consensus have the potential to improve the stability of trust relationships in the WoT paradigm. As we work through this integration, the synergy between blockchain and WoT emerges, providing a strong foundation for establishing confidence in the large and linked web of digital identities.

Going further, the incorporation of blockchain technology represents a paradigm shift in how trust is generated and maintained inside the WoT architecture. By exploiting a blockchain's tamper-resistant characteristics, the integrity of trust claims is strengthened and the possibility of trust manipulation is considerably reduced. This fusion not only overcomes the issues with subjective trust evaluations but also prepares the door for more robust, transparent, and decentralized interactions within the WoT. This section investigates the opportunities created by combining blockchain technology with the fundamental principles of the WoT paradigm:

- **Decentralization and distributed consensus**: The capacity of blockchain technology to facilitate decentralization and distributed consensus is one of its key features. Decentralization guarantees that confidence is dispersed among participants rather than centralized in a central authority under the WoC concept. Blockchain does this by keeping a distributed ledger in which each participant has a copy of the whole transaction history. PoW and PoS consensus procedures guarantee that transactions are confirmed and agreed upon by the network. The WoT paradigm becomes more impervious to assaults by including blockchains, as it takes the agreement of a majority of network participants to violate trust.
- **Immutable and tamper-resistant data storage**: Blockchain data storage's immutability and tamper resistance bring substantial value to the WoT paradigm. Trust-related information in classic WoT systems might be tampered with or manipulated. All trust-related transactions and data are recorded in a decentralized and cryptographically secure way using blockchain. Once a transaction is posted to the blockchain, it is very difficult to change or tamper with it, maintaining the data's integrity and reliability. This element improves the WoT model's dependability since participants may have faith in the authenticity and immutability of trust-related information.
- **Transparency and auditing**: Transparency is an important feature of the WoT paradigm because it allows members to assess the trustworthiness of others. Blockchain technology offers a transparent and auditable platform in which all transactions are recorded and available to all participants. The blockchain's decentralized structure ensures that no single party has control over the transaction history. Participants may validate the integrity and correctness of transactions, increasing network trust. Furthermore, blockchain transparency enables audits and accountability by allowing for the traceability and verification of trust-related interactions.
- **Trustless interactions**: Within the WoT concept, blockchain technology allows trustless interactions. Individual trust must be established between members in conventional systems. However, blockchain-based WoT solutions enable the formation of trust without the need for pre-existing ties. Participants may check the validity and reputation of others using cryptographic methods and smart contracts based on their interactions recorded on the blockchain. Trust may be built by objectively evaluating previous transactions, which eliminates the requirement for a centralized authority to confirm trustworthiness. The WoT model's trustless nature improves its efficiency and scalability.

- **Scalability and performance**: Both the WoT and blockchain systems have significant scaling challenges. Blockchain technology improvements, such as sharding, sidechains, and layer-2 solutions, have the potential to overcome scalability difficulties. Blockchain-based WoT systems can achieve increased transaction throughput and improved speed by utilizing these strategies. Scalability solutions allow for parallel transaction processing, lowering the load on the main blockchain and boosting the overall efficiency of the WoT paradigm.

Decentralization, immutability, transparency, trustless interactions, and scalability are significantly enhanced by integrating blockchain infrastructure into the WoT model, providing a solid foundation for establishing trust and security within decentralized systems. The following section will explore real-world use cases that demonstrate the practical application of blockchains in improving trust and security within the WoT.

Real-world use cases

As we explore real-world environments, we will see how the WoT concept transcends theory and transforms it into practical solutions. The examination of real-world use cases reveals the WoT model's practical applicability, illustrating its effect across a variety of industries. From safe online transactions and decentralized identity management to collaborative ecosystems that build trust, these use cases serve as beacons for the WoT model's promise and agility in tackling the complex difficulties of our digital society. In this section, we explore the concrete manifestations of trust, dependability, and security via the dynamic interaction of the WoT model with real-world events.

Supply chain management

The WoT approach is being used in the supply chain management business to ensure product authenticity and traceability. Counterfeit items offer serious dangers to both customers and producers, resulting in financial losses, reputational harm, and severe health problems. Stakeholders can construct a decentralized trust network by including a WoT model in the supply chain, providing transparent and verifiable product provenance from the source to the end customer.

Each player in the supply chain, such as raw material suppliers, manufacturers, distributors, and retailers, operates as a node in the WoT network in this use case. Trust endorsements are recorded on a blockchain when items pass through various phases of the supply chain, attesting to the authenticity and quality of the product.

The following are key components of this WoT model use case:

- **Trust endorsements**: Participants at each level of the supply chain testify to the validity and provenance of the goods they handle. This might include checking the source of raw materials, the production process, quality inspections, and regulatory compliance.
- **Decentralized ledger**: The blockchain is a decentralized and unchangeable database that records all trust endorsements and transactional data. This provides openness and prevents product information from being tampered with or manipulated.

- **Smart contracts**: Smart contracts are used to automate and enforce trust-related agreements, such as supplier quality standards compliance or distributors ensuring correct product handling and storage.
- **Consumer verification**: Consumers can have access to the WoT information by scanning QR codes on product packaging or via mobile applications. They can validate the product's legitimacy and track its path through the supply chain, giving them more trust in their purchasing decisions.

Companies may develop trust and credibility with consumers, decrease the dangers of counterfeit items, simplify product recalls if necessary, and maintain compliance with industry standards by utilizing the WoT model in supply chain management. Furthermore, the WoT network's decentralized design improves data security and eliminates reliance on centralized agencies for product verification, resulting in a more resilient and efficient supply chain ecosystem.

Identity verification and authentication

A real-world application of the WoT paradigm in the identity verification and authentication sector is the development of a decentralized and dependable identification system. Traditional identity verification systems frequently rely on centralized databases, which are prone to data breaches and privacy problems. Individuals may have more control over their identities while still guaranteeing that their information is safely vetted and validated by deploying a WoT architecture.

In this application, the WoT model serves as a trusted network in which individuals attest to the legitimacy of one another's identities. Each user keeps track of their own identification data and can seek trust endorsements from other users they trust. The more trust recommendations a person obtains from credible sources, the better their reputation score rises, increasing their total identity credibility within the network.

The following are key components of this WoT model use case:

- **Self-sovereign identity**: Individuals have self-sovereign control over their identification data under the WoT approach. They have the ability to choose which information to share with others while maintaining control of their personal data, lowering the danger of data abuse or breaches.
- **Trust endorsements**: Users in the network can validate the identities of others by endorsing them with a series of verifiable attestations. These trust endorsements are stored on a blockchain, which ensures transparency and immutability.
- **Reputation-based authentication**: Reputation ratings, which are based on the amount and quality of trust endorsements, are becoming increasingly important in validating an individual's identity. This reputation score may be used by service providers and reliant parties to assess the amount of confidence they place in a user.
- **Multi-factor authentication** (**MFA**): To provide an extra degree of protection, the WoT concept may be used with MFA mechanisms. For example, in order for a user to access sensitive information or services, a service provider may demand a combination of a reputation score, biometric data, and a **one-time password** (**OTP**).

Individuals may benefit from a more secure and privacy-focused identity management solution by implementing the WoT concept in the identity verification and authentication business. It eliminates dependency on centralized databases, increases transparency, and enables individuals to govern their own identities while allowing service providers to make educated and trust-based decisions when authenticating users. This use case has the potential to revolutionize the identity verification environment by providing a more decentralized, safe, and user-centric approach to identity management.

Voting systems

A real-world application of the WoT paradigm in voting systems in the public sector is for improving election security, transparency, and integrity. Voter fraud, coercion, and a lack of faith in the election process are all common problems for traditional voting systems. Governments may develop a decentralized and verifiable system that guarantees every vote is validated and recorded properly by employing a WoT model in voting.

In this application, the WoT architecture serves as a trusted network for voters, election officials, and other stakeholders. Each voter keeps track of their own identification and trust data, and they may vouch for the legitimacy of other voters they know or trust. Trust endorsements are stored on a blockchain or distributed ledger, which ensures transparency and prevents unauthorized access to or manipulation of voting data.

The following are key components of this WoT model use case:

- **Voter authentication**: Voter authentication may be increased by utilizing the WoT concept by using trust endorsements from other trusted voters. The amount and quality of trust endorsements received by a voter can be used to establish their eligibility and degree of trust within the voting network.
- **Verifiable voting records**: Each vote is recorded on the blockchain, which makes it tamper-resistant and transparent. Voters may independently confirm that their vote was accurately counted, increasing faith in the democratic process.
- **Election oversight**: To promote fairness and openness, election authorities and observers can join the WoT network. They may authenticate and audit trust endorsements, adding another level of scrutiny to the voting process.
- **Secure and decentralized system**: The WoT model's decentralized structure ensures that no single entity has complete influence over the voting process. This decreases the possibility of election result hacking, fraud, or manipulation.

Governments may establish a more secure, transparent, and trust-based election process by incorporating the WoT paradigm into voting systems. It boosts voter trust, minimizes the possibility of election fraud, and develops a democratic climate in which individuals actively engage and certify the voting system's integrity. This use case has the potential to transform how public elections are conducted, encouraging fair and accountable democratic practices.

Intellectual property rights

A real-world application of the WoT paradigm in the **intellectual property rights** (**IPR**) management business is the creation of a decentralized and transparent platform for creative work copyright and ownership authentication. In businesses such as music, literature, art, and software, correct attribution and intellectual property protection are critical. The WoT paradigm may be used to create a trust network among authors, distributors, and consumers, allowing for safe and verifiable IPR management.

Each creator or copyright holder keeps their own identity and trust data in this use case, and they may obtain trust endorsements from other credible institutions such as recognized publishers, copyright agencies, or other artists. These endorsements serve as verifiable attestations to the creative works' ownership and validity, and they are maintained on a blockchain or distributed ledger, assuring transparency and immutability.

The following are key components of this WoT model use case:

- **Copyright validation**: Creators can use the WoT approach to demonstrate the legitimacy and ownership of their creative property. Trusted entity endorsements lend legitimacy to their copyright claims, lowering the danger of copying and unauthorized use.
- **Royalty tracking and payment**: The WoT paradigm allows for the creation of transparent and automated royalties monitoring and payment systems. Smart contracts may be used to enforce licensing agreements and deliver royalties to copyright holders based on how their works are utilized.
- **IP dispute resolution**: In the event of a disagreement or ownership claim, the WoT paradigm can support a decentralized dispute resolution procedure. Verified trust endorsements can be used as evidence in settling disputes and confirming intellectual property ownership.
- **Consumer trust**: Before acquiring or employing creative works, consumers may use the WoT information to verify their legitimacy and ownership. This aids in the development of customer trust and confidence in the authenticity of the items they consume.

The industry may establish a more efficient, safe, and egalitarian environment for creators, publishers, and consumers by applying the WoT paradigm in IPR management. It guarantees adequate intellectual property rights attribution and protection while providing a decentralized and trust-based method for maintaining and certifying ownership of creative works. This use case has the potential to transform how intellectual property is managed, creating a more equitable and accountable environment for both producers and consumers.

Decentralized social networks

A practical application of the WoT paradigm in decentralized social networks is the creation of a trust-based and privacy-focused platform for social interactions. Traditional centralized social media networks frequently experience problems with data privacy, content control, and algorithmic biases. Users may have more control over their data and create trust connections with others in the network by implementing the WoT paradigm in decentralized social networks.

Each member in the decentralized social network retains their own identification and reputation data in this use case. They may attest to the authenticity and integrity of those they know or trust, endorsing their network connections. Trust endorsements are recorded on a blockchain or distributed ledger, guaranteeing transparency and prohibiting centralized institutions from manipulating data.

The following are key components of this WoT model use case:

- **Privacy and data control**: Users have self-determination over their personal data and may select which information to share with others. This decreases the possibility of data breaches and unauthorized data mining by centralized systems.
- **Content moderation**: Rather than depending exclusively on centralized algorithms to monitor information, the WoC approach empowers people to control their own feeds based on their confidence in other users. Users may choose who they want to follow and what material they want to see first, resulting in a more personalized and relevant social media experience.
- **Trust-based interactions**: Users may discover and communicate with reliable and trustworthy folks thanks to trust endorsements. This aids in the reduction of disinformation and promotes polite and meaningful dialogues inside the social network.
- **Community governance**: Community-driven governance methods can be implemented in decentralized social networks based on the WoT architecture. Users may vote on platform rules, content regulations, and feature additions collectively, resulting in a more democratic and inclusive social media network.

The industry can solve privacy issues, battle disinformation, and build a more authentic and engaging social media experience by embracing the WoT concept in decentralized social networks. It enables users to take control of their data and activities, resulting in a more transparent, user-centric, and socially responsible online social interaction platform. This use case has the ability to transform the social media environment and pave the way for a more decentralized and trust-based online community.

Finally, the sources of truth are critical to the effectiveness and legitimacy of a WoT architecture. Because a decentralized trust network relies on members vouching for the authenticity and integrity of others, it is critical to develop dependable and verifiable sources of truth. The accuracy and trustworthiness of these sources have a direct impact on the model's level of trust endorsements and overall performance. We strengthen the WoT's resilience against harmful actors and disinformation by anchoring trust data to trustworthy and transparent sources. Furthermore, ensuring that these sources are well-maintained and constantly updated ensures the trust network's endurance and relevance, establishing a safe and trustworthy environment in which members can create meaningful connections and make informed decisions.

Recognizing and validating the sources of truth will remain an essential foundation for the WoT model's widespread adoption and successful implementation in various domains, revolutionizing how trust is established and maintained in our digital interactions.

Summary

The chapter focused on the crucial role of sources of truth in identity data verification, highlighting their relevance in maintaining data integrity, improving identity assurance, and allowing effective identity management. It emphasized the importance of authoritative and credible references in establishing and sustaining trust in an individual's identification, preventing fraud, and assuring safe access to systems and resources. The chapter also discussed the issues and considerations that come with integrating with sources of truth, such as data privacy and security, data sharing and interoperability, scalability and performance, identity data governance, and auditability. It also examined the WoT paradigm and its relationship to trust anchors in identity data verification, highlighting the need for thorough assessments, reputation systems, and community participation in determining the trustworthiness of trust anchors. It also analyzed the possibilities for incorporating blockchain technology to improve the WoT paradigm, namely in terms of decentralization, transparency, and trustless interactions. Overall, the publication emphasized the importance of sources of truth and the WoT paradigm in assuring the reliability and security of identity verification procedures, as well as the problems and potential solutions to successfully applying these principles. In the following chapter, we'll look at the fundamentals of historical authority. We'll trace the development of identity verification from antiquity to the digital era, revealing how historical writings, legislative decrees, and institutional records shaped current authentication and verification procedures.

5

Historical Source of Authority

The internet flung us into the digital world, and over the previous several decades, our identities have become more redundant and splintered with each new service provider and authority. Data associated with our identities, as well as instances of our identities, have proliferated to uncontrollable proportions. Consider your most recent relocation. In how many places did you have to disclose your address? Consider how accepted you would be in the current culture if you denied yourself any sort of identification. As digital forms of identification become more prevalent in our daily lives, it is clear that the current approach to identity and identity proofing is out of step with how individuals trade and interact across diverse physical and digital realms. The required operations for verifying, validating, and managing the identity throughout its life cycle are repetitive and time-consuming, necessitating human data reconciliation and validation activities in the background.

Many of the critical mechanisms were developed for early systems in the late 1960s and 1970s, and they remain the same today: as the concept of digital identity evolved, such primitives remained the building blocks with which new systems should be integrated, and to which they may be either well-suited or not at all.

In this chapter, we will cover the following topics:

- The practical use of and controlling access to sources of authority
- Securing and leveraging access control lists
- Best practices for implementing access control lists

Practical uses of historical sources of authority

Historical sources of authority play an important role in proving the validity and legitimacy of an individual's identity in traditional centralized digital identity verification systems. These sources are historical documents and databases that are created and maintained throughout time by recognized institutions and organizations. Centralized identity verification systems may assure accurate and trustworthy identity identification by using these trusted repositories, offering a secure basis for diverse online services and transactions. This section investigates the practical uses of historical authority sources and their significance in traditional centralized digital identity verification systems.

Historically, several sources of authority have been used to validate identities. Here are a few examples:

- **Government**: Governments have always been a major source of authority for certifying IDs. They are responsible for issuing official identification papers such as passports, driver's licenses, and national identity cards, as well as maintaining official records of citizens and residents.
- **Institutions**: Institutions such as banks, educational institutions, and businesses have also served as authoritative sources for identity verification. They may ask for confirmation of identification as part of the application or enrollment process, and they may have internal databases of confirmed identities.
- **Trusted third parties**: Traditionally, trusted third parties such as notaries, attorneys, and other professionals are used to authenticate identities. A notary, for example, may be required to confirm the identity of someone signing a legal document.
- **Community and social networks**: Community and social networks have played a part in identity verification in several civilizations. A person's reputation within their community or social network, for example, may be used to determine their identity.
- **Personal knowledge**: Personal knowledge has been used to authenticate identification in several circumstances. A bank teller, for example, may recognize a regular customer and be able to confirm their identification based on the personal knowledge of the individual.

These historical authority sources inspired the creation of current identity verification systems, which frequently rely on a combination of government-issued identification documents, institutional databases, and trusted third-party verification services. Emerging technologies such as blockchain and self-sovereign identification, on the other hand, have the potential to revolutionize the way identities are confirmed and certified in the future.

Government-issued documents, such as passports, national identity cards, and driver's licenses, are one of the key sources of authority in identity verification. These documents act as official identification documents and are supported by government institutions in charge of preserving and validating citizen data. Individuals are required to provide physical copies of these papers for verification in conventional systems, and their information is cross-referenced with central government databases to authenticate their legitimacy. This historical source of authority creates a solid relationship between an individual's physical identity and their digital representation, instilling trust in the identity verification process.

Financial organizations, such as banks and credit agencies, play an important role in identity verification as historical sources of authority. Individuals' financial histories, credit ratings, and transaction records are essential in determining creditworthiness and general trustworthiness. Traditional centralized identity verification systems work with financial institutions to validate identification data and financial histories, ensuring that the individual being verified has a consistent and dependable financial track record. This adds another layer of protection and lowers the danger of fraudulent financial transactions.

Educational institutions act as historical authoritative sources for validating academic qualifications and accomplishments. Transcripts, degrees, and certificates from approved institutions and universities provide important information about a person's educational past. Traditional centralized identity verification systems can work with educational institutions to validate the legitimacy of these papers, assuring the validity and trustworthiness of users' educational claims. This is especially crucial when individuals need to demonstrate their qualifications for professional reasons or job applications.

Electricity, water, and telecommunications companies all keep historical records of their customers' accounts and billing information. These firms can act as authorities in verifying a person's residential address and utility consumption history. Traditional centralized identity verification systems can cross-reference utility company data to check the correctness of an individual's reported residential address, hence enhancing the identity verification process.

Traditional centralized digital identity verification systems rely heavily on historical sources of authority to prove the validity and legitimacy of an individual's identity. These systems can assure accurate and reliable identity identification by using data from government-issued papers, financial organizations, educational institutions, and utility businesses. The utilization of historical authority sources improves the credibility of identity verification procedures, allowing for secure access to online services, e-commerce transactions, and digital interactions. However, as technology improves and decentralized identification solutions arise, a balance must be struck between using historical sources of authority and guaranteeing user privacy and data protection. As we continue to investigate the evolution of digital identity verification, combining the best of traditional and novel techniques will be critical to developing safe, user-centric, and trustworthy identity ecosystems.

Now that we've looked at the practical uses of historical sources of authority, we'll move on to the essential topic of regulating access to them. In the following section, we will look at techniques and processes for guaranteeing secure and efficient control of access to the source of authority while maintaining its integrity and dependability.

Controlling access to the source of authority

Controlling access to sources of truth for identity verification is a vital part of guaranteeing the verification process's confidentiality, privacy, and correctness. These sources must be protected against unauthorized access and potential exploitation as historical custodians of sensitive data such as government-issued papers, financial information, educational qualifications, and utility consumption histories. A comprehensive approach to access control entails the use of a variety of techniques and technologies to prevent unauthorized disclosure, manipulation, or theft of personal information. Some of the major tactics and best practices for controlling access to sources of truth for identity verification are as follows:

- **Multi-factor authentication** (**MFA**): Using MFA is an efficient technique to improve access control to sources of truth. The system provides an additional layer of protection by demanding various types of authentications, such as passwords, biometrics (fingerprint or facial recognition), and one-time passcodes. This guarantees that only those with the proper credentials have access to critical identifying data. Even if one of the authentication factors is compromised, MFA considerably minimizes the danger of unauthorized access.
- **Role-based access control** (**RBAC**): RBAC is a popular access control architecture in which permissions are assigned based on predetermined roles and responsibilities. This enables administrators to provide particular permissions to people or groups inside an organization. Employees in a financial institution, for example, may have access to financial records for verification purposes, while others may only have access to non-sensitive information. RBAC guarantees that only authorized individuals have access to sensitive sources of truth, lowering the risk of data breaches or insider threats.
- **Data encryption**: Encrypting identity-related data during storage and transmission offers an extra degree of security. Robust encryption techniques guarantee that even if unauthorized access occurs, the data remains unreadable in the absence of the necessary decryption keys. End-to-end encryption for data transmission ensures that the data is safe throughout its trip from the source to the verification system.
- **Access auditing and monitoring**: Auditing access to sources of truth regularly and monitoring data consumption can help discover and respond to any security breaches quickly. Access logs and activity monitoring can assist administrators in detecting any suspicious actions or unauthorized attempts to access sensitive data, allowing them to take necessary action and tighten security measures.
- **Secure APIs and integrations**: APIs are used by many identity verification systems to gain access to external sources of truth. It is critical to ensure that these APIs are safe and verified properly. Implementing techniques such as **Open Authorization** (**OAuth**) for API access and secure tokenization aids in preventing unauthorized access via API endpoints.
- **Consent management**: Obtaining express user agreement is critical in instances where identity verification entails accessing personal information from third-party sources. Users must be notified about the data being accessed, the reason for which it is being used, and how long it will be kept. Implementing strong permission management practices empowers individuals by giving them greater control over their data and increasing trust in the identity verification process.

- **Secure communication protocols**: It is critical to use strong encryption protocols (for example, SSL/TLS) to protect communication routes between the identity verification system and the sources of truth. Secure communication protects data integrity and confidentiality by preventing eavesdropping and man-in-the-middle attacks.
- **Data minimization**: Adhering to the idea of data reduction aids in limiting access to sources of truth. Only the information that's needed for identity verification should be accessible and retained. Minimizing data collection decreases the potential effect of a data breach and reduces the danger of sensitive information being exposed.
- **Regular security assessments**: Regular security audits, vulnerability testing, and penetration testing are required to uncover any flaws in access control measures. These evaluations assist organizations in proactively addressing security holes and ensuring continual progress in sources of truth protection.
- **Blockchain for data integrity**: In some cases, blockchain technology can improve data integrity and access management. The decentralized and tamper-resistant structure of blockchain means that once data is stored on the blockchain, it cannot be changed or erased without network consensus. Using blockchain to store cryptographic hashes or attestations of identification data can give sources of truth an additional degree of trust and security.

To summarize, restricting access to sources of truth for identity verification is a difficult and multidimensional task. Organizations can establish a robust and trustworthy system for identity verification by implementing a comprehensive approach that includes MFA, RBAC, data encryption, access auditing, secure APIs, consent management, secure communication protocols, data minimization, regular security assessments, and potentially blockchain technology. These safeguards are critical for securing sensitive data, ensuring individual privacy, and fostering confidence among users and stakeholders. Access control systems must adapt and develop as technology evolves to keep ahead of possible security risks and protect the integrity and trustworthiness of sources of truth for identity verification.

While all of the aforementioned control points are crucial, access control is the most important since it allows security rules and procedures to be implemented across all security and control verticals.

Now that we've discussed the significance of managing access to the source of authority, let's look at one of the primary strategies for implementing access control: **access control lists** (**ACLs**). In the next section, we'll look at how ACLs provide granular control over resource access, ensuring that only authorized entities may successfully interact with the source of authority.

ACLs

Controlling access to sensitive information is critical in the realms of identity management and data security. Organizations require strong systems to guarantee that only authorized persons have access to and can interact with sensitive identification data and records. ACLs emerge as a critical answer to this problem. ACLs are a granular and flexible access control technique that allows or denies certain individuals or groups access to resources such as data sources of truth that carry essential identification information.

An ACL is a collection of access control entries, each of which specifies the user or group and the accompanying rights they have on a certain resource. An ACL would establish which person or entity is permitted to read, edit, or query specific identity information in the context of safeguarding access to data sources of truth for identity data and records. Organizations may modify access rights depending on user roles, responsibilities, and the sensitivity of the data involved with this degree of granularity.

Access control entries (**ACEs**) are ACL components that describe the exact rights or limits that are allowed or denied to individuals or groups for a certain resource, such as files, folders, or system objects. ACEs play an important role in deciding who has access to or manipulates certain resources, as well as the amount of access allowed to each individual or organization.

Implementing ACLs to safeguard access to data sources of truth entails two major steps:

- **Defining ACEs**: The first step is to establish the ACL's ACEs. Each ACE is made up of the following parts:
  - **Subject**: The individual or group to whom the ACE applies. This might be a single-user account or a group of users with comparable access needs.
  - **Object**: The ACE regulates access to the resource or data source of truth. This might refer to a specific database, file, or data repository that contains identity records.
  - **Permissions**: The collection of object permissions that are granted to the subject. Read, write, edit, delete, and execute are all common permissions. When it comes to identity data, read access is frequently used to see information, while write access is necessary to make adjustments.
- **Evaluating access control decisions**: Following the definition of the ACL, access control choices are made based on the permissions indicated in the ACEs. When a user seeks to access a data source of truth, the system looks for a corresponding ACE that gives the necessary rights. If an ACE matches, the access request is granted; otherwise, it is rejected.

ACLs have various advantages when it comes to safeguarding access to data sources of truth for identification data and records:

- **Granular control**: Granular control over access permissions is provided by ACLs, which enable organizations to adapt access privileges to individual users or groups depending on their roles and responsibilities. This fine-grained control guarantees that users have just the data they require, reducing the danger of unauthorized access.

- **Flexibility and scalability**: Because ACLs are very adaptable and scalable, they are ideal for regulating access to big databases and complicated identification information. As access needs change over time, organizations may quickly amend and update ACEs.
- **Auditability**: ACLs offer an audit record of access control decisions, allowing organizations to trace and assess access attempts. This auditability improves openness and accountability in managing identification data and records.
- **Data privacy and compliance**: ACLs contribute to data privacy and compliance with data protection rules by limiting access to sensitive identification information. Organizations can show compliance by instilling safe access restrictions using ACLs.

ACLs are critical in ensuring access to data sources of truth for identification data and records. ACLs help organizations impose strict access controls, maintain data privacy, and ensure compliance with data protection standards by establishing granular access control entries and assessing access decisions based on permissions. ACLs are a core tool for identity management systems, offering the required security and control in maintaining vital identity information due to their flexibility and scalability.

Now that we've demonstrated the capabilities of ACLs for restricting resource access; let's look at how they might be used to secure access to sources of truth. In the subsequent section, we'll look at how ACLs may help enforce strict access controls, protecting the integrity and confidentiality of crucial sources of truth.

Securing access to sources of truth with ACLs

Data is a precious asset in the digital age that must be protected against unauthorized access and possible exploitation. Access to data sources must be secured at all times, especially when dealing with sensitive information such as personal information, financial records, and proprietary knowledge. ACLs provide a strong and versatile tool for successfully controlling and managing access privileges to data sources. This section delves into the ideas and best practices for utilizing ACLs to safeguard data sources, ensuring that only authorized users have access to the information they require while guarding against unauthorized or malicious activity.

ACLs are granular and adaptable access control mechanisms that restrict permissions to resources such as files, databases, and directories. An ACL is made up of a list of ACEs, each of which defines the access permissions for a certain user or group on a specific resource. ACEs generally contain data on the subject (user or group), the object (resource), and the precise rights to be granted or denied. ACLs go beyond normal file or resource permissions to provide an additional layer of access control, allowing organizations to modify access privileges depending on user roles and responsibilities.

Implementing ACLs for data sources

ACLs for data sources are required for a variety of reasons. ACLs aid in enforcing security measures by indicating which people or organizations are permitted to access or alter specified data sources.

This guarantees that only authorized users may interact with critical information, lowering the risk of unauthorized access and data breaches. You need to keep the following in mind when implementing ACLs:

- **Defining ACEs**: ACEs must be properly specified in the context of safeguarding access to data sources to represent the appropriate level of access for individuals or groups. For example, an ACE may offer a certain person read-only access to a financial database, while another ACE may grant write access to a group of data analysts. Understanding the access needs of different users, as well as the sensitivity of the data they need to access, is required to define these ACEs.
- **Access control decisions**: The system checks the ACEs in the associated ACL when a user seeks to access a data source. Access is granted if a matching ACE is located that gives the relevant permissions. If no matching ACE is found or if the user's access permissions are expressly prohibited, the access request is refused. This procedure guarantees that users only receive access to data sources based on the access privileges they have been granted, lowering the danger of unauthorized data disclosure.
- **Managing ACLs**: ACL management becomes increasingly important as organizations evolve and access requirements vary. Administrators must evaluate and update ACLs regularly to ensure that they comply with the organization's security policies and data access requirements. Implementing correct version control and access control governance mechanisms can help with ACL management.

In summary, ACLs are an important part of data security and access management because they provide a systematic and organized way to limit, monitor, and safeguard access to data sources.

Advantages of ACLs for securing data sources

In the ever-changing world of data security, ACLs play an important role by providing a standardized way to control access rights for data sources. ACLs have several benefits, including protecting sensitive information, maintaining compliance, and strengthening organizational defenses against unwanted access and data breaches. Let's look at some of the advantages:

- **Granular control**: ACLs provide organizations granular control over access permissions, allowing them to set access privileges down to individual individuals or groups. This degree of detail guarantees that users only see the information they need, reducing the risk of data breaches.
- **Flexibility and scalability**: ACLs are very adaptable and scalable, making them ideal for controlling access to large databases and varied user groups. To react to changing access needs, organizations may quickly adjust and update ACEs.
- **Auditability and accountability**: ACLs enable organizations to track and assess access attempts by providing an audit trail of access control decisions. This auditability improves data access management transparency and accountability.
- **Data privacy and compliance**: ACLs contribute to data privacy and compliance with data protection rules by restricting access to sensitive data sources. By adopting strong access restrictions using ACLs, organizations may show compliance with data privacy requirements.

In conclusion, the adoption of ACLs emerges as an essential method that offers not only a strong defense against unwanted access but also supports compliance, privacy, and effective resource management in the ever-changing data security context. The numerous benefits of ACLs contribute to a robust and regulated data environment, which is critical for the integrity and confidentiality of modern companies.

Now that we've discussed the importance of ACLs in ensuring access to sources of truth, let's focus on the recommended practices for increasing their efficacy. In the next section, we will look at key tactics and principles for deploying ACLs to provide strong security measures surrounding data sources, hence improving overall data safety and integrity.

Best practices for securing data sources with ACLs

There are various best practices to consider when you're securing data sources with ACLs. First, you should apply the concept of least privilege, allowing users just the access required to undertake their responsibilities. This reduces the risk of data disclosure as you can avoid offering unneeded or excessive access privileges.

You should also review and update ACLs regularly to ensure they are in line with the organization's developing security policies and access requirements. Also, implement continuous ACL management and version control mechanisms.

Another thing to consider is enforcing the separation of tasks by requiring several levels of authorization for key actions such as giving or changing access privileges. This decreases the possibility of access rights being misused or abused.

To make access control administration easier, use groups. Assign permissions to groups rather than individual individuals to make managing access privileges for several users in similar roles easier.

ACLs are a useful tool for safeguarding data sources and preventing unauthorized access to sensitive information. Organizations may impose strict access controls and secure their precious data assets by designing granular access control entries and executing frequent reviews and changes. ACLs improve data privacy, ensure compliance with data protection standards, and add to an organization's data management practices' overall security posture. Organizations may develop a strong and trustworthy data access management system by following best practices and exploiting the flexibility and scalability of ACLs.

ACLs and capabilities are the two most prevalent operating-system-level paradigms. The ACL model is widely used, but the latter's potential has yet to be completely recognized and utilized. The following diagram depicts the network points where ACLs need to be implemented:

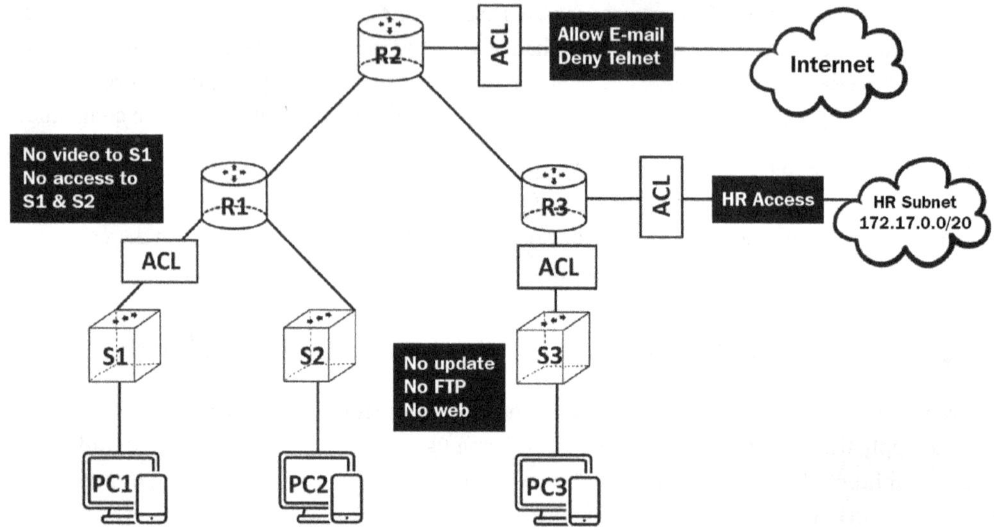

Figure 5.1 – Access control lists for system access

A documented form of identity holds inherent value for the identity holder. Usually, it's human-centric and facilitates functions such as trust and ownership. It also facilitates moving goods, people data, funds, and other resources. When entities must create or validate the identity to move resources, what is accepted and the framework by which an identity can be validated vary widely across different cultures, countries, contexts, and industries. The following figure provides examples for which identity proofing is needed:

Identity Examples	People	Organizations	Governments & Public Sector	Connected Devices	Things	Virtual Entities
Identity Proof Requirement Examples	Employment (i.e. background checks)	Banking and Insurance (i.e. KYC)	Getting an ID (i.e. driver's license)	Social benefits and welfare	Goods in supply chain	Social media
	Healthcare services	Making payments	Paying and collecting taxes	Trade finance	Forests and wildlife tracking	Workforce management
	Border control	Telecom (i.e. getting a mobile phone contact)	Travel and hospitality	Paying and collecting taxes	Processes	Machine to machine

Figure 5.2 – Identity proofing cheat sheet

State-issued ID systems, for example, follow a similar reasoning. However, the United States considers **social security numbers (SSNs)** and other identifiers that are not primarily used for identification. A passport, an identification document that's designed to represent the user across international boundaries, is nothing more than a collection of information that's been gathered and validated by a central authority. Many state-run digital identity solutions simply repackage and securely transport the data contained in a passport in digital form.

In summary, following the best practices for safeguarding data sources using ACLs creates a strong barrier against unwanted access and ensures data integrity, compliance, and effective resource management. Implementing these practices may help firms strengthen their data security posture, proactively minimizing risks and cultivating a robust and regulated environment in the ever-changing field of information technology.

Now that we've detailed the recommended practices for safeguarding data sources with ACLs, let's look at the possible limitations of using historical sources of truth for identity verification. In the next section, we will look at the limits and issues that come with this method, emphasizing the need for alternate solutions to improve identification data verification procedures.

Cons of using historical sources of truth for verification of identity data

Using historical data sources of authority for identity verification has its downsides, and organizations must be aware of these issues to establish a strong and safe identity verification procedure. The disadvantages of depending only on historical data sources are data accuracy, data privacy, security risks, possible biases, and the requirement for constant updates. To address these issues, integrating complementary techniques such as data validation, MFA, data anonymization, and frequent data audits can considerably improve the identity verification process' dependability and trustworthiness. Let's take a closer look:

- **Data accuracy and currency**:
    - **Con**: Historical data sources may have mistakes as a result of obsolete or stale information. People's addresses, names, and other personal information change with time, rendering previous data less dependable for proper identification verification. Relying on old information without frequent updates might result in identification mismatches, perhaps rejecting genuine users or granting access to unauthorized ones.
    - **Mitigation**: Organizations should develop frequent updates and validation processes for historical data sources to overcome data accuracy challenges. Periodic data audits can assist in identifying and correcting old or incorrect information. Furthermore, where possible, incorporating real-time data sources can enhance previous data and give more up-to-date information for verification reasons.
- **Data privacy and compliance**:
    - **Con**: Historical data sources frequently contain sensitive personal information, raising privacy problems. Data breaches or unauthorized access to centralized data repositories might jeopardize people's privacy and violate data protection legislation.
    - **Mitigation**: Organizations should put strict data privacy and security safeguards in place. Encrypting sensitive data, limiting access to authorized individuals, and performing regular security audits can all assist in reducing privacy threats. Compliance with data protection rules and regulations, such as GDPR, emphasizes responsible data processing.

- **Security vulnerabilities**:
  - **Con**: Malicious actors looking to steal important identification information will be drawn to centralized historical data sources. Identity theft, fraud, and other cybercrimes might result from a data leak from one of these sources.
  - **Mitigation**: Organizations should use MFA when accessing historical data sources to improve security. MFA offers an additional layer of security, making it more difficult for attackers to breach user accounts. Furthermore, implementing strong access restrictions, intrusion detection systems, and frequent security upgrades reduces the danger of security breaches.
- **Potential biases and discrimination**:
  - **Con**: Historical data sources may contain biases or outmoded social conventions that might affect the identity verification process accidentally. Using previous address data, for example, may result in mistakes in establishing present residence for people who move frequently.
  - **Mitigation**: Using a varied variety of data sources and validation methodologies can assist in mitigating potential biases. To prevent continuing biased practices, organizations should aim for inclusion and diversity in their data sources. Furthermore, regular monitoring and analysis of verification data can aid in identifying and correcting any discriminating trends.
- **Data aggregation challenges**:
  - **Con**: Integrating data from numerous historical sources can be difficult and time-consuming, resulting in data aggregation problems. Inconsistencies or inconsistencies in data from diverse sources might influence the verification process's accuracy.
  - **Mitigation**: Implementing data cleaning and matching algorithms can aid in addressing data aggregation issues. Validating and cross-referencing data from several sources can aid in identifying and resolving conflicts, boosting the overall accuracy of the verification process.
- **Lack of user control**:
  - **Con**: Using just historical data sources may limit consumer control and transparency over their personal information. Users may be unable to change outdated or erroneous information, thereby causing difficulty in accessing services or verifying their identity.
  - **Mitigation**: Users get more control over their identity data when self-sovereign identification systems are used. Individuals may maintain and distribute their own verified identity data using self-sovereign identification, decreasing dependency on centralized history data sources and placing users in control of their identity verification.

- **Lack of real-time verification**:
  - **Con**: The most recent changes to an individual's identification may not be reflected in historical data sources. Because real-time modifications may not be reflected in prior data, this might cause delays in gaining access to services or goods.
  - **Mitigation**: Real-time identity verification methods can be applied to supplement past data. Real-time verification APIs and services may perform fast identification checks, eliminating delays and assuring access to services based on the most recent data.

In conclusion, while historical authority data sources play an important role in identity verification, they are not without their drawbacks. Organizations must be cautious in handling issues such as data accuracy, privacy, security, bias, and real-time verification. Organizations may improve the reliability and efficacy of identity verification procedures by introducing mitigation methods such as data validation, MFA, data anonymization, continuous monitoring, and self-sovereign identity solutions. Building a strong and safe identity verification ecosystem requires striking a balance between historical data sources and complementary current verification methodologies.

Summary

This chapter covered the historical sources of authority that are used to validate identities, including government-issued documents, educational institutions, utility corporations, community and social networks, and individual knowledge. It underlined the significance of restricting access to various sources to preserve confidentiality, privacy, and accuracy in the identity verification process. This chapter also discussed the benefits of utilizing ACLs to secure data sources and recommended best practices for maintaining ACLs. It also discussed problems and mitigation measures for data accuracy, privacy, security risks, and data aggregation concerns in historical authoritative sources. Finally, it proposed a balanced strategy that blends historical data sources with contemporary verification approaches to create a safe and dependable identity verification environment.

In the next chapter, we'll look at the relationship between trust and risk as it relates to managing sensitive **personally identifiable information** (PII) regarding identity data.

6

The Relationship between Trust and Risk

The delicate balance between trust and threat is the cornerstone of security and functionality in the complicated arena of ultramodern access operation systems. This chapter digs into the complicated network of links that strengthens the dynamic interplay between trust and risk within these systems. Understanding the complexities of this symbiotic connection becomes critical as we negotiate the ever-changing topography of digital security, ensuring the integrity of sensitive data and the perfect execution of access restrictions. Join us on a journey into the complex world of access management, where trust and risk combine to build the future of safe and effective information management. This chapter focuses on enabling you to understand the difference between, and the impact of, trust and risk in enterprises.

The chapter will cover the following topics:

- The impact of trust and risk and the risks arising from compromised identity
- Various forms of risks and their threat vectors for online identity theft
- Risk management principles, as well as assessment and mitigation strategies

The impact of trust and risk

Trust and risk are closely related concepts that have a significant impact on individual and organizational decision-making.

Trust is a belief that an individual or organization will act in a reliable, responsible, and ethical manner. Trust is essential for building strong relationships between individuals and organizations, and for promoting cooperation and collaboration.

Risk, on the other hand, refers to the potential for harm or loss associated with a particular decision or action. Risk can come from many sources, including financial, legal, reputational, and physical.

The relationship between trust and risk is complex. On one hand, trust can help mitigate risk by providing a sense of security and confidence that an individual or organization will act responsibly and ethically. For example, if an individual trusts a financial institution to manage their investments, they are more likely to take on higher levels of risk because they believe that the institution will act in their best interest.

Risk, on the other hand, may weaken trust by introducing uncertainty and unpredictability. For example, if a person has a bad experience with a product or service, they may lose trust in the company that gave it and be less inclined to do business with them again. The link between trust and risk is, in general, a balancing act.

Individuals and organizations must consider the possible risks and advantages of a certain choice or action and determine whether their degree of confidence in the other party is sufficient to reduce those risks. Building and sustaining trust is critical for successful risk management and the development of solid relationships between individuals and organizations.

As the internet becomes more participatory and social, new threats to online trust emerge. Aside from viruses, these threats attack the link between online services and users by penetrating the core of their relationship, namely identification. While customers are becoming more aware of online threats, they should trust the notion of an interactive internet to support web service changes. Passwords are seen to be the foundation of trust, but trust requires an infrastructure for transmission and delegation. The simplest basic concept of trust is a user's faith in a website. This ultimate model depicts the human concept of trust, but it can't be applied to the complicated internet world, where the bulk of transactions are concealed. There are several organizational and technological levels between the service provider and the user: the internet service provider, the browser, and the systems and standards that ensure interoperability.

The concept of risk is currently one of the most vibrant scientific fields, attracting the interest of academics from a wide range of disciplines. Studies on the likelihood of an event occurring and its potential repercussions are followed by responses that inform concerns about measures affecting the health and safety of communities and the environment. This raises a basic concern about the human perception of risk.

There is a good consensus that the age of the Renaissance, discoveries, and long-distance maritime commerce should be seen as milestones in the earliest consolidation of risk analysis. In a risk-averse society, increased risk awareness eventually leads to risk assessment and the quest for answers to help with adopting proper risk management. Whether the threats are environmental, technological, or otherwise, including perception research in risk management procedures has become a top concern for government policymakers. We can already imagine how we would have to deal with today's and tomorrow's difficulties without these operating tools.

Risk is associated with the idea of vulnerability, which describes the potential for loss resulting from the effect of hazardous occurrences on a specific asset. The credibility or validity of a quantitative risk analysis based on estimated probabilities becomes difficult in particular settings and conditions of ambiguity. In certain cases, vulnerability analysis and management may be a suitable and successful option.

Risks arising from compromised identity

When an online identity is compromised, it causes actual harm. While internet services spend a lot of money to make their systems safe, compromising the identity procedures has an impact on the entire system. One prominent example is the banking industry, which is an appealing target since a breach might allow an attacker to steal cash. Losing the keys to systems that secure sensitive information, such as medical data, can have serious ramifications for users:

Figure 6.1 – Compromised identity risks

The issue with the web's trust source and identity procedures is that credentials act as keys to all the values stored in the systems. A social security number, for example, can be a unique identifier for all US citizens, but it is also widely used as a validation secret, something that only the user knows and can thus be used to authenticate their identities. Furthermore, dangers exist across the systems. When a popular social networking website was hacked and users' credentials were compromised, experts cautioned that many users would use the same passwords for other sites. To mitigate this risk, several other websites proactively reorganized their consumers' passwords. Because of this value concentration, identification information is an accessible target.

Attacks made on online identity break trust

A wide range of dangers, aimed at the users' property and privacy, can undermine trust online. These dangers might arise from various factors in the internet's ecosystem. Some attack vectors include **phishing** attacks, **man-in-the-middle** (**MiTM**) attacks, **social engineering**, and **brute-force** attacks, which target the customers' relationship with web services. Each abuses a different aspect of the customer's online experience.

Establishing trust in online identities is critical in the digital age as online interactions and transactions have become a fundamental part of our lives. Accessing services, completing financial transactions, and participating in social networks all rely on online identification. However, various cyberattacks that aim to compromise online identities can significantly harm trust and confidence in online networks. These threats jeopardize personal data integrity, jeopardize privacy, and diminish consumers' faith in online platforms and services.

Phishing is one of the most common online identity threats. Phishing attacks entail tricking people into disclosing sensitive information such as usernames, passwords, and personal information by impersonating a genuine company via bogus emails, websites, or texts. These attacks focus on human weaknesses, tricking unwary users into freely sharing their credentials. Once an attacker has gained access to a user's online identity, they can breach sensitive accounts and commit fraud, resulting in financial loss and reputational harm.

Identity theft is another serious hazard to online identity. Malicious actors steal personal information to mimic someone else, resulting in unauthorized access to the victim's accounts and data. Identity thieves might use stolen information to make money, apply for loans or credit cards illegally, and participate in unlawful actions under the victim's identity. Identity theft erodes confidence between individuals and the internet platforms they utilize, producing anxiety and aversion to participating in online activities.

Furthermore, credential-stuffing attacks endanger online identity security. Here, attackers employ automated tools to test a huge number of login and password combinations, taking advantage of the fact that users frequently repeat passwords across various platforms. When attackers acquire unauthorized access to a user's account, they might harvest sensitive data, change personal information, or conduct malevolent actions, eroding user faith in online service providers and platforms.

DDoS assaults, while not directly attacking online identity, can have a huge influence on user trust in online services. DDoS assaults exhaust a system's resources, resulting in service interruptions and downtime. These assaults disrupt user access to vital services, causing annoyance, discomfort, and the feeling that the platform is not secure or trustworthy enough.

Organizations must use a multi-layered strategy to identity security to counteract these assaults and retain trust in online identity. Strong authentication solutions, such as **multi-factor authentication** (**MFA**), can lower the likelihood of successful phishing assaults dramatically. MFA provides an extra layer of security for regular usernames and passwords, making it more difficult for attackers to breach online identities.

Data encryption is critical in protecting personal information against identity theft threats. Organizations may ensure that sensitive data stays unreadable and useless even if attackers get access to it by encrypting it at rest and in transit.

Furthermore, using artificial intelligence and machine learning to detect odd user behavior can aid in the prevention of credential-stuffing assaults. These technologies can detect suspect login patterns and prohibit criminal activity in real time, preserving user identities and increasing trust in online networks.

To protect against DDoS assaults, strong intrusion detection and prevention systems must be implemented. These systems are capable of detecting and blocking fraudulent traffic, assuring the continued availability and dependability of online services, which is critical for retaining user confidence.

Finally, online identity assaults pose considerable risks to user trust in online platforms and services. Phishing, identity theft, credential stuffing, and **distributed denial of service** (**DDoS**) attacks can compromise sensitive information, weaken privacy, and destroy trust in the security of online identities. To tackle these threats, organizations must have a multi-layered identity security strategy that includes strong authentication techniques, data encryption, AI-based anomaly detection, and powerful DDoS protection. Organizations may maintain user trust and confidence in their digital services and platforms by adopting proactive actions to safeguard online identities.

Local network risks

Most people assume that when they communicate over a landline, the data leaving their computer via a data connection to a line linked to the **internet service provider** (**ISP**) is secure. Users have a well-established relationship with their service provider and a good awareness of the physical dependability of the cables, if not the database or information systems. It is difficult for a normal cybercriminal to disrupt data on a telephone modem cable or internet connection that they do not physically choose. The same cannot be said for wireless network communications. Nasty actors can benefit greatly from unsecured Wi-Fi networks in their neighborhoods.

Local networks, often known as **local area networks** (**LANs**), are an essential component of modern businesses and homes, enabling seamless communication, resource sharing, and data access among linked devices. Local networks, on the other hand, are not immune to security challenges, and a range of hazards lie inside these supposedly safe environments. Understanding and managing these local network risks is critical for safeguarding sensitive data, protecting privacy, and maintaining the network's overall integrity.

Unauthorized access is one of the most serious local network threats. Local networks can be vulnerable to attackers seeking to obtain unauthorized access to linked devices and resources if they are not adequately secured. This unauthorized access might lead to data breaches, which expose sensitive information such as bank records, personal data, and proprietary company information to unscrupulous actors. Unauthorized access can also result in the unauthorized usage of network resources such as bandwidth and processing power, which can have a substantial impact on network performance.

Local networks are frequently vulnerable to internal attacks as well. Insider threats, whether deliberate or unintentional, represent a substantial danger to network security. Employees who have access to the local network may unintentionally install malware or incorrectly configure network settings, potentially resulting in data loss or interruptions. Furthermore, dissatisfied workers or insiders with malevolent intent might destroy the network or steal vital information, inflicting significant harm on the organization.

The availability of susceptible or poorly configured devices is another prevalent danger in local networks. Many devices on a local network, such as **Internet of Things (IoT)** devices, may have obsolete firmware or inadequate security mechanisms, making them ideal targets for exploitation. Attackers can utilize these devices to gain access to the larger network or use them as stepping stones to more valuable assets.

Local networks are also vulnerable to MITM attacks. In an MITM attack, an attacker intercepts network traffic and gains access to sensitive data carried across the network. When sensitive information, such as login passwords or financial data, is delivered without sufficient encryption, this assault can be especially harmful.

Furthermore, local networks may be vulnerable to unauthorized network monitoring and surveillance. Local networks can be scanned by cybercriminals to detect weaknesses and possible entry points for exploitation. Once vulnerabilities have been identified, attackers can execute targeted attacks to compromise devices and acquire network access.

Organizations and individuals can use a variety of security techniques to reduce local network hazards. To begin, it is critical to secure the network with strong passwords and to install appropriate access controls. To address known vulnerabilities, network equipment should be constantly updated with the most recent firmware and security updates. Network segmentation can help mitigate the effect of a possible breach and prevent lateral movement inside the network.

Furthermore, adopting **intrusion detection and prevention systems (IDS/IPS)** can assist in detecting and blocking unauthorized network activity. Network monitoring and recording are critical for detecting unusual activity and responding to any threats promptly.

It is also critical to educate network users on security best practices, such as recognizing phishing attempts and avoiding suspicious links, to prevent social engineering attacks that target network users. Regular security awareness training can help consumers become more alert and less vulnerable to social engineering approaches.

To summarize, local networks are not without their hazards, and addressing these vulnerabilities is critical to ensuring the network's security and integrity. Unauthorized access, internal threats, susceptible devices, MITM attacks, and network scanning are just a few of the major hazards that businesses and people must be aware of and manage. Local network risks may be addressed by establishing robust security measures, conducting frequent security assessments, and developing a culture of cybersecurity awareness. This protects sensitive data while also retaining trust and confidence in the network's security.

Online surveillance

Online surveillance demonstrates the fundamental component of trust: the separation of contexts such that activities in areas that are intended to be unrelated do not impact one another. Unfortunately, it is relatively easy to monitor people as they use the web in different ways. Cookies, which are little tokens that are recorded in your computer's browser, may monitor users across many websites. Under some settings, a cookie set by an advertisement on one website can be read by adverts on another website, letting the advertiser know that the visitor visited both sites.

The process of systematically monitoring, tracking, and gathering individuals' digital activities, conversations, and behavior on the internet is referred to as online surveillance. Governments, companies, and other groups use this practice for a variety of goals, including law enforcement, intelligence collecting, marketing, and data analysis. While internet monitoring has important uses, it also raises serious privacy and ethical problems.

Gathering metadata, which contains information about people's online activities, such as the websites they visit, the duration of their visits, and the communication patterns they display, is one of the most common ways of online surveillance. Even if the actual content of the communication is not collected, metadata may be quite revealing since it gives insights into individuals' habits, hobbies, and social relationships.

Intercepting and monitoring communications, such as emails, instant messaging, and social media activities, is another common kind of internet surveillance. Governments and intelligence services, in particular, have been known to conduct large-scale surveillance programs to monitor internet conversations. Concerns have been raised regarding potential abuses of privacy rights and misuse of authority by those performing the monitoring.

Furthermore, the usage of monitoring technology such as facial recognition systems and location tracking is becoming more common. These technologies allow for real-time surveillance and identification of people based on physical characteristics or geographical data. While they can be utilized for security and public safety, they also offer considerable privacy hazards since individuals may be unaware that they are being watched or have no influence over the information that's being collected about them.

Online surveillance also includes tracking people's online behavior commercially. Companies amass massive quantities of information on customers' online behaviors, interests, and purchasing histories to develop detailed profiles for targeted advertising and personalized content delivery. This type of surveillance raises concerns about the monetization of personal data, as well as the possibility of individual manipulation and exploitation based on their digital footprints.

Various techniques can be applied to alleviate the privacy and ethical problems that are raised by internet surveillance. Individuals can be protected by privacy laws and regulations that control data collection, usage, and sharing. Strong data protection frameworks, such as the **General Data Protection Regulation** (**GDPR**) in the European Union, establish explicit criteria for how organizations must handle personal data.

Encryption and anonymization methods can also assist in safeguarding people's privacy by making it more difficult for surveillance agencies to access or identify specific people's data. Encrypted communications guarantee that messages are safe and only available to the intended recipients, whereas anonymization methods remove **personally identifiable information** (**PII**) from data, making it less likely to be tracked back to an individual.

Furthermore, raising awareness about online monitoring and encouraging digital literacy can enable individuals to make educated decisions about their digital activities and make efforts to preserve their privacy. Understanding the risks and implications of online monitoring might inspire people to use privacy-enhancing tools and exercise caution when sharing information online.

Finally, cyber spying has become a common practice with both useful and troubling uses. While it has its uses in security, law enforcement, and business, it also presents substantial privacy and ethical concerns. Online surveillance tactics include metadata gathering, communication interception, surveillance technology, and commercial tracking. To defend people's privacy and rights, strong data protection legislation, encryption, anonymization, digital literacy, and awareness campaigns are required. Striking a balance between acceptable monitoring for public safety and respecting people's right to privacy is a difficult subject that needs to be discussed continually and the ethical consequences must be evaluated thoroughly.

Browser-based web risks

Browsers must be able to comprehend and execute scripts that are delivered by websites. A new type of security vulnerability has emerged that allows unauthorized code to be injected into the browser side of a web application.

Browser-based online risks are security threats and vulnerabilities that occur as a result of utilizing web browsers to access and interact with the internet. Web browsers, being the major entryway to the online world, are vulnerable to a variety of threats, including virus assaults, phishing efforts, browser extension vulnerabilities, and **cross-site scripting** (**XSS**) attacks.

Malware dissemination is one of the most common browser-based online hazards. Malicious actors employ web browser vulnerabilities to inject malicious code into legitimate websites, resulting in the unintentional download and execution of malware on users' devices. This malware can take the shape of viruses, ransomware, spyware, or adware, and it can compromise data integrity, steal sensitive information, or impair system functionality.

Web users are also in danger from phishing assaults. Phishing attempts sometimes entail the creation of bogus websites that seem exactly like authentic ones to trick users into submitting personal information such as usernames, passwords, and credit card information. Unsuspecting users may submit their credentials on these bogus sites unintentionally, allowing attackers to capture their information for unlawful purposes.

While browser extensions improve user experience and usefulness, they can also present security issues. Browser extensions that are malicious or poorly developed might jeopardize user privacy by gathering and transferring sensitive information without the user's knowledge or agreement. When installing extensions, users should take caution and make sure they come from trusted developers and recognized software marketplaces.

As mentioned previously, another common issue in browser-based online systems is XSS. XSS attacks leverage web application vulnerabilities to insert malicious scripts into web pages seen by other users. These scripts can steal critical data or undertake unauthorized activities on the user's behalf, potentially resulting in serious security breaches.

Furthermore, the large range of plugins, media players, and Java applets that web browsers allow magnifies browser-based online dangers. These new features can provide new attack vectors and broaden the surface area for possible vulnerabilities. Outdated plugins and unsupported software versions can be especially dangerous since they may include known security holes that attackers can exploit.

To reduce browser-based online dangers, users should follow best practices and utilize updated security technologies. It is critical to keep web browsers and their plugins up to date with the latest security updates to avoid known vulnerabilities from being exploited. Regular security upgrades also help browser developers handle new risks and keep up with growing attack strategies.

Users should also be cautious while visiting strange websites and clicking on links. Verifying website URLs is critical for avoiding phishing attempts, especially when completing sensitive actions such as inputting login credentials or completing financial transactions.

Using browser security extensions and add-ons can provide an additional layer of protection against web threats. These security solutions may block malicious material, alert users about potentially dangerous websites, and analyze web pages in real time to identify and mitigate dangers.

Organizations can also deploy web application security measures to protect themselves against XSS attacks and other online-based dangers. This involves input validation, output encoding, and the use of security frameworks to prevent dangerous scripts from being injected into online applications.

Finally, browser-based online dangers are a major worry in today's digital world. Malware dissemination, phishing efforts, browser extension vulnerabilities, and XSS attacks are just a few of the hazards that internet users come across. To mitigate these dangers, users must be watchful, update their browsers and plugins, and utilize security extensions and tools. To protect consumers from possible vulnerabilities, organizations must deploy effective web application security safeguards. Users and organizations may improve the overall security of their web browsing experiences and limit the effect of browser-based web hazards by taking a proactive and security-conscious approach.

Social engineering

Social engineering directly infiltrates the target users, convincing them to readily hand over identification information or actively execute an action to assist the attacker. In terms of online identity, the most common kind of attack is **phishing**, in which the attacker forces the victim to expose their online identity data to a website controlled by the attacker.

Social engineering assaults, which use psychological manipulation to deceive and exploit victims, pose major threats to both individuals and organizations. These assaults do not rely on advanced technological vulnerabilities, but rather on the human factor, using trust, curiosity, fear, and other emotions to obtain unauthorized access to sensitive data, systems, or resources. Data breaches, financial loss, reputational harm, and compromised network security are some of the major hazards that are connected to social engineering assaults.

Data leaks are one of the most serious threats of social engineering attempts. Attackers may mimic trustworthy entities, such as coworkers, service providers, or technical support employees, to fool people into disclosing sensitive information, such as login credentials, passwords, or personal information. Attackers with this information can acquire unauthorized access to systems, email accounts, or financial platforms, possibly resulting in data theft, identity theft, or privacy breaches.

Another serious risk that's posed by social engineering assaults is financial loss. Attackers trick victims into transferring cash or making payments to phony accounts in schemes such as **business email compromise** (**BEC**) or CEO fraud. Attackers fool those responsible for financial transactions by impersonating high-ranking executives or trusted partners, resulting in significant financial losses for organizations.

Social engineering attacks can potentially inflict significant reputational harm. Phishing assaults, in which attackers send misleading emails or messages to lure people into clicking harmful links or installing software, can jeopardize an organization's image. Customers, clients, and partners who are victims of these assaults may lose faith in the afflicted organization, resulting in unfavorable publicity and eventual commercial loss.

Furthermore, social engineering attacks can weaken network security. Attackers may utilize social engineering techniques, acting as workers or contractors, to obtain physical access to an organization's facilities. Once inside, attackers can take advantage of the chance to install malware, steal sensitive information, or mess with network equipment, thus jeopardizing the security of the entire network.

Furthermore, targeted spear-phishing assaults, in which attackers personalize their communications to specific persons, can be difficult to detect, particularly when attackers perform considerable research to make the messages believable. These assaults, if successful, can provide attackers with access to extremely sensitive information, intellectual property, or trade secrets, inflicting considerable damage on organizations.

To prevent social engineering attacks, a mix of technological restrictions and security awareness training is required. To prohibit harmful emails and websites related to social engineering efforts, organizations should adopt comprehensive email filtering and web filtering systems. MFA can offer an extra layer of security by demanding additional verification beyond passwords to access critical accounts or systems.

Employee security awareness training is critical for educating individuals about the numerous social engineering strategies that are employed by attackers. Employees should be educated to identify social engineering warning flags such as unsolicited demands for sensitive information, urgent or threatening messages, and questionable email attachments or links.

Regularly simulating phishing exercises can also assist organizations in evaluating the efficiency of their security awareness training programs. These drills entail sending simulated phishing emails to workers and measuring their answers to identify areas for improvement and providing targeted training when necessary.

To summarize, social engineering assaults pose major hazards to individuals and organizations by taking advantage of human psychology to deceive and manipulate victims. Among the key hazards connected with social engineering attacks, there's data breaches, financial loss, reputational harm, and compromised network security. To counteract these threats, organizations must install technological measures such as email filtering and MFA, as well as provide extensive staff security awareness training. Organizations can better protect themselves against social engineering attacks by cultivating a culture of cybersecurity knowledge and vigilance.

Risk management principles and assessments

Risk management is a key process that's used by organizations to detect, analyze, and reduce possible risks to their operations, assets, and goals. Businesses may make more informed decisions, allocate resources more efficiently, and increase their resilience to uncertainties and threats by using solid risk management concepts. This section examines the major risk management concepts and their importance in developing a solid risk management framework.

Proactive approach

Taking a proactive approach is the core of good risk management. Rather than waiting for risks to manifest, organizations must identify and manage possible hazards before they become major difficulties. This principle requires detecting and comprehending risks, vulnerabilities, and potential repercussions to implement timely and suitable risk mitigation solutions.

Risk identification

Identifying possible hazards is the first stage in the risk management process. This entails identifying internal and external elements that may have an impact on the organization's goals. Methods for identifying risks may include brainstorming meetings, data analysis, scenario planning, and studying historical data. Organizations can better prepare for risk management by compiling a detailed inventory of possible hazards.

Risk assessment and analysis

Once hazards have been identified, they are subjected to a detailed risk assessment and analysis to determine the likelihood and possible effect of each risk. Risks are evaluated using qualitative and quantitative methodologies, allowing organizations to prioritize and focus their efforts on high-risk areas. This procedure includes analyzing the likelihood of the risk occurring and evaluating the potential ramifications for the organization.

Risk evaluation

Risk evaluation entails comparing measured hazards to predetermined risk criteria, as well as the organization's risk appetite. This risk appetite reflects the amount of risk an organization is ready to take to achieve its goals. Risks that lie within the organization's risk appetite are deemed acceptable, but those that surpass the threshold necessitate additional mitigation measures or other techniques.

Risk mitigation strategies

Organizations establish risk mitigation plans that are adapted to their unique needs and objectives after they have a thorough grasp of the risks. Risk avoidance, risk reduction, risk sharing, and risk transfer are examples of risk management practices. Risk mitigation measures must be implemented to reduce the possibility or effect of prospective hazards and strengthen the organization's resilience.

Monitoring and review

Risk management is a continual activity that needs constant monitoring and evaluation. Organizations should examine the success of risk mitigation measures regularly and adapt their policies accordingly. Performing risk assessments, internal audits, and management reviews frequently serves to ensure that risk management practices are current and up to date.

Communication and collaboration

Clear and open communication among stakeholders is required for effective risk management. This includes disseminating risk information, raising risk awareness, and motivating employees to report possible hazards. Collaboration across departments and organizational levels improves the ability to integrate risk management practices into diverse business processes.

Crisis preparedness

While the goal of risk management is to prevent or reduce hazards, organizations must also be prepared to deal with any crises. Crisis management strategies should be prepared, including the procedures to be undertaken if a major danger materializes. Regular crisis drills and simulations help the organization prepare to respond quickly and effectively to unforeseen circumstances.

Risk assessment

Risk assessment is an important part of the risk management process because it provides organizations with information about potential risks and vulnerabilities. This section considers the complexities of risk assessment, its methodology, and the need to undertake a complete and systematic risk assessment.

Risk identification methods

Various strategies are used to detect possible dangers during the risk identification phase. Brainstorming meetings, in which stakeholders discuss and share their ideas on possible hazards, aid in gathering a diverse variety of perspectives. Data analysis, which includes reviewing historical data and doing statistical analyses, can reveal trends and patterns that suggest possible threats. Furthermore, scenario planning entails developing hypothetical events to analyze risks in diverse circumstances.

Qualitative risk assessment

Qualitative risk assessment is a subjective technique that depends on expert judgment and criteria. This strategy entails evaluating risks based on their effect and likelihood, as well as frequently utilizing qualitative scales such as low, medium, or high. When accurate data is lacking or when dealing with emergent threats, qualitative risk assessment is important.

Quantitative risk assessment

In contrast, quantitative risk assessment assesses hazards using objective data and numerical analysis. Risks are assigned values based on likelihood, impact, and exposure. Monte Carlo simulations and other quantitative tools assist organizations in quantifying and prioritizing risks based on their potential financial effect.

Risk heat maps

Risk heat maps are visual tools that depict the effect and likelihood of hazards. These graphical representations give organizations a quick overview of the most critical risks, allowing them to focus their resources on high-impact, high-likelihood concerns. By showing complicated risk data simply and succinctly, risk heat maps aid in communication and decision-making.

Risk tolerance and assessment

Organizations determine their risk tolerance and acceptance levels during risk assessment. The utmost level of risk that an organization is prepared to take is referred to as risk tolerance. Risks that fall within this tolerance level are deemed acceptable. Risks that surpass the tolerance limit, on the other hand, necessitate more investigation and risk mitigation activities.

Risk scenarios and sensitivity analysis

Risk assessment entails assessing multiple risk scenarios, each with its own set of hazards and potential effects. Sensitivity analysis is used to examine the impact of changes in risk variables on total risk exposure. Organizations acquire a better grasp of potential outcomes and can establish contingency plans as a result of evaluating alternative risk scenarios.

Risk ranking and prioritization

Risks are ranked and prioritized based on their possible effect and likelihood after risk assessment. This technique enables organizations to focus on high-priority issues and allocate resources properly. Risk rating aids decision-making and ensures that risk is managed effectively.

Finally, risk management concepts and risk assessment are essential components of a successful risk management framework for identity-based networks. Organizations can improve their resilience, protect sensitive information, and achieve their goals in an ever-changing risk landscape by taking a proactive approach, identifying and assessing risks, implementing risk mitigation strategies, and continuously monitoring and reviewing risk management practices. A comprehensive and systematic risk assessment gives useful insights into possible risks and vulnerabilities, allowing organizations to make educated decisions and effectively prioritize risk mitigation activities.

Risk mitigation strategies

Risk mitigation methods are critical components of a complete risk management strategy that aim to reduce or eliminate possible hazards and vulnerabilities that may affect people, organizations, or systems. These techniques help with discovering, analyzing, and handling possible risks to avoid or mitigate their negative consequences on many elements, such as security, operations, finances, and reputation. Individuals and organizations may proactively protect themselves against possible risks, improve resilience, and preserve continuity in the face of uncertainty and hardship by employing risk mitigation methods.

A systematic and proactive strategy for detecting possible risks, analyzing their impact and likelihood, and creating methods to minimize their effects is required for effective risk mitigation techniques. To get a thorough understanding of prospective threats and their potential repercussions, you may have to conduct risk assessments, vulnerability studies, and scenario modeling. Specific mitigation actions are undertaken once the risks have been recognized to lessen the likelihood or severity of the hazards.

Depending on the nature of the hazards and the context in which they arise, risk mitigation methods can take numerous shapes. Implementing security measures such as firewalls and encryption to protect against cybersecurity threats, establishing redundancy and backup systems to ensure business continuity in the event of disruptions, or conducting employee training to raise awareness about potential risks and how to mitigate them are some examples of these strategies.

Collaboration and communication among stakeholders, including leadership, workers, suppliers, customers, and partners, is critical to the effectiveness of risk mitigation measures. Organizations may guarantee that risk mitigation policies are adopted and followed properly throughout the organization by developing a culture of risk awareness and accountability.

Continuously monitoring and frequently appraising risk mitigation methods is critical in this dynamic and growing context, where threats can appear or alter fast. Individuals and organizations may respond proactively to emerging risks and challenges by being aware and adaptable, boosting their resilience and preparedness to face future threats effectively.

To summarize, risk mitigation methods are an important component of risk management initiatives because they enable individuals and organizations to identify, analyze, and handle possible hazards proactively. Individuals and organizations may decrease vulnerabilities, safeguard assets, and maintain continuity in the face of uncertainty by applying these measures. Risk mitigation solutions remain adaptive and successful in tackling new and emerging hazards with continual monitoring and evaluation, ensuring a robust and resilient approach to risk management.

Risk and trust management roadmap

Setting risk mitigation goals and objectives is critical for establishing a clear path for controlling possible risks and vulnerabilities. Organizations may determine the desired objectives of their risk mitigation initiatives by creating clear and quantifiable targets. Goals should be reasonable and feasible and should be in line with the organization's risk tolerance and resources. Clear goals allow risk mitigation initiatives to be prioritized, ensuring that resources are directed toward the most essential threats. Furthermore, having well-defined objectives enables organizations to monitor and evaluate risk mitigation measures effectively, allowing them to quantify their success and make required modifications as needed.

The following diagram depicts the roadmap toward establishing a strong trust and risk model:

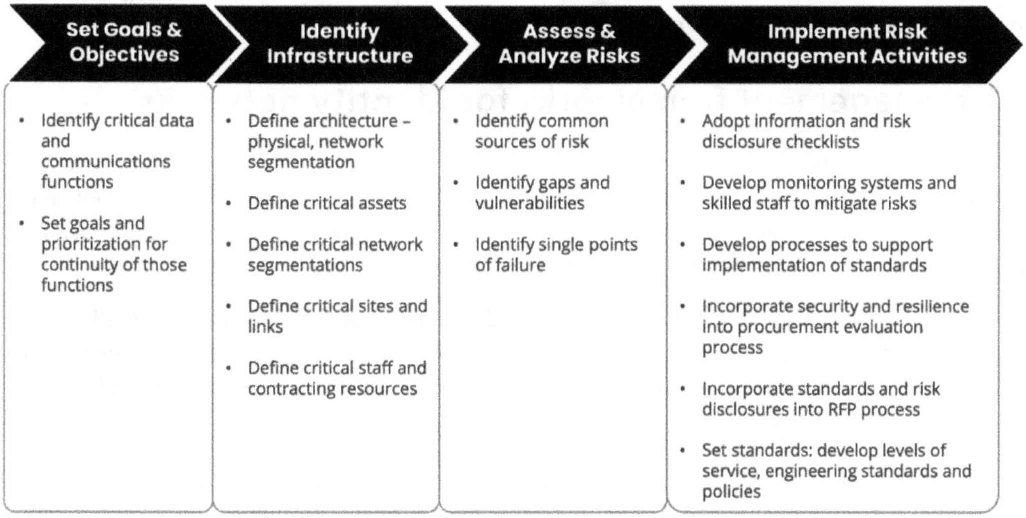

Figure 6.2 – Risk management roadmap

It is critical to identify infrastructure security components to mitigate risk effectively. These components include firewalls, IDS, encryption, access restrictions, and network segmentation. Firewalls protect the network by screening and monitoring incoming and outgoing traffic. IDS identifies unusual network activity and possible threats. Encryption protects the confidentiality and integrity of data during transmission and storage. Unauthorized access to sensitive resources is prevented via access restrictions. Critical assets are isolated through network segmentation, lowering the effect of a possible compromise. Organizations may increase their infrastructure security by recognizing and integrating these components, eliminating vulnerabilities, and managing possible hazards.

Risk access and analysis are critical steps in the risk management process. It entails detecting possible risks and vulnerabilities, as well as the impact they may have on an individual, organization, or system. To estimate the possibility and severity of each risk, risk assessment approaches such as risk matrices and qualitative or quantitative analysis are used. Decision-makers may prioritize mitigation measures and allocate resources more efficiently if they understand and evaluate risks. Regular risk access and analysis enables proactive risk management, prompt reactions to new risks, and the development of robust risk mitigation measures.

Putting the strategies, plans, and measures discovered throughout the risk assessment and analysis process into practice is what risk management activities entail. It involves incorporating risk-mitigation measures into the organization's activities, such as security protocols, contingency plans, and disaster recovery processes. To guarantee constant adherence to risk management practices, effective implementation necessitates collaboration across stakeholders, including employees, management, and external partners. Risk management operations must be monitored, evaluated, and adjusted regularly to ensure they remain relevant and successful in an ever-changing risk landscape. Organizations may improve their resilience, preserve assets, and reduce the effect of possible hazards by making proactive efforts to execute risk management operations.

Risk management frameworks for identity networks

Identity-based networks are critical in giving safe access to resources and services in today's linked digital world. However, these networks are not without danger. To secure sensitive information and preserve the integrity of identity-based networks, strong risk management frameworks must be implemented due to the possibility of unauthorized access, data breaches, and identity theft.

A risk management framework is a methodical methodology that's used by organizations to discover, analyze, and reduce any hazards that are connected to their identity-based networks. These frameworks provide a systematic approach to assessing the risk picture, making informed decisions, and allocating resources to protect network assets and user identities efficiently:

- **Identifying risks**: Identifying possible hazards is the first step in any risk management approach. In the case of identity-based networks, this entails identifying flaws in authentication techniques, access controls, and user identity management procedures. Threat modeling and risk assessments are critical methods for identifying and prioritizing threats while considering criteria such as data sensitivity, network asset value, and the effect of prospective breaches.
- **Risk assessment and analysis**: Once hazards have been identified, they are subjected to a detailed risk assessment and analysis to determine the likelihood and possible effect of each risk. Risks are evaluated using quantitative and qualitative methodologies, allowing organizations to prioritize and focus their efforts on high-risk areas. Organizations may make data-driven risk mitigation decisions by understanding the risk landscape.

- **Risk mitigation strategies**: Organizations can design risk mitigation methods that are suited to their identity-based networks after they have a good grasp of the threats. Instilling MFA to improve access restrictions, installing encryption protocols to safeguard data in transit and at rest, and deploying **identity and access management** (**IAM**) systems for centralized user identity management are examples of these tactics.
- **Continuous monitoring and evaluation**: Risk management is a never-ending process. Continuous monitoring and assessment are essential for ensuring that risk reduction measures remain successful and relevant. Regular security audits, vulnerability assessments, and incident response exercises aid in identifying new threats and adapting risk management procedures.
- **Compliance and regulatory alignment**: Identity-based networks frequently handle sensitive personal information, necessitating adherence to data protection laws and industry standards. Risk management frameworks should verify that network security measures comply with any legal and regulatory standards. This involves following GDPR, HIPAA, and PCI DSS regulations, as well as industry-specific norms.
- **Training and awareness**: Human error continues to be a substantial contributor to security vulnerabilities. Comprehensive training and awareness programs are required to educate staff, administrators, and users on network security and identity protection best practices. Employees should be educated so that they can identify social engineering attempts, prevent phishing frauds, and use safe authentication practices.
- **Incident response and recovery**: Security problems may occur despite attempts to mitigate them. An incident response plan, which defines the procedures to be implemented in the case of a breach, should be included in an effective risk management framework. Drills and rehearsals for incident response serve to ensure a quick and effective reaction, reducing the impact of security events and enabling recovery.

To summarize, identity-based networks are vital assets that need strong risk management frameworks so that they can be protected against possible threats and vulnerabilities. Organizations may increase network security and protect sensitive identities and data by methodically identifying risks, analyzing their effect, and applying suitable risk mitigation techniques. A complete risk management approach must include continuous monitoring, compliance alignment, and user training. Organizations may establish a safe and trustworthy identity-based network environment via proactive risk management, encouraging confidence among users and stakeholders.

Summary

This chapter examined the important link between trust and risk in ultramodern access control systems, focusing on the impact of social engineering attacks on individuals and organizations. It emphasized the necessity of technological solutions such as email filtering and MFA, as well as extensive security awareness training, to minimize these dangers. Furthermore, this chapter discussed the importance of risk management concepts and assessments, infrastructure security components, risk identification methodologies, and risk mitigation strategies in securing local and online networks. It also underlined the importance of proactive risk management, ongoing monitoring, compliance alignment, and user training in creating a safe and trustworthy identity-based network environment. In the next chapter, we will take a look at informed consent and why it matters.

7

Informed Consent and Why It Matters

Informed consent is a fundamental premise in modern healthcare, research, and other fields where people's autonomy and rights are vital. It symbolizes the ethical obligation to respect people's autonomy in making decisions about their own bodies, health, and involvement in research or therapies. Beyond its importance in clinical contexts, informed consent pervades many facets of our lives, including digital privacy agreements and consumer contracts. Its significance rests not just in legal compliance but also in sustaining essential ethical ideals such as transparency, autonomy, and respect for human dignity. In this chapter, we will look at the core of informed consent, including its historical roots, ethical underpinnings, practical uses, and long-term importance in current society. Through a multidisciplinary perspective, we examine the complexity, problems, and subtleties of informed consent, shedding light on why it is more important than ever. Understanding the complexities of this symbiotic connection is critical for protecting the integrity of sensitive data and the smooth functioning of access restrictions as we traverse the ever-changing environment of digital security. Join us as we explore the complex realm of informed consent, where trust and risk combine to define the future of safe and efficient information management.

In this chapter, we will cover the following topics:

- The fundamentals of informed consent
- The re-purposed data problem and privacy-by-design concepts
- The role of privacy in various jurisdictions

What is informed consent?

Informed consent is a process by which individuals are fully informed about the risks and benefits of a particular decision or action and then make a voluntary decision based on that information. In healthcare, informed consent is required before any medical procedure or treatment is performed, but it is also important in many other areas, such as research, data privacy, and online services.

Informed consent matters because it is a fundamental principle of ethics and human rights. It ensures that individuals have the right to make their own decisions about their bodies and their personal information and that they are not subjected to any unwanted or harmful interventions. Informed consent is particularly important in situations where there is a power imbalance between the individual and the person or organization requesting consent, such as in healthcare or research.

From a legal perspective, informed consent is also required in many jurisdictions. For example, the **General Data Protection Regulation** (**GDPR**) in the European Union requires organizations to obtain explicit and informed consent from individuals before collecting, processing, or sharing their personal data.

Informed consent is important for building trust between individuals and organizations and for promoting transparency and accountability. By providing individuals with clear and accurate information about the risks and benefits of a particular decision or action, organizations can build trust and credibility with their stakeholders.

Overall, informed consent is a critical principle that helps ensure that individuals have control over their own bodies and personal information and that they are treated with respect and dignity. It is essential for promoting ethical and responsible behavior in healthcare, research, data privacy, and other areas and for building trust between individuals and organizations.

Informed consent is an ethical, and sometimes legal, mechanism that ensures that individuals provide information voluntarily with complete knowledge and understanding of pertinent risks. In the context of digital identity, it implies that individuals are fully aware of their use of their digital identity, as well as all related data trails, and are able to comprehend the decisions made regarding the use of their data. Indeed, the main answer is closely related to ethical considerations and ethical development practices. As we would with a set of personal data, we would confirm we have 100% consent from the individual since the data belongs to them.

Informed consent matters due to its accountability and transparency. If we—being a humanitarian or development organization—are gathering personal data, informed consent facilitates us in our objective to be truthful with beneficiaries and self-assured that we are answerable to our choices and actions. As the World Bank principles state, guaranteeing user literacy concerning legal identification systems may foster an environment of understanding and mutual trust. But we can also spot a gap in the clear need for informed consent and claims about the difficulty of obtaining it, especially in humanitarian organizations. So, if this is the scenario, how can we bridge this gap? The answer lies in educating the user about informed consent and the benefits they can get out of it.

Educating the user about informed consent

Educating the digital identity recipients on the what, who, when, where, and why of their data gathering is the most important and first step. But where does the responsibility of the data collector end? Is it sufficient to give someone information and then hope that they read, listen to, and know it?

Think about a person's reaction when they are alerted about a new update on their iPhone and Apple gives them tens of pages of terms and conditions that are inexplicable. Do users accept them because not doing so would take away a resource that is valuable to them? Do they accept them within reading because they're excessive? Or do they take the time to read it fully and understand it?

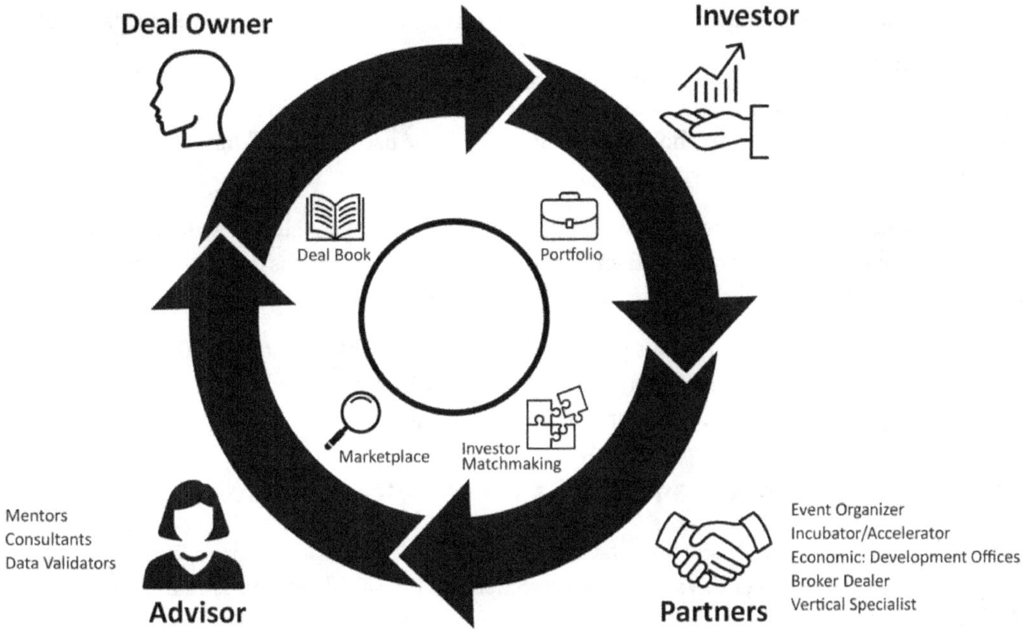

Figure 7.1 – Informed consent role model

This should have given you an idea of how the project beneficiaries can feel when presented with tons of information on data collection opportunities and risks. Often, they are at a vast disadvantage, maybe not having an idea of what data is, let alone why they must be aware of how the data is shared.

Maybe it is up to digital identity service providers and project executors to ensure that the information is stated and comprehended and not only provided. This would be a logical step, given that only by fully understanding the ramifications of sharing their data can people provide genuinely informed permission.

Indeed, providing evocative education is an actual challenge in contexts where the recipients are not digitally literate or are in a state that makes it tough for them to say no. Also, informed consent cannot be obtained in humanitarian emergencies because of technology illiteracy and urgency, as well as the complexity of digital identity data. Imagine how tough it would be for a user to give informed consent if they didn't have any experience with or exposure to the technology but are asked to comprehend what biometric data is and what the consequences are of it being shared with diverse organizations. If that gets in the way of you getting aid, you can envision where your priorities might lie.

Understanding informed consent

Informed consent is a crucial element in the management and control of user data. It refers to people's free and informed consent to enable an organization or service provider to collect, handle, and utilize their personal data. In the digital age, where large quantities of personal information are gathered and processed, informed consent is critical to protecting users' privacy and giving them control over their data.

Several crucial factors are included in the idea of informed consent. To begin, individuals must be fully informed about the purpose and extent of data collection and processing. Users should be given clear and intelligible information about how their data will be used, who will have access to it, and any potential dangers connected with giving it.

Secondly, informed consent requires that people provide their consent willingly and without compulsion. Individuals should not be forced to share their data or face severe repercussions for denying consent, and consent should be freely granted. To get consent, organizations must avoid any dishonest or misleading practices.

Third, informed consent is a continuous process. Users should be able to withdraw their permission at any moment without incurring repercussions. Organizations must make it simple for users to withdraw their permission and prevent further data processing.

Legal and regulatory frameworks for informed consent

Informed permission is not only an ethical factor in user data management; it is also a legal necessity in many places throughout the world. Various data protection and privacy regulations require organizations to seek individuals' explicit and informed consent before collecting and processing their personal data.

The GDPR of the European Union is one of the most important legislative frameworks governing informed consent. According to the GDPR, organizations must seek specific, informed, and unambiguous permission. Individuals have the right to withdraw their consent at any time, and consent requests must be made in clear and unambiguous terms.

Similarly, in the United States, the **California Consumer Privacy Act (CCPA)** guarantees customers the right to know what personal information is being gathered about them and to opt out of the sale of their data. The CCPA emphasizes the importance of open data practices and giving individuals ownership over their data.

Many other nations and areas have established data protection legislation that includes measures for informed consent. These regulatory frameworks stress the significance of gaining meaningful permission from users, as well as honoring their data privacy and control rights.

Challenges and limitations of informed consent

Despite the importance of informed consent in user data management, various obstacles and limits must be addressed in order for it to be effective. One big issue is *consent fatigue*, which occurs when consumers are bombarded with consent requests from many websites and services. As a result, users may mechanically click to accept without fully comprehending the ramifications of their agreement.

Furthermore, the complexities of privacy rules and terms of service agreements might make it difficult for consumers to give informed permission. These contracts are frequently long and dense with legal language, making it difficult for consumers to understand the ramifications of giving their data.

Furthermore, the power differential between organizations and people might have an influence on the validity of informed consent. Large organizations may hold enormous power over users, perhaps resulting in coercion or a lack of true choice in obtaining permission.

Improving informed consent

Organizations and governments can employ a variety of ways to overcome the issues and constraints of informed consent in user data management. One way is to build consent mechanisms with a more user-centric perspective. This includes presenting clear and concise consent forms that are simple to comprehend, as well as giving consumers a granular choice over the sorts of data they provide.

Another critical step is to simplify privacy rules and terms of service agreements. Organizations should attempt to communicate information in simple language and utilize visuals, such as infographics or symbols, to help people understand data practices.

Additionally, organizations might experiment with novel means of collecting consent, such as employing interactive consent dashboards or using new technologies such as blockchains to develop more transparent and auditable consent procedures.

The future of informed consent

Technological breakthroughs, dynamic regulatory environments, and changing user expectations are likely to affect the future of informed consent in user data management. The broad usage of artificial intelligence and the **Internet of Things** (**IoT**) will provide new issues and possibilities for getting informed consent.

Furthermore, as indicated by continuing talks about the need for a uniform worldwide approach to data protection, regulators may impose increasingly strict standards for getting consent.

Furthermore, user knowledge and privacy rights advocacy will continue to affect how organizations approach informed consent. As consumers grow more aware of their data rights, they will most likely expect more clear and user-friendly consent practices from the organizations with which they engage.

Finally, informed consent is an important element in user data management and control. It enables individuals to make informed decisions regarding the gathering and use of their personal data, and it guarantees that organizations follow ethical and lawful data practices. Organizations may cultivate more trust with their users and establish a more transparent and responsible data environment for the future by addressing difficulties and constraints and implementing user-centric methods.

The re-purposed data problem

The repurposed data problem is a crucial issue in data management and privacy. It occurs when data obtained for one reason is reused or repurposed for new—often unanticipated—goals without the necessary authorization of the persons whose data is involved. The repurposed data dilemma raises important ethical, legal, and security problems in today's data-driven society, where massive volumes of information are created and processed.

The question of informed permission is central to the repurposed data problem. Informed consent is a key data privacy principle that allows individuals to make informed decisions regarding the collection, use, and sharing of their personal data. Individuals may lose control over their information if it is repurposed without their explicit authorization, potentially resulting in privacy violations and breaches of trust.

In cases involving large data analytics and **artificial intelligence** (**AI**), the reused data problem gets much more difficult. The potential for data repurposing grows as organizations collect enormous amounts of information from numerous sources. The interconnectedness of data streams from social media, internet surfing, IoT devices, and other sources allows data to be repurposed for purposes other than its original goal.

Furthermore, the lack of defined data reuse norms and laws exacerbates the problem. Many governments' data protection rules may not expressly cover data repurposing, allowing the opportunity for ambiguity and potential misuse. As a result, organizations may interpret data usage permissions more widely than consumers intended, thus leading to unforeseen consequences and infringement of privacy rights.

Data anonymization and pseudonymization are frequently employed as mitigation techniques for the reused data problem. Anonymization is the process of deleting or encrypting **personally identifying information** (**PII**) from datasets, making it difficult to attribute data to specific people. Pseudonymization, on the other hand, is replacing direct identifiers with fictitious identifiers, guaranteeing that data cannot be directly linked to people.

Even anonymized or pseudonymized data, when joined with other datasets or given to advanced data analysis tools, might pose hazards. Even with these safeguards in place, the process of re-identification, in which anonymous data is connected back to the identities of people, highlights the difficulties in securing perfect data privacy.

To address the issue of reused data, organizations must take a privacy-by-design strategy. Privacy by design focuses on incorporating privacy issues into all stages of the data lifecycle, from collection and storage through to sharing and disposal. Organizations can reduce the dangers associated with data repurposing by employing privacy-enhancing technology and practices such as differential privacy.

Furthermore, strong data governance frameworks must be implemented to enable transparent and responsible data practices. Organizations must explicitly explain the objectives for which data is gathered and get users' explicit agreement for each use scenario. Consent management systems and data access controls can help track and manage user consent choices while also allowing users to cancel or alter their consent at any moment.

In addition to technology and administrative solutions, increasing user awareness of data repurposing is critical. Transparency in data practices and open communication about data usage are critical for building trust between organizations and users. Individuals can be empowered through user education to make educated decisions about their data and understand the ramifications of providing consent.

Regulatory organizations are also crucial in tackling the reused data issue. Strengthening data protection regulations to include specific rules on data repurposing can offer organizations clarity and set clear limitations for data usage. Penalties for failing to comply with data protection legislation can serve as a deterrent to data abuse and urge organizations to prioritize data privacy.

Finally, in the data-driven world, the repurposed data problem is a complicated and diverse difficulty. It poses ethical, legal, and security issues regarding personal privacy and data security. Organizations can mitigate the risks associated with data repurposing by adopting a privacy-by-design approach, implementing robust data governance frameworks, raising user awareness, and strengthening data protection regulations. This will foster a data ecosystem that respects individual privacy and empowers users with control over their data.

Privacy by design

Privacy by design (**PbD**) is a fundamental approach to data security that emphasizes the incorporation of privacy issues into system, process, and technology design and development. It tries to handle privacy concerns proactively by incorporating privacy safeguards throughout the whole data processing lifecycle rather than installing them as an afterthought. The notion of informed consent is a major pillar of PbD since it ensures that individuals have control over their personal data and understand the ramifications of its usage.

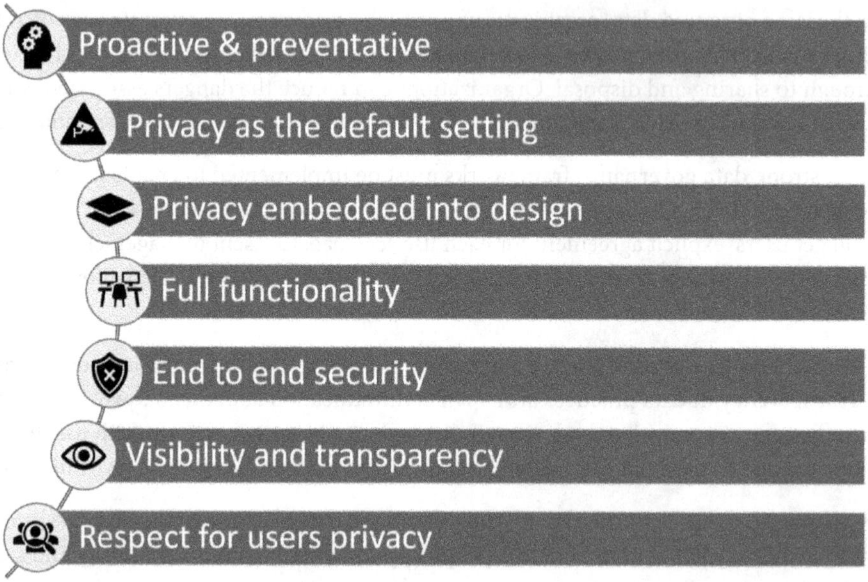

Figure 7.2 – The pillars of PbD

As a component of PbD, informed consent refers to individuals' voluntary and informed acceptance of an organization's or service provider's collection, processing, and sharing of their personal data. It is a crucial component of data protection laws and regulations across the world, notably the European Union's GDPR and the United States' CCPA.

The informed consent principle includes many critical factors that contribute to its efficacy in protecting user privacy and control over personal data. To begin, organizations must offer individuals clear and transparent information about the goal and extent of data collection, the exact data categories being gathered, and how the data will be handled and shared. This information must be stated concisely and clearly, avoiding elaborate legal language that may confuse or mislead people.

Organizations must gain meaningful and legitimate permission by deliberate acts, such as selecting a checkbox or signing, rather than depending on pre-selected alternatives. This means that persons must actively express their consent; silence or inactivity do not constitute consent.

Furthermore, informed permission must be freely provided without force or pressure. Individuals should not face negative repercussions or be refused services if they refuse to grant permission. Organizations must avoid misleading practices, such as employing ambiguous wording or burying consent requests in extensive privacy policies, which might weaken consent's voluntariness.

PbD urges organizations to build granular consent systems that allow consumers to offer explicit and distinct consent for different types of data processing. This strategy gives users a choice over how their data is used and shared, enabling them to provide consent for specific reasons while denying consent for others.

Furthermore, under PbD, the idea of *purpose limitation* is intimately linked to informed permission. Data shall only be collected and processed for the precise objectives expressly indicated to the subject at the time of data collection, according to purpose limitation. If an organization chooses to use the data for reasons other than those previously revealed, it must get further agreement from the individual.

Incorporating informed consent into PbD entails taking the user's point of view into account and ensuring that data processing practices are user-centric. Individuals should be provided with user-friendly interfaces that clearly explain the options accessible to them, including the opportunity to simply grant or withdraw consent.

Transparency is a key tenet of informed consent in PbD. Organizations must maintain clear records of when and how consent was obtained, including the relevant information provided to the individual at the time of consent. This transparency not only reinforces trust between organizations and individuals but also helps demonstrate compliance with data protection regulations.

As technology advances, so do the difficulties and opportunities associated with informed consent in PbD. Artificial intelligence and machine learning, for example, are introducing new complications into data processing and decision-making. In these cases, organizations must take further precautions to ensure that users understand the possible implications of data usage and are given meaningful permission options.

Furthermore, as big data analytics and data-sharing platforms have grown in popularity, data flows across organizations have surged. Obtaining and managing informed permission becomes more complicated in this setting, as organizations must consider data transfers and third-party data processors while respecting user preferences and consent choices.

Organizations can use privacy-enhancing technology to improve the efficacy of informed consent under PbD. Cryptographic approaches, such as *secure multi-party computation*, for example, allow data processing without disclosing raw personal data, maintaining privacy while gaining useful insights.

In addition, privacy-enhancing user interfaces and user-friendly consent management platforms are critical components of putting informed consent into practice. These interfaces should be intended to provide information in a clear and intelligible manner, allowing users to make more educated data decisions.

Education and awareness campaigns are crucial to ensuring that people understand the value of informed consent and their data privacy rights. Organizations may help educate users on their data rights, the importance of informed consent, and the measures they can take to successfully manage their data preferences.

Finally, informed consent is a fundamental element of PbD, giving individuals control over their personal data and allowing them to make informed decisions regarding data collection, processing, and sharing. Organizations may instill confidence in users and guarantee compliance with data protection laws and regulations by including informed consent in the design and development of data processing systems and technologies. In order to properly defend informed consent and protect data privacy in the ever-changing digital ecosystem, user-centric interfaces, clear data practices, and privacy-enhancing technology must be implemented.

The Personal Information Protection and Electronic Documents Act (PIPEDA)

The **Personal Information Protection and Electronic Documents Act** (**PIPEDA**) is a key piece of law in Canada that controls the private sector's acquisition, use, and disclosure of personal information. PIPEDA, enacted in 2000, is intended to protect individuals' privacy by regulating how organizations handle personal data and requiring them to get informed permission for data processing operations. Informed consent is critical under PIPEDA because it allows individuals to make informed decisions about their personal information and gives them control over its usage.

The fundamental goal of PIPEDA is to safeguard personal information during commercial operations such as data gathering by enterprises, banks, merchants, and other commercial organizations. The act applies to federally controlled organizations as well as private-sector organizations operating in provinces that do not have significantly equivalent privacy legislation.

The notion of informed consent is at the heart of PIPEDA. Organizations must notify individuals of the purpose for collecting their personal information before or at the time of collection according to this principle. Individuals must understand how their data will be utilized, and the aim must be precise. This ensures that individuals understand the possible repercussions of disclosing personal information.

Individuals' informed consent must be gained via meaningful and active dialogue. Organizations must offer clear and straightforward explanations of their data management practices in a language that a typical person can comprehend. This is critical to avoiding any deceit or confusion that may lead to people granting permission unwittingly without fully knowing the ramifications of their decision.

Furthermore, PIPEDA requires that consent be given willingly, without compulsion or undue pressure. Organizations must not require permission as a condition of service unless it is required to execute a contract or legal duty. This rule guarantees that individuals are not coerced into disclosing personal information against their choice.

The act also acknowledges that permission can be revoked at any moment. Individuals have the right to cancel their permission if they no longer want the organization to process their personal information. Organizations must abide by this ruling and immediately discontinue further data processing operations.

When personal information is to be used for a new purpose that has not previously been specified, PIPEDA requires organizations to get additional consent from the people affected. This criterion, called *purpose limitation*, guarantees that data is only used for the reasons for which it was obtained or for compatible uses that the subject would reasonably anticipate.

Organizations must implement strong privacy policies and practices in order to comply with PIPEDA's informed consent requirements. These policies should explain the organization's data management practices, such as the sorts of personal information gathered, the reasons for collecting it, how it will be used, and how long it will be maintained.

Organizations must also establish clear protocols for acquiring and managing informed consent. This may entail employing permission forms or online consent methods that explicitly explain the data processing activities and seek individuals' express approval.

Informed consent is essential not just for regulatory compliance, but also for establishing trust and sustaining strong relationships with consumers and clients. Organizations that show a commitment to protecting individuals' privacy and offering meaningful choices about their personal information are more likely to acquire the trust of their stakeholders.

Non-compliance with the informed consent requirements of the PIPEDA can have substantial ramifications for organizations. The **Privacy Commissioner of Canada** (**OPC**) has the jurisdiction to investigate complaints and issue compliance orders. Organizations found in breach of the PIPEDA may face financial fines as well as reputational harm.

As technology advances and new data processing methods develop, obtaining informed permission remains a difficulty. Organizations must stay up to date on technological advances and change their consent practices accordingly. Addressing the intricacies of data analytics, artificial intelligence, and IoT technologies, which might present unique obstacles in acquiring meaningful permission from individuals, is part of this.

Finally, the PIPEDA is a critical component of Canadian privacy law that controls the processing of personal information in the private sector. Informed consent is a cornerstone of PIPEDA, allowing individuals to make informed decisions about their personal data and control over its usage. Organizations subject to the PIPEDA must follow clear and transparent consent procedures, ensure voluntary and meaningful consent, and respect individuals' right to withdraw consent at any time. Organizations may create trust with their consumers and clients while complying with the core elements of PIPEDA by following the principles of informed consent.

The role of consent in other jurisdictions

User consent is a critical and ever-changing part of data privacy and security, delicately woven into the fabric of geographically dispersed governments. In our increasingly interconnected world, where data easily crosses borders, it is critical to recognize that the meaning, interpretation, and enforcement of user permissions can vary greatly from state to jurisdiction. This chapter digs into the complex landscape of user consent needs across geographically separated areas, offering insight into the many legal and regulatory regimes that impact users' relationships with their data. Understanding the subtle interactions of these elements is critical for businesses and organizations attempting to negotiate the complicated landscape of global data compliance while honoring user rights and expectations.

European Union

Consent is an important part of data protection and privacy rules in the **European Union** (EU). The GDPR lies at the heart of the EU's data protection architecture, enshrining consent as a vital pillar for guaranteeing the legitimate and fair processing of personal data. The GDPR establishes rigorous requirements for acquiring, managing, and utilizing consent, emphasizing the significance of giving individuals control over their personal information.

Consent is defined in the GDPR as "*any freely given, specific, informed, and unambiguous indication of the data subject's wishes by which he or she, by a statement or a clear affirmative action, signifies agreement to the processing of personal data relating to him or her.*" This definition describes the key elements of valid permission that organizations must consider when collecting and processing personal data.

First and foremost, consent must be freely granted. Individuals must have a true option and must not face pressure or negative repercussions for declining permission. Organizations must prevent any power imbalances that might impair an individual's capacity to grant voluntary consent.

Second, permission must be expressed and informed. Individuals must be fully informed about the reasons for which their data will be processed, the types of personal data being collected, and any third parties with whom the data may be shared. This criterion guarantees that individuals are aware of the entire extent of data processing and may make informed consent decisions.

Third, consent must be unequivocal and need a clear affirmative action. Consent cannot be granted based on pre-ticked boxes or silence. Individuals must instead actively take an affirmative move, such as checking a box or clicking a button, to indicate their acceptance of the data processing operations.

Furthermore, the GDPR requires that permission be tailored to each processing purpose. Organizations cannot get broad or ambiguous permission to engage in a wide variety of unconnected activities. Instead, permission must be granular, which means that people must be given different consent or refusal alternatives for each particular processing purpose.

In addition, the GDPR recognizes that permission is not the sole legal basis for processing personal data. There are various legal bases for data processing, including the need for processing for the fulfillment of a contract, compliance with a legal duty, protection of vital interests, and the data controller's or a third party's legitimate interest.

When relying on permission as the lawful basis for processing, organizations must adhere to the GDPR's rigorous standards. For the processing of sensitive personal data, such as health information or biometric data, expressed consent must be obtained. Consent for non-sensitive personal data must nevertheless be clear and affirmative.

Obtaining valid consent is a continuous process, not a one-time occurrence. Organizations must guarantee that individuals have the ability to withdraw their permission at any moment and without repercussions. Withdrawing consent should be as simple as granting it, and organizations should halt further processing as soon as a withdrawal request is received.

Organizations bear a considerable weight of duty under the GDPR to show compliance with consent obligations. This entails keeping meticulous records of consent, including the date and time of consent, the information offered to the subject, and the technique used to get consent.

The GDPR's approach to data protection is not limited to consent. Transparency, purpose limitation, data reduction, accuracy, storage limitation, and security are also emphasized in the rule. Together with permission, these principles create a complete framework for safeguarding individuals' rights and guaranteeing the responsible and ethical use of personal data.

Furthermore, the GDPR imposes severe fines for failure to comply with consent requirements and other data protection duties. Organizations that violate the GDPR face fines of up to €20 million or 4% of their worldwide annual sales—whichever is greater. These sanctions demonstrate the EU's seriousness about data protection and the crucial role of consent in protecting individuals' privacy rights.

In the EU, the function of consent goes beyond the GDPR to additional rules and standards promoting data protection and privacy. The ePrivacy Directive, which will shortly be superseded by the ePrivacy Regulation, for example, particularly addresses permission in the context of electronic communications and cookies. Unless strictly essential for the provision of a service specifically requested by the user, this law demands explicit agreement for the use of cookies and similar tracking technology.

In conclusion, the function of consent in the EU is critical in protecting individuals' privacy and data protection rights. Consent must be freely provided, explicit, informed, and unequivocal under the GDPR and other relevant rules. It gives people control over their personal data and allows them to make educated decisions about how it is used. Organizations must comply with tight standards for acquiring and managing consent or else face harsh fines. Organizations may develop a culture of privacy and responsible data management in the EU and beyond by adhering to the principles of consent and data protection.

United States of America

The role of consent in data protection and privacy in the United States of America is a diverse and developing environment controlled mostly by a patchwork of federal and state laws, industry self-regulation, and individual corporate rules. Unlike the EU's GDPR, which emphasizes explicit and specific consent as a primary legal basis for data processing, the United States takes a more nuanced approach to consent, with multiple legal frameworks and sector-specific regulations shaping consent's role in data privacy.

The United States does not have a comprehensive data privacy regulation similar to the GDPR at the federal level. Instead, sector-specific legislation handles certain areas of data privacy and consent. The **Health Insurance Portability and Accountability Act** (**HIPAA**), which controls the use and sharing of **protected health information** (**PHI**) by healthcare providers, health plans, and other entities covered by the act, is one of the most important federal laws. Individuals are required to obtain written authorization for certain uses and disclosures of personal health information under HIPAA, with exceptions for treatment, payment, and healthcare operations.

Similarly, the **Children's Online Privacy Protection Act** (**COPPA**) is a federal legislation that requires parental approval to collect and process personal information from children under the age of 13. Before collecting, using, or disclosing personal information from minors, online businesses must get verified parental approval under COPPA.

State-level data protection rules also have an impact on the function of consent in the United States. Several states have established their own data privacy laws, which may include permission requirements. The CCPA and the **Virginia Consumer Data Protection Act** (**VCDPA**), for example, both give customers the ability to refuse the sale of their personal information. This entails gaining individuals' prior authorization before selling their data to third parties.

In addition to particular regulations, industry self-regulation initiatives influence consent practices in the United States. The **Digital Advertising Alliance** (**DAA**) and the **Interactive Advertising Bureau** (**IAB**), for example, have self-regulatory frameworks in place for online behavioral advertising. These frameworks frequently call for gaining customer consent before using cookies and tracking technology for targeted advertising objectives.

Furthermore, specific corporate rules and terms of service agreements impact consent practices in the United States. Companies may provide individuals with the choice to opt in or opt out of specific data processing activities, and consent requests are frequently included in privacy policies or user agreements.

However, the United States' absence of comprehensive federal data protection legislation has resulted in certain restrictions with regard to the role of consent in data privacy. In contrast to the GDPR, there is no general requirement for express and specific permission as the basic legal foundation for data processing. Instead, many laws and regulations in the United States focus on informing people about data practices and giving them the option to opt out of particular activities.

Furthermore, the complexity and variety of data privacy rules across different states might provide issues for national firms. Compliance with a patchwork of state regulations may necessitate organizations adopting different consent practices depending on where their clients or users are located.

Furthermore, as new technologies and data processing practices arise, the role of permission in data privacy is always expanding. Emerging technologies such as artificial intelligence, machine learning, and the IoT pose new issues in getting meaningful permission from individuals, particularly when data is gathered and processed in complicated and automated ways.

Finally, in the United States, the role of consent in data protection and privacy is determined by a combination of federal and state legislation, industry self-regulation, and corporate policy. While specialized statutes such as HIPAA, COPPA, CCPA, and VCDPA address permission requirements for certain industries, there is no overall federal data protection legislation like the GDPR. Because there is no consistent approach to permission, there is a diverse landscape of consent practices among businesses and jurisdictions. As technology advances, so will the importance of permission in data privacy, demanding continual efforts to achieve a balance between empowering individuals with data ownership and enabling responsible data practices for businesses and organizations.

India

In India, the function of informed consent is crucial to data security and privacy, as governed by the country's data protection framework and several sector-specific rules. The **Personal Data Protection Bill (PDPB)** is the principal legislation governing data protection in India, with the goal of establishing a comprehensive and effective data protection framework in the country. Informed permission is critical in the PDPB because it gives individuals control over their personal data and guarantees that data processing activities are carried out publicly and responsibly.

Informed consent, according to the PDPB, is "*any manifestation of a data principal's free, specific, informed, and unambiguous expression of will by which he or she, through a statement or by a clear affirmative action, signifies agreement to the processing of personal data.*" This definition is consistent with worldwide best practices and emphasizes the requirement for consent to be explicit, precise, and gained by the data principal's unambiguous and positive action.

The PDPB requires organizations to give individuals clear and comprehensive information regarding the reasons for which their personal data will be processed in order to get informed permission. The data principal must be fully informed about the types of personal data that will be collected, the data processing methods, the data retention periods, and any intended data sharing with third parties. This extensive disclosure is required to ensure that persons have a thorough awareness of the data processing activities and may make informed consent decisions.

The PDPB requires that permission be freely granted, which means that persons must be able to make an informed decision without coercion or undue influence from the data fiduciary (the organization collecting and processing the data). Organizations are forbidden from making the supply of services or benefits conditional on getting consent unless the data processing is required for contract performance.

Furthermore, the PDPB recognizes the data principal's right to withdraw permission at any moment. Withdrawing consent must be as simple as obtaining consent, and organizations must stop data processing operations as soon as a withdrawal request is received. This clause emphasizes the data principal's sovereignty over their personal data and guarantees that they may successfully exercise their data choice management rights.

To acquire informed permission, the PDPB requires organizations to use plain and straightforward language that the ordinary data analyst can comprehend. Complex legal jargon or extensive privacy rules that might confuse or mislead people are avoided. Organizations are instead urged to implement user-friendly interfaces and consent procedures that allow for easy comprehension and explicit declaration of consent.

In addition, the PDPB recognizes that permission is not the only legal basis for data processing. The PDPB, like the GDPR, allows alternative legal reasons for data processing, including contract performance, legal obligation fulfillment, and legitimate interests pursued by the data fiduciary or a third party. When relying on consent as the lawful basis for processing, organizations must follow the PDPB's rigorous standards for acquiring and managing consent.

Aside from the PDPB, India has a number of sector-specific legislations and regulations governing the role of consent in data protection. The Information Technology (Reasonable Security Practices and Procedures and Sensitive Personal Data or Information) Rules, 2011, for example, require organizations handling sensitive personal data or information to obtain prior consent from data subjects before collecting and processing such data under the Information **Technology Act (IT Act)**, 2000 .

Furthermore, in the healthcare sector, the Medical Council of India (MCI) has developed rules on informed consent. The standards for gaining informed permission from patients prior to medical operations, treatments, or the exchange of medical information are outlined in these recommendations.

Case law and court interpretations also have an impact on the function of informed consent in India. In situations involving medical treatments and clinical trials, Indian courts have recognized the necessity of informed consent as a basic right.

In India, however, there are obstacles to properly implementing informed consent practices. Individuals' knowledge of consent and the capacity to offer meaningful consent might be impacted by language hurdles, limited computer literacy, and various cultural norms. Furthermore, the fast-changing technological landscape, such as the growth of artificial intelligence and big data analytics, adds significant difficulties to gaining informed permission for data processing operations.

Finally, the function of informed consent in India is an important component of the country's data protection and privacy policy. Informed permission, guided by the PDPB and sector-specific legislation, gives individuals control over their personal data while also ensuring openness and responsibility in data processing practices. Organizations must implement simple and easy-to-use consent processes while preserving the data principal's ability to withdraw consent at any time. As India's data protection regime evolves, obtaining meaningful informed consent will remain a problem, necessitating constant efforts to find a balance between individual liberty and responsible data processing.

Challenges to meaningful informed consent

The idea of informed consent poses considerable problems in an era of fast-increasing technology and altering economic models, particularly in connection to new technologies and human behavior. As innovative technologies such as artificial intelligence, the IoT, biometric identification, and big data analytics become more prevalent in our everyday lives, the complexity of data processing activities raises questions regarding the sufficiency of getting informed permission. Furthermore, the emergence of novel business models that depend heavily on data-driven decision-making presents further hurdles to ensuring that individuals have a thorough awareness of how their data is being utilized. Furthermore, the digital era and continual contact change people's perceptions and willingness to grant meaningful permission. Addressing these numerous issues is critical to protecting individual privacy and preserving confidence in the digital environment.

New technologies and business models

Because of their data-intensive nature and frequently opaque data processing processes, new technologies present particular obstacles to getting informed consent. Artificial intelligence algorithms and machine learning models, for example, may require complicated data processing and decision-making processes that are difficult for people to completely grasp. As a result, individuals may be unsure about the dangers and repercussions of consenting to the use of personal data in such technologies. Furthermore, many emerging technologies, such as IoT devices, may gather data in real-time, demanding continuing and dynamic consent methods to guarantee that users are always aware and in control of their data.

Innovative business models that rely on data as a primary asset, such as targeted advertising, personalized services, and data monetization, pose difficulties in getting informed permission. These models frequently entail considerable data sharing and processing across numerous organizations, which may result in a lack of transparency about data flows and goals. When data is shared and used for diverse purposes throughout a network of firms, individuals may struggle to comprehend the entire scope of data processing. Furthermore, business models that rely on lengthy and complicated privacy rules or permission methods might overwhelm consumers and result in *consent fatigue*, where people assent without fully comprehending the ramifications.

The rise of new technologies and novel business models, such as big data and the IoT, creates significant issues in getting consumers' informed permission. The term "big data" refers to huge amounts of data created by numerous sources such as social media, sensors, and internet activities. The IoT consists of networked devices that collect and share data to allow smart and seamless interactions. Both of these technologies provide significant opportunities for organizations to improve their goods and services, but they also create privacy and data protection issues. Because of the sheer scope and complexity of data processing in big data and IoT contexts, gaining genuine informed consent is a difficult issue.

The difficulty in providing users with complete information about data processing activities is one of the key obstacles to getting informed consent in the context of big data and IoT. Data is acquired from numerous sources with these technologies, frequently without direct interaction with the data subject. The obtained data may be merged, consolidated, and analyzed to yield important insights. However, due to the volume and diversity of data, specifying the actual aims and potential repercussions of data processing at the moment of data collection becomes difficult. Because of this complexity, consent forms may be ambiguous or confusing, making it difficult for individuals to make informed judgments regarding their data.

Furthermore, the decentralized nature of data collecting in IoT devices makes gaining explicit and granular consent difficult. IoT devices, such as smart home appliances or wearable gadgets, frequently gather data in an ongoing and automated manner. Because of the impracticality of obtaining agreement for each data point, consent may be gained on a wider level. Such broad permission, however, may fail to sufficiently inform consumers about the numerous ways their data is used across multiple devices and services, potentially leading to inadvertent data sharing and uses that users did not anticipate.

Furthermore, in big data and IoT contexts, the idea of *secondary uses* of data creates significant issues for informed permission. The practice of using data obtained for one reason for unrelated objectives is referred to as secondary usage. For example, data generated to improve the operation of a product may be repurposed for marketing or profiling reasons. Individuals may not have explicitly consented to such secondary uses in these circumstances, raising issues about data control and transparency.

Another major issue is the possibility of data anonymization and re-identification. Data is frequently anonymized in big data and IoT environments to protect individuals' identities. However, developments in data analytics and re-identification techniques have cast doubt on anonymization's usefulness. Individuals may unintentionally consent to data sharing due to the mistaken notion that their identities are sufficiently safeguarded, yet the danger of re-identification may jeopardize their privacy.

The fast growth of these technologies adds to the difficulties of informed consent. Data processing methods and business model innovations can outstrip regulatory frameworks and best practices. This volatile climate makes it difficult for organizations to maintain compliance and transparency in their consent practices, and it may also result in variations in how informed consent is gained across industries and countries.

Addressing these issues necessitates a proactive and multifaceted strategy. Transparency in data collecting and processing practices must be prioritized by organizations, with individuals receiving clear and intelligible information about how their data will be used and shared. Adopting granular consent options and implementing user-friendly consent processes can allow individuals to make meaningful decisions about their data.

Regulators are critical in ensuring that data protection laws and standards keep up with technological advances. Clear and up-to-date legislation can help organizations acquire valid informed consent while protecting individuals' privacy rights. Furthermore, authorities and industry stakeholders should work together to create best practices and standards for consent in big data and IoT contexts.

Furthermore, educating people about their data rights and the repercussions of data sharing is critical. Awareness campaigns may help people realize the value of their data and the significance of giving informed permission. Giving people control over their data can help to develop a culture of privacy and appropriate data practices.

To summarize, acquiring informed permission in the context of big data and the IoT is a complex task. The complexity of data processing, decentralized data gathering, secondary uses, and developing technology all present substantial challenges to providing users with transparent and meaningful consent alternatives. To address these difficulties, organizations, authorities, and people must work together to create a balance between technological innovation and data protection, ensuring that privacy stays at the forefront of data-driven breakthroughs.

Human behavior

Human behavior, which is impacted by the frequent use of digital platforms, can also have an impact on the efficacy of informed consent. Because digital material is fast-paced and easy to consume, people may scan over permission forms or privacy rules without paying close attention to the contents. This behavior, termed *click-through consent*, may result in people accidentally providing consent without fully comprehending it. Individuals' willingness to grant informed consent might also be influenced by their sense of the value exchanged between sharing data and obtaining services or rewards. Individuals may be hesitant to engage in complex permission processes if they do not see any real advantage of data sharing.

Addressing these issues necessitates a multifaceted strategy including stakeholders from numerous disciplines. To begin, technology developers and data processors should prioritize user-centric design by displaying information in a clear and intelligible manner and implementing dynamic permission mechanisms that allow users to make granular data choices. Transparent data processing procedures should be established so that people may understand how their data is being utilized in emerging technologies.

Second, firms must prioritize data openness and responsibility. Simplifying privacy rules and permission forms, as well as making information regarding data processing practices widely accessible, can encourage more informed consent. Furthermore, organizations should investigate alternative data governance models such as data trusts or data cooperatives, which provide individuals with collective ownership over their data.

Regulators have a significant impact on the landscape of informed consent. In the context of new technology and commercial models, they should create clear norms and criteria for getting informed consent. Furthermore, regulatory agencies have the authority to enforce compliance and hold organizations accountable for getting valid permission and maintaining data security.

It is critical to educate people about the value of their data and the significance of informed consent. Awareness campaigns and digital literacy efforts may assist people in understanding their data rights and urge them to think critically about consent processes.

Finally, the difficulties in getting informed consent in the context of new technology, economic models, and human behavior need a comprehensive and collaborative strategy. To protect individual privacy rights, the combination of breakthrough technologies, data-driven business models, and developing human behavior necessitates clear, user-centric, and dynamic permission practices. By resolving these issues, society may strike a balance between technical progress and data security, encouraging trust and confidence in the digital era.

Alternatives to consent

Obtaining informed permission from individuals for data processing activities is frequently regarded as the gold standard in the digital identity area for safeguarding privacy and giving users control over their personal information. However, as the digital world advances and data processing grows more sophisticated, it is becoming clear that depending exclusively on consent may no longer be practicable or adequate. As a result, different methods of gaining consent are being investigated in order to solve the difficulties of getting meaningful permission in particular settings while still protecting human rights and privacy.

The idea of *legitimate interests* as a valid basis for data processing is an alternative to consent. Legitimate interests enable organizations to handle personal data without the data subject's explicit consent where there is a genuine and compelling basis for doing so and the data subject's rights and freedoms are not violated. This method is recognized in the GDPR of the EU and provides a flexible and balanced framework for data processing. It allows organizations to process data for legitimate objectives such as fraud protection, network security, and direct marketing without requiring express agreement for each usage.

Another option is to use the data *anonymization* and *pseudonymization* concepts. Anonymization is the process of removing identifiers from data in such a way that the individual cannot be identified directly or indirectly. Pseudonymization, on the other hand, is the process of replacing identifying information with pseudonyms, which makes the data less clearly traceable. Organizations can reduce privacy concerns and perhaps process data for research or statistical purposes without obtaining individual authorization by using strong anonymization or pseudonymization processes. To guarantee privacy, it is critical to verify that the data is really anonymized or effectively secured against re-identification.

In the context of digital identification, blockchain technology provides another option for permission. DLT enables the establishment of self-sovereign identities, in which people maintain complete control over their personal data without relying on a centralized authority. Individuals can choose to disclose certain qualities or data points to service providers or other parties in self-sovereign identification systems, removing the requirement for continuous and wide consent for data sharing. Instead of requesting consent for each transaction, consumers may have more granular control over their data, increasing privacy and security.

Furthermore, contextual integrity is a privacy concept that emphasizes data practices that are aligned with the environment in which information is exchanged. Contextual integrity, as opposed to a one-size-fits-all approach to permission, takes into account the unique context of data collection and usage, such as the connection between the data subject and data collector, as well as the societal norms and expectations of the circumstance. This paradigm enables more nuanced and situation-specific privacy practices, eliminating superfluous data gathering or processing when it is not essential.

PbD is a larger approach that proposes incorporating privacy issues into system and process design and development. Organizations may proactively integrate privacy safeguards into their goods and services by embracing PbD principles, minimizing their reliance on permission as the primary privacy precaution. To enhance privacy from the start, PbD emphasizes data reduction, purpose limitation, and data protection measures.

Furthermore, as alternatives to standard permission models, data sharing agreements and data trusts are emerging. Data trusts are organizations or legal entities that operate as data custodians on behalf of data subjects, ensuring that data is utilized for agreed-upon objectives and in the data subjects' best interests. Organizations may build collaborative and responsible data-sharing practices by utilizing data trusts without requiring explicit consent for each data transfer.

To summarize, although informed consent remains a critical pillar of privacy and data security, the digital identity sector is experimenting with new ways to solve the issues provided by emerging technologies and data processing practices. Anonymization, self-sovereign identities, contextual integrity, PbD, data trusts, and data sharing agreements are all possible alternatives to relying entirely on permission. These methods offer a more nuanced and adaptable framework for handling data in a privacy-conscious way, enabling creative and responsible data practices while respecting individual rights and maintaining privacy in the digital era.

Enforcement models in informed consent

Informed consent and digital identity enforcement mechanisms are critical in ensuring that individuals' privacy rights are safeguarded and that organizations follow data protection legislation. These enforcement methods include means for monitoring compliance, investigating suspected violations, and penalizing organizations that fail to secure valid and meaningful permission or mishandle digital identity data. The success of enforcement models can have a substantial influence on establishing responsible data practices and generating confidence in the digital ecosystem.

Regulatory monitoring by government agencies is one of the key enforcement models in the informed consent and digital identity sector. Many nations have data protection authorities or supervisory organizations in place to ensure that data protection laws and regulations are followed. These authorities have the power to investigate complaints, perform audits, and levy fines or punishments on organizations that violate data protection regulations. Each member state of the EU, for example, has its own data protection authority, and the GDPR authorizes these authorities to enforce data protection regulations and levy fines of up to 4% of a company's global annual sales for major violations.

Self-regulation efforts and industry rules of conduct are another type of enforcement mechanism. To encourage best practices and accountability in data processing, industry associations and organizations may adopt voluntary codes of conduct. Companies who comply with these guidelines demonstrate their commitment to proper data processing and may benefit from improved customer trust. However, the efficacy of self-regulation models is dependent on organizations' desire to engage, as well as the presence of robust systems to monitor compliance and remedy possible breaches.

Civil action is another means of enforcing informed consent and digital identity. People or groups of people who think their privacy rights have been infringed may sue the data controller or processor. Civil actions can result in monetary compensation for the people involved and can serve as a deterrence for other organizations to follow data privacy requirements. Civil litigation, on the other hand, may be a time-consuming and expensive procedure, and many people may lack the finances or experience to seek legal action.

Furthermore, **data protection impact assessments** (**DPIAs**) are crucial in implementing informed consent and digital identity practices, particularly for high-risk processing operations. DPIAs are systematic analyses of the possible risks and consequences of data processing on the privacy of users. Certain forms of data processing require organizations to do DPIAs, and authorities may monitor and evaluate these assessments to guarantee compliance with data protection regulations. If a DPIA identifies major privacy concerns that have not been sufficiently managed, the regulator may step in and mandate adjustments to data processing operations.

In addition to existing enforcement strategies, new technologies are being investigated in order to increase enforcement in the digital identity domain. For example, blockchain and distributed ledger technologies enable the creation of immutable and transparent records of data processing processes. Regulators and people may get better insight into how data is used and shared by employing blockchain-based systems to record consent transactions and data-sharing events, boosting accountability and trust.

Artificial intelligence and machine learning can also help with enforcement in the area of digital identification and consent. By analyzing data processing processes at scale, AI-powered solutions can assist in identifying possible infractions or non-compliance with consent laws. Regulators may prioritize investigations and spend resources more effectively by employing AI to spot abnormalities or trends suggestive of privacy violations.

Despite the relevance of enforcement methods, they are difficult to establish and sustain in the informed consent and digital identity domain. The worldwide nature of data transfers, as well as the possibility of jurisdictional conflicts, is a key concern. International organizations must traverse various data protection laws and regulatory obligations, making it difficult to maintain uniform enforcement across borders.

Furthermore, the constant evolution of technology and data processing practices necessitates the adaptation of enforcement strategies in order to keep ahead of new dangers. To effectively address innovative privacy problems, regulators and authorities must be nimble and well-informed. Collaboration between regulatory agencies, industry stakeholders, and technology specialists is essential for designing responsive enforcement procedures that can keep up with technological advances.

Furthermore, limited resources might limit the breadth and efficacy of enforcement actions. Data protection agencies may be constrained by limited finances and personnel, making it difficult to address an increasing number of data protection complaints and investigations. Adequate financing and capacity-building programs are critical for bolstering enforcement skills and sustaining a vibrant privacy enforcement environment.

In conclusion, informed consent and digital identity enforcement mechanisms are critical to protecting individuals' privacy rights and guaranteeing responsible data practices. Data protection obligations are enforced by governmental monitoring, self-regulation efforts, civil lawsuits, data protection impact assessments, and developing technologies. However, enforcement attempts must overcome obstacles such as jurisdictional complications, technological improvements, and limited resources. Collaboration between regulators and industry and technology specialists will be critical in building effective and agile enforcement mechanisms that preserve privacy in the digital era. By supporting and enforcing privacy rights, enforcement models may establish a trustworthy and privacy-centric digital environment that empowers individuals while also encouraging responsible data innovation.

The future of privacy

As technology advances and society grapples with the delicate balance between convenience and data protection, the future of privacy in the digital identity realm is poised to undergo profound adjustments. With the rising prevalence of digital interactions, the necessity for strong privacy safeguards in maintaining digital identities becomes more apparent. Innovative solutions that give people more control over their personal information while providing smooth and safe digital experiences are expected to emerge in the future.

The growth of **self-sovereign identities** (**SSI**) is one of the important themes affecting the future of privacy in the digital identity field. Individuals have full ownership and control over their digital identities in a decentralized identity concept that is SSI. Without relying on centralized authority, people may securely maintain their identity traits, selectively release information, and establish their identity using SSI. Blockchain technology is critical to the implementation of SSI, providing a tamper-resistant and verifiable platform for identity management. SSI claims to improve privacy, eliminate identity fraud, and encourage user autonomy in the digital arena by removing the need for middlemen and central databases.

Furthermore, advancements in zero-knowledge proofs and cryptographic techniques will be critical in improving privacy in the digital identity field. Individuals can use zero-knowledge proofs to verify certain truths about themselves without exposing the underlying data. Users, for example, can show they are above a specific age without providing their precise birthday. These strategies allow for *minimal disclosure* while protecting privacy, which is especially useful in situations when data disclosure is required for verification or access.

In the future, privacy-enhancing technologies will be combined with AI and machine learning to create more secure and user-centric digital identities. Identity verification systems driven by AI can detect and prevent fraudulent activity while protecting sensitive personal data. Furthermore, AI algorithms may be constructed with privacy in mind, allowing data processing and analysis while maintaining user anonymity and data security.

Furthermore, the rising relevance of user permissions and granular data control will be a characteristic of the digital identity space's future. Organizations and service providers must prioritize data openness and provide user-friendly consent processes. To coincide with expanding data processing activities, the use of dynamic consent, in which individuals may adjust their choices and permissions in real time, will become more common. Organizations will recognize that gaining meaningful permission is critical to instilling confidence in consumers and fostering a privacy-first culture as data becomes more valuable.

Furthermore, cross-border data flows and international data protection legislation will affect the future of digital identity privacy. Global data protection regulations, such as the EU's GDPR and equivalent legislation throughout the world, will affect how organizations handle digital identities. Countries will likely collaborate more in the future to harmonize privacy rules and promote secure and lawful data sharing across borders.

As the IoT and linked gadgets grow more common, so will privacy problems in the digital identity realm. To ensure data privacy in IoT contexts, strong security mechanisms, data encryption, and explicit user authorization for data collection and sharing will be required. To reduce the hazards associated with unauthorized access and identity fraud, secure identity verification technologies such as biometric authentication will become increasingly common.

Finally, the future of privacy in the digital identity area promises intriguing advancements and opportunities. Self-sovereign identities, zero-knowledge proofs, AI-powered identity verification, granular consent systems, and worldwide data protection standards will fundamentally alter how digital identities are maintained and safeguarded. Giving people more control over their personal data while ensuring seamless and safe digital experiences will be key to the future of privacy. The digital identity sector may build a more privacy-centric and trustworthy environment for users globally by adopting privacy-enhancing technology, supporting openness, and adhering to PbD principles.

Summary

This chapter looked at the difficulties and obstacles of acquiring informed permission in a variety of settings, including data privacy, digital identity, and humanitarian organizations. It highlighted the significance of user-centered design, transparent data processing protocols, and ongoing consent withdrawal choices in order to improve informed consent. The chapter also covered how new technologies, such as big data and the IoT, affect getting informed permission, as well as the legal and regulatory frameworks that regulate consent requirements, including the GDPR. It also emphasized the need for clear and user-friendly permission methods, as well as the possible penalties for non-compliance, including significant fines under the GDPR. In the next chapter, we will take an in-depth look at the security perspective of identity and access management systems.

8

IAM – the Security Perspective

IAM, or **identity and access management**, is a security framework that focuses on managing user identities and controlling access to resources and applications within an organization. IAM systems provide a way to manage user authentication, authorization, and user account provisioning and de-provisioning. From a security perspective, IAM plays a critical role in protecting an organization's digital assets by controlling who has access to what information, applications, and resources.

In this chapter, we will cover the following topics:

- IAM principles and frameworks
- The lifecycle of a strong IAM system
- Compliance and regulatory considerations

IAM security fundamentals

IAM is essential for maintaining the confidentiality, integrity, and availability of an organization's information systems. By providing centralized control over user accounts, IAM systems help prevent unauthorized access to sensitive data and resources. IAM systems also provide a way to monitor and audit user activity, which can help detect and prevent security incidents.

From a security perspective, IAM systems must be designed with several key considerations in mind. These include the following:

- **Strong authentication**: IAM systems must provide strong authentication mechanisms, such as multi-factor authentication, to prevent unauthorized access to user accounts
- **Authorization controls**: IAM systems must provide granular authorization controls, allowing administrators to manage user access to specific resources based on job roles, responsibilities, and other criteria
- **User account management**: IAM systems must provide tools for managing user accounts, including creating and de-provisioning user accounts, managing passwords, and enforcing password policies

- **Auditing and logging**: IAM systems must provide comprehensive logging and auditing capabilities, allowing administrators to monitor user activity and detect potential security incidents
- **Integration with other security controls**: IAM systems must integrate with other security controls, such as firewalls, intrusion detection systems, and **security information and event management** (**SIEM**) systems, to provide a layered defense against cyber threats.

Overall, IAM plays a critical role in securing an organization's digital assets, and it must be designed and implemented with security considerations in mind.

Multi-factor authentication and privilege-based controls are considered the kinds of identity and access management that may help security leaders address major vulnerabilities in their enterprises.

IAM is the umbrella term for the IT security frameworks, procedures, and practices aimed at managing digital identities. Identity management is centered around the procedures of provisioning and de-provisioning identities, as well as assuring identity security, authentication, and authorization for resource access or specified tasks. Even though people have several accounts linked to them, they only have one unique digital identity. The implementation of distinct access controls for every account is dependent on the particular resource and context.

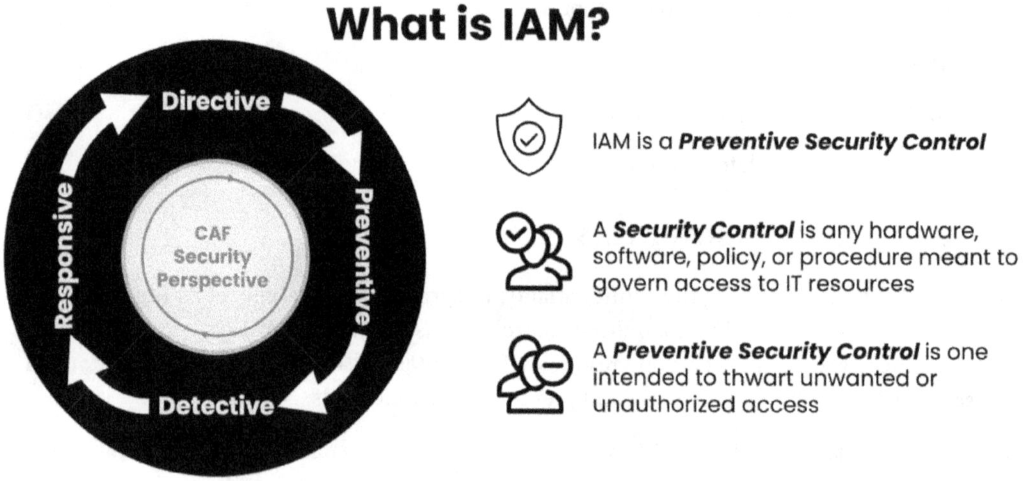

Figure 8.1 – IAM security perspective

The principal objective of IAM is to make sure that any specified identity has access to the correct resources (databases, applications, networks, etc.) within the correct context.

IAM principals

Identity management is the cornerstone of security that ensures that users are granted the appropriate access and that unauthorized users are prevented from accessing data, systems, and applications. IAM enterprise policies set rules for the following:

- Identifying users and giving them designated roles
- Recognizing the data, platforms, and other domains that IAM protects
- Defining specific security levels and access rights for systems, sensitive data, information, and places
- Overseeing the IAM system's insertion, deletion, and change of individuals
- Overseeing the IAM system's addition, deletion, and adjustment of access rights for roles

The following diagram depicts the flow between personal cloud-based applications and organizational services, be they public or private sector applications, that enable connectivity and interoperability between these services:

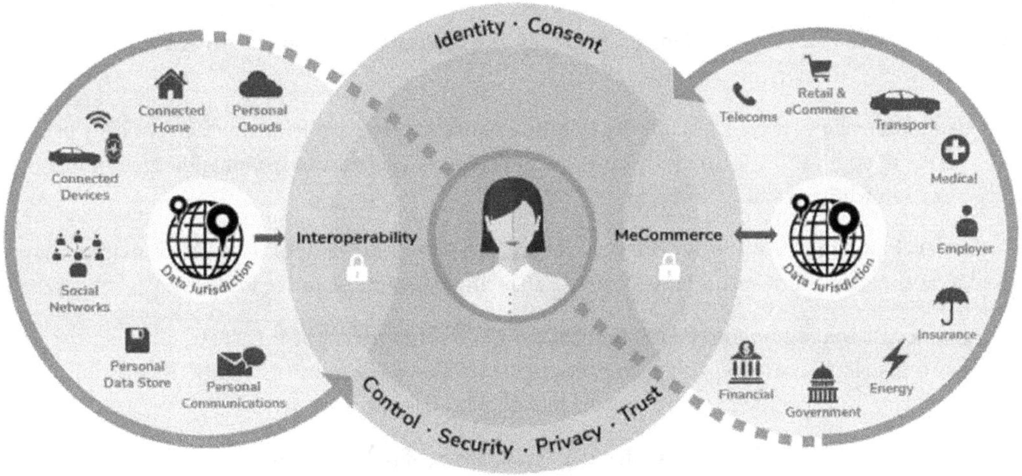

Figure 8.2 – Interoperability flow between cloud infrastructures

IAM is a critical component of every organization's security strategy. In today's digital geography, where operations and structure are critical, applying strong IAM principles is critical to safeguarding sensitive data and protecting against unauthorized access. This section delves into the key IAM concepts, emphasizing their use to improve security in both operational and structural settings.

IAM's underlying component is authentication. It entails verifying the identification of people or entities attempting to access systems, applications, or infrastructure. Passwords, fingerprints, and **multi-factor authentication** (**MFA**) are used as authentication mechanisms to guarantee that only authorized individuals obtain access. Robust authentication systems are required in application and infrastructure security to prevent unauthorized entry points.

Once authenticated, authorization decides which activities and resources a person or entity is permitted to access. Effective authorization guarantees that only those with the necessary permissions may carry out certain tasks. Granular authorization rules must be enforced for apps and infrastructure, restricting access to just what is required for a user's job or purpose.

The idea of least privilege states that users should be given just the access necessary to do their responsibilities. By using this idea at both the application and infrastructure levels, the attack surface is reduced, and possible harm is limited in the event of a breach.

RBAC is a popular technique in application security that allocates rights depending on the roles of the users. You may expedite access control and guarantee that users have the appropriate permissions for their job duties by using RBAC within apps.

SSO improves application security by allowing users to log in once and access different applications without having to re-enter their credentials. SSO may link centralized authentication and authorization with IAM systems, simplifying access control.

Identity federation allows users from one organization to access resources from another, typically using SSO. In terms of application security, this can extend trust and access control outside the bounds of the organization while retaining security.

With the development of APIs, it is critical to secure access to these endpoints. IAM principles may be used to authenticate and authorize API users, ensuring data and resource integrity.

Privileged access management (**PAM**) is a critical IAM component for infrastructure security. It controls and restricts access to privileged accounts such as system administrators. PAM solutions improve security by imposing rigorous constraints on who has access to sensitive systems and data.

Enforcing 2FA for privileged accounts offers an extra degree of protection to infrastructure security. Even if credentials are compromised, attackers must also have a secondary authentication mechanism, making unauthorized access more difficult.

Robust audit and monitoring capabilities are required for effective IAM. This includes tracking who accesses vital systems, when they do so, and what actions are performed. This information is critical for quickly recognizing and responding to security events.

IAM covers the whole user lifetime, from onboarding to offboarding. To avoid unauthorized access or lingering privileges, it is critical for application and infrastructure security to ensure that user access is granted and de-provisioned effectively.

Many sectors have strict compliance standards. It is critical for both application and infrastructure security that IAM practices adhere to these rules. Noncompliance can have serious legal and financial consequences.

IAM is a continuous process, not a one-time event. It is critical to examine and update access controls, authentication mechanisms, and monitoring practices on a regular basis in order to respond to emerging security risks.

Security breaches and data theft are continual risks in today's digital world. To mitigate these risks, strong IAM principles must be implemented. Organizations may secure sensitive data, preserve operational continuity, and develop a robust defense against hostile actors by following these principles at both the application and infrastructure levels. Understanding IAM concepts and how to use them is the first step toward a more secure digital future.

After establishing the fundamental concepts of IAM security, we will proceed to investigate alternative access control approaches and frameworks. In the subsequent section, we will look at the complexities of these models and frameworks, explaining their roles in improving security measures inside IAM systems.

Access control models and frameworks

Access control models and frameworks are important components of security systems because they regulate how people or entities are permitted or denied access to resources, systems, or data inside an organization's network or information systems. These models and frameworks specify the rules, regulations, and procedures for managing and enforcing access rights. They are critical for assuring the confidentiality, integrity, and availability of data and resources, as well as ensuring that only authorized users or entities have proper access.

Here's a brief explanation of these terms:

- **Access control models**: These are philosophical or theoretical approaches to access control that outline how it should be implemented. They offer an organized approach to thinking about and designing access control systems. The following are examples of common access control models:
  - **Discretionary access control** (**DAC**): Access choices in DAC are made at the resource owner's discretion. Users have complete control over who has access to their resources.
  - **Mandatory access control** (**MAC**): Access is enforced by MAC using security labels and preset security policies. Users have little influence over access selections.
  - **Role-based access control** (**RBAC**): RBAC grants permissions to roles, which are subsequently assigned to users. Roles, rather than individual identities, are used to make access choices. RBAC is the main technology behind IAM systems. It uses job roles that have been predefined to control access to particular systems and data. User access privileges are changed to reflect their new work roles when they join or leave the organization. Workers in the HR division, for instance, might have access to various systems and personnel information related to their particular job functions, as illustrated in the following table:

Business Role		Employee Data Access			
Role	**Task**	**Contact Details**	**Benefits Data**	**Salary Data**	**Performance Data**
HR Benefits Administrator	Manage employment benefits	Yes	Yes	No	No
HR Payroll Administrator	Manage payroll and salary	Yes	Yes	Yes	No
HR Talent Management	Manage training and promotions	Yes	No	Yes	Yes

Figure 8.3 – Role definition matrix

    - **Attribute-based access control** (**ABAC**): ABAC makes access choices based on attributes or characteristics (for example, user attributes, resource attributes, and environmental attributes).

- **Access control frameworks**: Access control frameworks are realistic implementations or systems that manage and enforce resource access using access control concepts. These frameworks provide access control tools, protocols, and technologies for use in real-world applications. Access control frameworks that are often used include the following:

    - **Access control lists** (**ACLs**): ACLs are lists connected with resources that indicate which people or entities have access to those resources and which do not.
    - **Security information and event management** (**SIEM**): SIEM systems gather, aggregate, and analyze security data, including access control events, by combining **security information management** (**SIM**) with **security event management** (**SEM**).
    - **Policy-based access control** (**PBAC**): PBAC imposes access control based on preset policies, rules, or conditions that determine who has access to resources and under what conditions.

Access control models and frameworks are important in safeguarding digital systems because they ensure that access is allowed only to authorized people or organizations in accordance with the organization's security policies and needs. They are an essential component of overall information security practices, assisting in the prevention of unauthorized access, data breaches, and other security concerns.

After exploring access control models and frameworks, we now turn our attention to another critical facet of IAM: **Identity Governance and Administration** (**IGA**). In the forthcoming section, we will look at the ideas and methods of IGA, emphasizing its critical role in guaranteeing compliance, enforcing regulations, and successfully managing identities across corporate ecosystems.

Identity governance and administration

IGA is a critical component of a company's cybersecurity strategy, concentrating on the administration of user identities, rights, and access governance. An IGA system is critical in protecting an organization's digital assets, guaranteeing regulatory compliance, and improving overall security posture in today's fast-expanding threat landscape. This chapter digs into the complexities of IGA, investigating its security implications, best practices, and the critical role it plays in ensuring data confidentiality, integrity, and availability.

Two of the service components of IGA are user onboarding and offboarding. However, the most critical component is the management of the user lifecycle in between the onboarding and offboarding process.

The practice of giving access to new workers and partners to ensure that only authorized personnel have access to an organization's systems and resources is known as user provisioning. Verifying user identities, providing suitable responsibilities, and regulating permissions are all security issues. Once a user is onboarded or provisioned within an enterprise, the next two factors that play a role are user authentication and authorization. The crux of IGA security is strong authentication. The use of MFA and biometric authentication technologies improves security against unauthorized access. This section delves into the complexities of authentication techniques and their roles in ensuring user access. Defining and enforcing user access regulations is what authorization implies. Within the framework of IGA security, RBAC, ABAC, and MAC models are considered. The principle of least privilege is determined by authorization models, which reduces the possibility of over-entitlement.

The following are the complexities of authentication techniques:

- **Password complexity**: It refers to the complexity of authentication methods and how they help to guarantee user access.
- **Multi-factor authentication** (**MFA**): By requiring users to present several pieces of identity, MFA improves security. Multiple authentication factor implementation and managing MFA, however, might be challenging.
- **Biometric authentication**: A strong degree of security is added by using distinguishing physical or behavioral characteristics, such as fingerprints or facial recognition. However, there can be technical complexity involved in integrating and maintaining biometric systems.
- **Token-based authentication**: Software- or hardware-based tokens offer transient access codes. On the other hand, handling and distributing tokens presents administrative challenges.
- **Risk-based authentication**: Potential dangers are identified by contextual factor analysis and user behavior analysis. Careful calibration is needed to strike a balance between causing consumers inconvenience and achieving an effective risk assessment.
- **Single sign-on** (**SSO**): Using a single set of credentials for entry simplifies user access with SSO. Centralizing access, however, might raise security issues, and it can be difficult to guarantee compatibility with different systems.

- **Certificate-based authentication**: An additional strong layer of security is added by using digital certificates to authenticate people and devices. However, controlling, rescinding, and giving certificates call for cautious management.

These complexities show that, in order to guarantee user comfort and security at the same time, authentication solutions must be carefully considered and calibrated.

Given the sensitive nature of identification and access information, IGA prioritizes data security. Encryption, data masking, and safe data transfer are critical components of the security architecture of an IGA system. Continuous auditing and monitoring of IGA operations is critical for detecting and responding to suspicious activity as soon as possible. The integration of SIEM, real-time alerts, and anomaly detection approaches are explored in order to strengthen IGA's proactive security posture.

IGA systems must comply with multiple regulatory frameworks, such as the **General Data Protection Regulation** (GDPR), the **Health Insurance Portability and Accountability Act** (**HIPAA**), and the **Sarbanes–Oxley Act** (**SOX**). This section covers how IGA aids in compliance and highlights the security needs and issues associated with these frameworks. User consent management is a critical component of data privacy and security. Granular control over user data and permissions must be enabled in IGA systems, allowing users to change their privacy preferences. The security procedures for data consent management are being examined. Insider attacks represent major security dangers, whether purposeful or inadvertent. IGA systems are critical for tracking user behavior, detecting abnormalities, and reducing insider risks. This section delves into the tactics used to combat these risks, such as behavior analytics and PAM.

Deprovisioning is also important when users depart an organization. It is critical to terminate access as soon as possible to prevent previous users from leveraging leftover rights or creating vulnerabilities. Effective de-provisioning strengthens an organization's resistance to insider threats.

IGA is an important component of an organization's security strategy because it ensures that the correct people have access to the right resources at the right time. IGA has far-reaching security concerns, ranging from user provisioning to data privacy and compliance. As technology advances, IGA's role in assuring data security, integrity, and availability will remain critical in protecting an organization's digital assets.

Identity management is a crucial component of modern cybersecurity, working hand in hand with IGA, which takes care of all security issues inside a company. It is a critical component of the overall security plan and concentrates on the creation and destruction of user identities. The next section will examine the identity management lifecycle, highlighting the critical role that it plays in protecting organizational resources and reducing security risks.

Identity lifecycle management

Identity management is a key component of contemporary cybersecurity, ensuring that only the proper people have access to organizational resources. The identity management lifecycle encompasses the establishment and termination of user identities. In this section, we will look at the identity management lifecycle from a security standpoint, with a particular emphasis on the vital role it plays in protecting an organization's assets and minimizing security threats.

Identity management is a complicated process that includes creating, providing, maintaining, and de-provisioning user accounts inside an organization. The identity management lifecycle is critical from a security standpoint for various reasons:

1. **Security controls**: Identity management effectively enforces security policies, lowering the risk of unauthorized access and data breaches
2. **Compliance**: To secure sensitive data and maintain compliance, several regulatory frameworks need comprehensive identity management practices
3. **Risk mitigation**: Insider threats, unauthorized access, and account misuse may all be mitigated with effective identity management
4. **Data confidentiality**: It contributes to the secrecy of organizational data by ensuring that only authorized users have access to it
5. **Data integrity**: The integrity of user data is maintained by appropriately representing and maintaining user identities
6. **Availability**: Reliable identity management guarantees that users have access to the resources they require, which promotes data availability

Let's look at some of the critical phases of the identity management lifecycle from a security standpoint.

During the user provisioning process, security regulations should be implemented. This covers password complexity requirements, MFA configuration, and account lockout procedures. Using RBAC during provisioning guarantees that users are given suitable rights, depending on their job duties, in accordance with the concept of least privilege. This security precaution reduces the possibility of over-entitlement. To prevent fraudulent account creation, implement secure registration and validation methods. CAPTCHAs, email verification, and identity verification procedures are all included.

From a security standpoint, authentication and authorization are at the heart of identity management. Strong authentication systems, such as biometrics, smart cards, and MFA, improve security by reliably confirming user identities. What activities users may do are governed by the creation of rigorous authorization policies, which are preferably implemented using ABAC. As work positions vary, authorization rules should be evaluated and updated on a regular basis.

The continuous maintenance of users is critical for maintaining security throughout the identity management lifecycle. Regular access audits ensure that users only have the appropriate permissions. Periodic access checks that are automated reduce the possibility of over-entitlement and unauthorized access. Password rules that are strict, such as password rotation, complexity requirements, and safe password storage, lower the likelihood of password-related security issues. Account lockout policies should be implemented to prevent brute-force attacks and unauthorized access.

The security of data is a critical component of the identity management lifecycle. To reduce the risk of data breaches, sensitive user data should be encrypted both at rest and in transit. Create and follow data retention rules to safely handle user data throughout its lifespan, taking into account both legal obligations and security concerns.

Deprovisioning, or terminating user accounts, is critical for security and risk reduction. Terminating user access as soon as an employee leaves reduces the danger of unauthorized access by former workers. It is vital to maintain data security by ensuring that user data is correctly cleared away after account termination. If required, this involves removing access and securely archiving data.

In terms of security, the identity management lifecycle is critical in preventing insider risks, whether purposeful or inadvertent. Implementing behavior analytics and user activity monitoring aids in the detection of abnormal behavior patterns, allowing for early action. PAM systems that control and monitor privileged access lessen the danger of insider misuse of high-privilege accounts.

Within the identity management lifecycle, compliance with legislative frameworks is a critical security factor. To secure personal and sensitive data, compliance with data protection requirements (for example, GDPR, HIPAA, etc.) needs tight identity management practices. Strong auditing and reporting skills guarantee that an organization can show compliance and effectively respond to security problems.

The identity management lifecycle is a critical component of a company's cybersecurity strategy. A security-focused viewpoint guarantees that this procedure is aligned.

Threat detection and IAM security

Identity management systems (**IDMS**) are critical for ensuring that only authorized users have access to a company's resources. They are, nevertheless, excellent targets for cybercriminals looking for unauthorized access. Threat detection is essential for detecting and minimizing these risks. This section will go into the topic of threat detection in identity management systems, examining techniques, best practices, and the most recent technology used to protect these systems from a variety of security issues. IDMS are critical components of an organization's security strategy because they manage who has access to what resources. Threats to these systems, such as data breaches, unauthorized access, and insider threats, can have serious implications. Effective threat detection is critical to IDMS's integrity and security. This section delves into the fundamentals of threat detection in IDMS:

- **Common threats**: One of the most common dangers is unauthorized access. Attackers try to obtain access to systems by compromising weak passwords, circumventing authentication methods, or exploiting flaws. By misusing their privileged access, stealing data, or disrupting systems from within, malicious insiders can represent major hazards. To mimic genuine users, attackers target user credentials using numerous methods such as phishing assaults, credential stuffing, or social engineering. Account hijacking is the process of gaining control of valid user accounts, usually by using stolen passwords or exploiting flaws. Identity theft is the fraudulent acquisition and exploitation of another person's personal information, usually for monetary benefit. Misconfigured access restrictions might expose sensitive data unintentionally, leaving systems open to abuse.

- **Detection strategies**: Anomaly detection entails watching for departures from recognized patterns in user behavior and system activity. Unexpected events might set out signals for additional research. Machine learning algorithms are used in behavior analytics systems to uncover trends and abnormalities in user behavior, assisting in the identification of insider threats and unauthorized access. Continuous real-time monitoring of IDMS activity aids in the detection and response to suspicious situations. Integration with threat intelligence feeds offers information about known threats and vulnerabilities, allowing for proactive threat detection. **User and entity behavior analytics** (**UEBA**) systems monitor both user and entity activity within the IDMS, providing full threat detection capabilities.
- **Best practices**: RBAC ensures that users have just the rights required for their responsibilities, lowering the attack surface and limiting possible risks. Use the least privilege concept to guarantee that users only have access to what is necessary for their respective jobs. Implement MFA to improve authentication security, making it far more difficult for attackers to get unauthorized access. Conduct frequent access audits to ensure that user permissions are appropriate for their jobs and responsibilities. To limit the risk of phishing attempts and social engineering, educate users and workers on security best practices.
- **Technical solutions**: SIEM systems gather and analyze data from a variety of sources, allowing for the identification of security incidents such as unauthorized access and unusual behavior. Advanced analytics are used in UEBA solutions to detect anomalous behavior patterns, allowing for the identification of insider threats and unauthorized access. Integrate threat intelligence systems to learn about known threats and vulnerabilities, allowing for proactive threat detection. Scalability and real-time threat detection capabilities are provided by cloud-based threat detection services for cloud-based IDMS. Deception technology uses misleading assets and lures to fool attackers and identify their existence in the IDMS.

A major part of current cybersecurity is threat detection in IDMS. Because cyber threats are becoming more sophisticated, organizations must adopt advanced strategies, best practices, and technical solutions to protect their IDMS. Organizations can safeguard their data, preserve operational continuity, and ensure the integrity of their identity management systems by successfully recognizing and mitigating risks. Robust threat detection skills are critical in defending digital identities and organizational resources in an ever-changing threat scenario.

After discussing the significance of IGA in managing identities within organizational frameworks, we now shift our focus to SIEM solutions. In the forthcoming section, we will look at how SIEM systems complement IGA by offering complete monitoring, analysis, and response capabilities that improve overall security posture and threat detection.

Security information and event management

SIEM systems are critical components of modern cybersecurity, acting as the eyes and ears of an organization's digital landscape. To detect and respond to security issues, these systems collect and analyze massive volumes of data from multiple sources. Failure to adopt or underutilize SIEM systems, on the other hand, might have serious security consequences. In this section, we will look at the repercussions of inadequate implementation or non-implementation of SIEM systems, as well as the vital role these systems play in protecting digital assets.

Organizations are responsible for protecting their digital assets, sensitive data, and user identities in an era of more sophisticated cyberattacks. SIEM systems provide a holistic approach to security by gathering, analyzing, and correlating security-related data from a variety of sources. SIEM systems enable real-time threat detection, incident response, and compliance monitoring capabilities when correctly integrated and configured. However, failing to adopt SIEM or failing to fully utilize its capabilities can expose organizations to a slew of security vulnerabilities.

Inadequate visibility into an organization's digital environment is one of the key security implications of poor SIEM deployment. This lack of visibility can be caused by neglecting to incorporate crucial data sources or incorrectly configuring the SIEM system. Because security personnel lack total visibility, they are unable to properly monitor, identify, and respond to security problems across the whole infrastructure. Weak or underutilized SIEM systems frequently result in incident detection delays. This lag may be disastrous, since sophisticated threats can swiftly increase, causing serious harm before they are ever detected. It is critical to notice security events as soon as possible in order to minimize and mitigate any damage. SIEM systems are intended to detect as well as respond to incidents. Inefficient incident response methods may come from poor SIEM setup or underutilization. Organizations may fail to effectively coordinate responses, prioritize problems, and minimize hazards. Inadequate SIEM configuration might result in a large number of false positive alarms. This alert fatigue can overload security professionals, causing them to overlook crucial notifications in the midst of the cacophony. False positives impair security operations' efficiency and can lead to serious threats remaining undetected. Many organizations are required to comply with regulatory obligations that demand extensive security monitoring. Weak or non-implementation of SIEM systems can result in compliance problems, which can lead to legal ramifications, penalties, or reputational harm.

Repercussions of a weak SIEM system

Inadequate data source inclusion is one of the fundamental reasons for poor SIEM deployment. SIEM systems rely on a diverse set of logs and data inputs from a variety of devices and apps. The system's capacity to connect events and detect threats is hampered by a lack of comprehensive data sources. The appropriate configuration of rules and correlation mechanisms is required for effective SIEM functioning. Inadequate rule sets or incorrectly designed correlation settings may result in missing occurrences or a large number of false positives. Bad log management practices frequently accompany bad SIEM installation. Inconsistent log collection, retention, and storage might result in missing historical data or restrict the system's capacity to investigate previous problems. SIEM systems can be resource-intensive, both in terms of hardware and software. Inadequate resource allocation may lead to system slowdowns, missed events, or unattended alerts.

While poor implementation is one side of the coin, the non-installation of SIEM systems is just as bad for an organization's security posture. Organizations that use weak SIEM systems suffer serious consequences. Without SIEM, organizations have significant gaps in their security monitoring. Attackers can operate unnoticed within these spaces, causing significant harm. SIEM systems save historical data, allowing organizations to review previous occurrences and discover trends. In the absence of implementation, organizations are deprived of vital historical data required for forensic investigation and incident response. Non-compliance with numerous regulatory and industry-specific regulations can result in non-compliance with SIEM, exposing organizations to legal risks and significant fines. Organizations without SIEM systems may have much longer incident reaction times when security issues occur. Due to a lack of automated threat detection and investigation technologies, incident response processes may be manual and time-consuming.

To reduce the security consequences of poor or non-implementation, organizations must deploy SIEM systems in accordance with best practices:

- **Comprehensive log collection**: In SIEM systems, include a diverse set of data sources to ensure that logs from all essential devices and applications are captured
- **Robust rule configuration**: Configure SIEM rules and correlation parameters to achieve a balance of precise threat detection and low false positives
- **Log management**: Implement best practices for log management, such as consistent log collection, retention, and storage. Make sure that past data is easily accessible for inquiry
- **Regular updates and monitoring**: Maintain SIEM systems with the most recent threat intelligence and regularly monitor system performance and alarms
- **Adequate resources**: Provide adequate resources, both in terms of hardware and manpower, to successfully maintain and run SIEM systems

The security consequences of the inadequate implementation or non-implementation of SIEM systems are significant and far-reaching. Organizations that do not fully utilize SIEM risk having insufficient visibility, delayed issue detection, ineffective incident response, compliance problems, and false positives. Additionally, organizations that use weak SIEM solutions suffer obstacles such as blind spots, a lack of historical data, regulatory issues, and longer incident response times. Effective SIEM implementation, which includes comprehensive data sources, robust rule configuration, sound log management, regular updates and monitoring, and sufficient resources, is critical to mitigating these security implications and maintaining a robust security posture in the face of evolving cyber threats.

Compliance and regulatory considerations

The requirement for effective IAM solutions is critical as organizations traverse an increasingly complex and linked digital ecosystem. A secure IAM platform not only enables effective access control but also plays an important role in meeting a plethora of regulatory obligations. This chapter digs into the essential compliance and regulatory issues that businesses must face when deploying a secure IAM platform. We'll look at major legislation and best practices to see how IAM might help organizations fulfill their compliance goals.

Importance of compliance in IAM

Because IAM systems store sensitive user information, they are essential to data privacy and protection rules. Compliance with data protection requirements, such as the GDPR, necessitates the implementation of tight safeguards for user data by organizations. IAM platforms protect an organization's resources. Compliance with security and confidentiality standards, such as ISO 27001, is critical to the protection of sensitive data and the retention of consumer confidence. Many legal frameworks demand that organizations keep extensive audit records and show accountability when it comes to regulating user access. By providing rigorous audit capabilities, IAM solutions assist organizations in meeting these criteria. IAM systems aid in access control and monitoring, which is critical in limiting insider risks. Measures to prevent unauthorized access and data breaches are required for regulatory compliance.

Key regulations and compliance frameworks

GDPR is a comprehensive data protection policy that applies to organizations that handle **European Union** (**EU**) citizens' personal data. GDPR compliance necessitates strong user consent management, data encryption, and the capacity to meet data subject demands. In the United States, HIPAA governs the security and privacy of healthcare information. Healthcare organizations require IAM solutions to govern access to electronic health records while remaining compliant with HIPAA. SOX imposes stringent controls on financial data and reporting. IAM solutions are crucial for implementing access rules and maintaining financial data integrity and confidentiality. PCI DSS applies to businesses that handle credit card information. IAM is critical for imposing rigorous access restrictions and ensuring PCI DSS compliance. ISO 27001 is a global standard that applies to **information security management systems** (**ISMS**). By guaranteeing limited access to sensitive information, IAM solutions help to satisfy the criteria of this standard.

Challenges and risks in IAM compliance

Compliance with many and sometimes overlapping requirements may be difficult. Organizations must invest in IAM solutions that can adapt to an ever-changing regulatory framework. Regulations evolve, and organizations must keep their IAM systems up to date in order to remain compliant. Adapting to new needs and ensuring that the IAM platform can handle changes without affecting operations are all part of this. Many organizations are limited in terms of both competent individuals and financial assets. Without appropriate resources, meeting compliance standards might be difficult. Third-party providers frequently require access to organizations' systems. It can be difficult to manage and monitor these external access privileges, especially in regulated businesses.

Compliance with IAM is essential for businesses looking to build strong security measures, but it also comes with a lot of possible hazards and difficulties. Several major compliance-related issues surface as companies depend more and more on IAM solutions to control user access to systems and sensitive data.

The dynamic nature of laws and standards is a major risk factor for IAM compliance. Compliance frameworks are updated and revised frequently by governments and industry associations to keep up with changing security concerns. In order to make sure that their IAM systems meet the most recent regulations, organizations need to remain alert. If this isn't done, it may be considered non-compliance, which could result in penalties, fines, and reputational harm for the company.

The intricacy of IAM systems presents another significant concern. Complex setups, system integrations, and user role administration are common tasks in the implementation and upkeep of IAM solutions. Because of the intricacy, there is a chance for errors, oversights, and vulnerabilities that could be exploited by hostile actors to obtain unauthorized access. Furthermore, complex IAM configurations can be hard for businesses to efficiently monitor and audit, which makes it hard to keep up continual compliance.

Risks associated with privileged access management also affect IAM compliance. Administrators and executives are examples of users with advanced privileges that present a serious security risk. Unauthorized system changes, jeopardized vital assets, and data breaches can result from improperly managed privileged access. To reduce the dangers associated with privileged access, organizations need to implement strict controls and monitoring systems that guarantee that only authorized workers can use elevated privileges.

An additional level of risk to IAM compliance is introduced by the expanding trend of remote work. It gets harder to guarantee safe and legal access when workers use different devices and places to access company systems. For an organization to maintain IAM compliance in a remote work environment, concerns such as device reliability, safe authentication techniques, and secure data transmission must be addressed. Organizations may be more vulnerable to security breaches if their IAM strategies are not adjusted to the realities of remote work.

IAM compliance also raises a lot of data privacy issues. IAM systems handle sensitive user data by default; improper treatment of this data can lead to privacy violations and legal repercussions. Providing streamlined access to authorized users while safeguarding personal data necessitates a sophisticated approach to IAM design and deployment. To guarantee compliance with data protection laws, organizations need to give priority to data encryption, put strong consent procedures in place, and audit their IAM systems on a regular basis.

Moreover, the more general problem of cybersecurity awareness and training is linked to IAM compliance. Because users are frequently the weakest link in the security chain, IAM compliance may unintentionally be compromised by their actions. The integrity of IAM systems can be compromised by phishing attacks, poor password policies, and social engineering techniques that compromise user credentials. Companies need to make investments in thorough cybersecurity education initiatives to equip users with the know-how and abilities required to thwart possible threats.

To sum up, even though IAM compliance is crucial for protecting corporate assets, there are dangers involved. It is critical to manage privileged access, navigate the changing regulatory landscape, handle the complexity of IAM configurations, manage remote work, protect data privacy, and raise cybersecurity awareness. In order to strengthen their defenses, uphold compliance, and protect the integrity of their digital ecosystems, organizations will be better positioned if they proactively address these risks and continuously improve their IAM policies.

Future trends in IAM compliance

Decentralized identification solutions, frequently based on blockchain technology, may offer new methods to satisfy compliance standards while giving people more control over their personal information. AI and machine learning are becoming more important components of IAM solutions, delivering improved threat detection and anomaly analysis capabilities that may help with compliance by recognizing anomalous user behaviors. **Regulation technology** (**RegTech**) solutions are being developed to assist organizations in automating compliance procedures, including IAM operations. These technologies have the potential to simplify compliance reporting and monitoring.

Compliance with regulatory regulations is not only a legal necessity; it is also critical to ensuring data security, user privacy, and confidence. A secure IAM platform is not just a way of managing access but also an important tool for meeting compliance goals. Organizations may use IAM to satisfy compliance responsibilities while strengthening their cybersecurity posture by knowing the regulatory landscape, following best practices, and staying up to date on emerging trends. As the compliance environment evolves, IAM will continue to be an important component in the goal of regulatory adherence and data security.

After investigating the features and benefits of SIEM systems, we now turn our attention to the changing environment of IAM security. In the following section, we will look at new technologies that are altering IAM security practices by providing novel solutions to increasing risks and difficulties in the digital ecosystem.

Emerging technologies in IAM security

IAM security is always evolving to keep up with the changing threat of environment and technological landscape. Emerging technologies are critical in boosting IAM security, improving user experience, and helping organizations adapt to the demands of a contemporary, interconnected world. We'll look at some of the most promising developing technologies that are transforming the IAM security environment:

- **Biometric authentication**: Biometric authentication systems, such as fingerprint recognition, face recognition, and iris scanning, have grown in popularity as extremely secure and convenient methods of confirming user identities. These methods enable MFA by confirming *who you are* based on distinct physiological or behavioral traits. Biometric authentication is rapidly being implemented into IAM systems, strengthening security while simplifying user login.

- **Behavioral analytics**: Machine learning and artificial intelligence are used in behavioral analytics to analyze user behavior trends. IAM solutions may create user behavior profiles by continually monitoring how users interact with systems and apps. Any deviations from established patterns might set up warnings, potentially revealing unauthorized access or compromised accounts. This technology assists organizations in detecting insider threats and sophisticated cyberattacks in advance.
- **Decentralized identity**: Decentralized identification solutions are gaining popularity as a means of giving individuals control over their digital identities. These systems make use of blockchain technology to generate self-sovereign identities, eliminating the need for centralized identity suppliers. Users may manage and distribute their identities and access credentials independently of any particular institution. This strategy empowers individuals while also improving privacy and lowering the danger of data breaches associated with centralized identity providers.
- **Continuous authentication**: At login, traditional IAM systems frequently rely on a single authentication event. In contrast, continuous authentication analyzes user behavior and applies authentication tests throughout a user's session. This strategy adjusts to the changing danger landscape, ensuring that users are verified and authorized depending on their continuing activity. Continuous authentication improves security while minimizing disruption to user processes.
- **Blockchain**: Blockchain technology has uses in IAM security in addition to cryptocurrency. Blockchains offer a secure, decentralized ledger for preserving user identities and access privileges. Blockchain-based IAM solutions provide increased security, transparency, and auditability. Users can have more autonomy over their digital identities, while businesses can rely less on centralized identity suppliers.
- **Zero trust architecture** (**ZTA**): Zero trust is a method of IAM security that trusts no one and checks everyone. It believes that threats might occur from both within and outside the network, necessitating ongoing user and device verification. Strong authentication, constant monitoring, and least-privilege access are used in ZTA to reduce the danger of unauthorized access and lateral movement by attackers.
- **AI-powered threat detection**: To improve threat detection capabilities, artificial intelligence and machine learning are being integrated into IAM solutions. By analyzing massive datasets and user behavior patterns, these technologies may detect abnormalities and possibly fraudulent activity. They help security teams notice security events earlier and deliver actionable insights.
- **Passwordless authentication**: Passwordless authentication is becoming more popular as a convenient and safe alternative to standard username and password combinations. It replaces readily compromised passwords with mechanisms such as biometrics, hardware tokens, or mobile device-based authentication.
- **API security**: With the development of web and mobile apps, it is critical to secure **application programming interfaces** (**APIs**). For API security in IAM, OAuth 2.0, an open protocol for access delegation, is commonly employed. It allows for delegated authority while protecting user credentials.

- **Quantum safe cryptography**: The advent of quantum computing puts standard cryptography techniques at risk. Quantum-safe cryptography, also known as post-quantum cryptography, is a developing topic aimed at developing cryptographic systems that are immune to quantum assaults. To preserve security in the quantum computing future, IAM systems will need to adapt to quantum-safe algorithms.

Emerging IAM security technologies are altering how organizations handle identities and access. Among the technologies altering the IAM landscape are biometric authentication, behavioral analytics, decentralized identification, continuous authentication, blockchain, zero trust, AI-powered threat detection, passwordless authentication, API security, and quantum-safe cryptography. These advancements are critical for reacting to a changing threat environment, safeguarding sensitive data, and improving user experiences. In an increasingly linked world, IAM is at the vanguard of preserving digital identities and access, and it will continue to grow with emerging technologies to solve tomorrow's security problems. Organizations that use these solutions will be better positioned to safeguard their digital assets and adapt to the ever-changing IAM security landscape.

After looking at the emerging technologies that are transforming IAM security, we will now focus on the problems and future directions in this dynamic industry. In the following section, we will look at the changing environment of IAM security, including rising risks, market trends, and possible areas for innovation in protecting digital identities and resources.

Challenges and future directions in IAM security

IAM is the foundation of contemporary cybersecurity, ensuring data privacy and regulatory compliance while safeguarding user access to digital assets. As the digital ecosystem changes, IAM security confronts new problems and must adapt to meet future expectations. In this section, we will look at the major difficulties that IAM security is now facing, as well as the future paths it must take to properly handle these challenges:

- **Evolving threat landscape**: As attackers utilize innovative tactics to breach user identities and obtain unauthorized access, cyber dangers are getting more complex. To successfully identify and respond to developing threats, IAM security must combine sophisticated threat detection, behavioral analytics, and AI-driven security solutions.
- **Legacy systems and silos**: Many organizations continue to rely on outdated IAM systems that are not connected with one another, resulting in identity data silos and wasteful operations. To break down barriers and build a single identity ecosystem, future IAM systems should prioritize integration and interoperability.
- **User experience versus security**: Finding an acceptable balance between user experience and security is a never-ending task. Complex security measures can impede usability, while too lax security might put organizations in danger. Future IAM systems should strive for a unified user experience while retaining a high level of security. Passwordless authentication and biometrics are two technologies that can assist in achieving this balance.

- **Regulatory compliance**: GDPR, HIPAA, and CCPA are just a few of the regulations that govern how user data is stored and secured. Organizations have a perpetual compliance problem. Future IAM systems should include compliance automation technologies, allowing organizations to expedite compliance operations and respond quickly to legislative changes.
- **Insider threats**: Insider threats, whether purposeful or inadvertent, continue to be a major problem. Users with legitimate access may abuse their rights. To detect atypical user behavior and reduce the risk of insider attacks, IAM systems should rely on behavioral analytics.
- **Resource constraints**: Many organizations confront resource limits, including experienced individuals and financial commitments, making comprehensive IAM systems challenging to adopt and maintain. More turnkey solutions, cloud-based options, and managed services that minimize the load on resource-constrained organizations should be included in the future of IAM.

IAM security is at a crossroads as it contends with a growing threat landscape and the need to strike a balance between security and usability. The development of creative IAM solutions is being driven by challenges such as growing risks, outdated systems, user experience, regulatory compliance, insider threats, and resource restrictions. The use of ZTA, biometric authentication, decentralized identification, AI-driven security, passwordless authentication, RegTech, quantum-safe cryptography, user behavior analysis, privacy-preserving technologies, and safeguarding APIs and microservices is the future of IAM security.

Organizations will be better positioned to preserve their digital identities, protect sensitive data, and face the ever-changing challenges of the IAM security landscape as they adopt these future trends in IAM security. The combination of modern technology, a user-centric strategy, and a strong commitment to regulatory compliance and data protection will be the key to future IAM security success.

Summary

This chapter delved further into SIEM systems and IAM concepts. It underlined SIEM's vital role in protecting digital assets, as well as the serious security ramifications of insufficient or non-implementation, such as delayed issue detection, compliance issues, and lengthier incident response times. It also emphasized the significance of strong rule setup, log management, and broad data sources for proper SIEM operation. On the other hand, IAM is defined as a security framework that focuses on maintaining user identities and limiting access to resources and applications inside an organization, with a heavy emphasis on compliance, access control models, and continuous monitoring and updates. The chapter also analyzed future trends in IAM compliance and security, highlighting the need for a user-centric strategy, cutting-edge technology, and a dedication to regulatory compliance and data protection for long-term success. In the next chapter, we shall take a more in-depth look at self-sovereign identity and what it brings to the table.

Part 3 - Digital Identity Era: The Near Future

This part examines emerging trends in identity management. *Chapter 9* introduces **Self-Sovereign Identity** (**SSI**), highlighting its benefits for privacy, data security, user control, data portability, interoperability, decentralization, and efficiency. *Chapter 10* discusses the importance of privacy by design in SSI, covering frameworks, user-centric controls, security best practices, and threat mitigation. *Chapter 11* explores the relationship between **Distributed Identifiers** (**DIDs**) and SSI, focusing on how DIDs enable verifiable credentials and enhance privacy and security. *Chapter 12* reviews protocols and standards for DIDs, including DID documents, methods, and verifiable credentials. *Chapter 13* covers DID authentication methods such as LDAP, Kerberos, OAuth2, OIDC, and SAML. *Chapter 14* addresses identity verification, emphasizing the extensive use of facial recognition, the fight against identity theft, and the advantages of digital identities.

This part has the following chapters:

9
Self-Sovereign Identity

Self-sovereign identity (**SSI**) is a revolutionary approach to digital identity management that gives individuals complete control over their personal data and how it is shared in the digital world. SSI is based on decentralized and cryptographic principles, utilizing blockchain and **distributed ledger technology** (**DLT**) to let users establish, maintain, and confirm their digital identities without depending on central authorities or identity providers. Verifiable credentials, cryptographic keys, and secure data storage are used by SSI systems to guarantee that individuals may establish their identities while maintaining privacy and security. This identity management paradigm change offers a more secure, user-centric, and interoperable solution for the digital age, in which people retain the keys to their own digital identities, and trust is built through cryptographic verification rather than by centralized intermediaries.

In this chapter, we will cover the following topics:

- What is SSI and why should it matter?
- Cryptographic methodologies and techniques in SSI
- The use of decentralized identifiers in the SSI space
- DID protocols in use in the SSI ecosystem
- Interoperability, standards, and protocols

Introduction to SSI

SSI is a new digital identification paradigm that empowers individuals to govern their own personal data. Individuals construct and control their own digital identities, which are kept on a decentralized network rather than depending on centralized authorities or organizations to handle their identities. SSI provides various advantages, including the following:

- **Privacy**: SSI gives individuals complete control over their personal data, which means they can choose what information to share and with whom. This helps protect their privacy and prevent unauthorized access to their personal information.

- **Security**: SSI uses strong encryption and decentralized storage to ensure that personal data is secure and protected from hacking and other cyber threats. This makes it more difficult for attackers to steal personal data or use it for fraudulent purposes.
- **Interoperability**: SSI is designed to be interoperable across different platforms and services, which means individuals can use their digital identities to access a wide range of applications and services. This helps reduce the need for multiple usernames and passwords and makes it easier to manage digital identity across different platforms.
- **Trust**: SSI is built on a foundation of trust, where individuals control their own personal data and establish trust relationships with other parties on a need-to-know basis. This helps build trust between individuals and organizations and reduces the risk of data breaches and other security incidents.
- **Flexibility**: SSI is a flexible and adaptable system that can be customized to meet the needs of different individuals and organizations. This makes it easier to implement and scale digital identity solutions that are tailored to specific use cases and requirements.

Overall, SSI brings a range of benefits to the table, including enhanced privacy, security, interoperability, trust, and flexibility. By giving individuals control over their own personal data, SSI has the potential to transform the way we manage digital identity and help build a more secure and trustworthy digital society.

Identity is exclusively a human concept. It represents the ineffable core of self-awareness and is widely recognized across cultures. Nonetheless, contemporary civilization has confused the conventional concept of identity. The following diagram depicts the visual differences between the current and the SSI model:

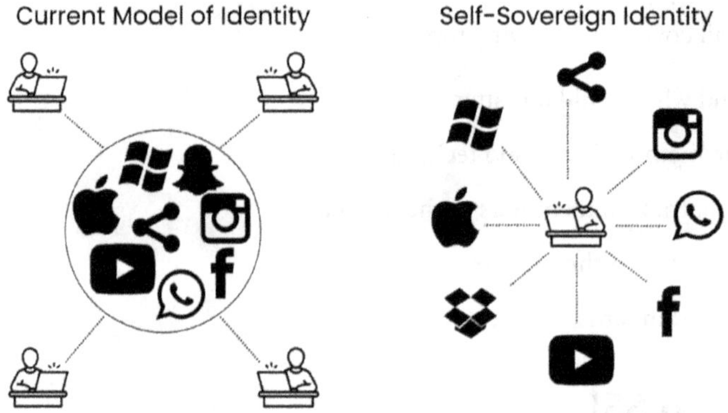

Figure 9.1 – SSI versus the current identity model landscape

Today, organizations and governments frequently associate driver's licenses, social security cards, and other state-issued credentials with one's identity. This is a dilemma since it implies that an individual's identification might be jeopardized if the state withdraws their credentials or if they merely cross state lines. User-centric designs changed centralized identities into interoperable merged identities under centralized management, allowing users to choose how and with whom to share their identities. While this was a significant step toward true user control of identification, it was only the beginning, with the next critical step needing user autonomy.

This serves as the foundation for SSI, a phrase that has grown in popularity and use in the 2010s. Rather than just highlighting that users should play a significant part in the identification process, SSI says that users should act as the rulers of their own identities.

SSI is the next step beyond user-centric identity, implying that it starts from the same place: putting the user at the center of identity management. This entails not only assuring the interoperability of a user's identity across many platforms with their agreement, but also allowing actual user control over their digital identity, hence encouraging user autonomy.

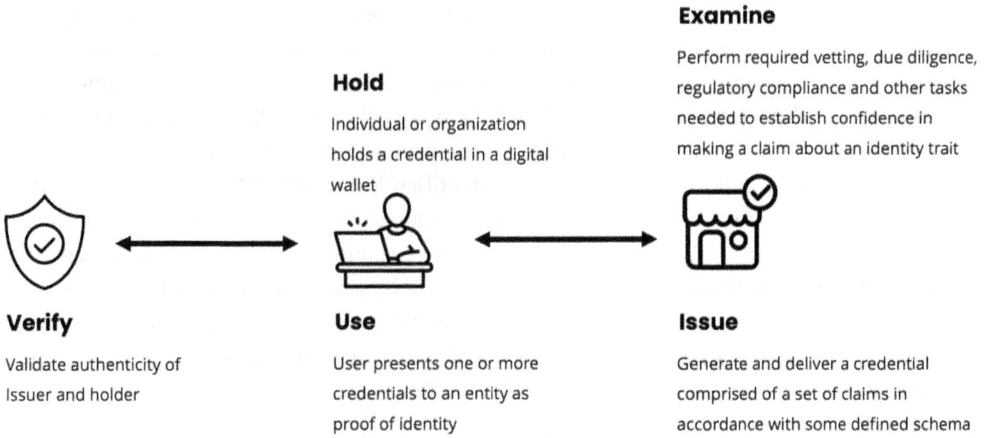
Examine
Perform required vetting, due diligence, regulatory compliance and other tasks needed to establish confidence in making a claim about an identity trait
Hold
Individual or organization holds a credential in a digital wallet
Verify
Validate authenticity of Issuer and holder
Use
User presents one or more credentials to an entity as proof of identity
Issue
Generate and deliver a credential comprised of a set of claims in accordance with some defined schema

Figure 9.2 – The SSI interaction landscape

For this to operate, an SSI must be mobile; it cannot be fixed in one location. It should also allow frequent users to provide information about themselves, such as personal facts or what they can do, and it may even incorporate information about the user that is supplied by other organizations or individuals.

An SSI system may be customized for a variety of applications, including banking, subsidies, microfinance, land management, and healthcare. The adaptability and compatibility of SSI can help prevent the development of separate identification systems tailored for specialized reasons, such as handling subsidies in a refugee camp. Separate digital identification solutions for each purpose sometimes place undue costs on marginalized communities, requiring them to handle additional passwords, cards, or bureaucratic processes.

We predict that numerous reasons will contribute to the broad acceptance of self-sovereign identification. For starters, the changing social scene, characterized by the increased digitalization of daily life, emphasizes the need for more secure and user-controlled identification solutions. The weaknesses of present systems expose consumers to a variety of threats, emphasizing the need for a more robust approach.

Second, the growth of supporting technologies, notably DLT, and the widespread use of cell phones play critical roles. DLT provides a safe and decentralized framework for maintaining identities, while smartphones are simple and powerful tools for individuals to govern and engage with their digital identities. These technological breakthroughs create a favorable environment for the emergence and integration of SSI.

Finally, the growing acceptance of new technologies, particularly in the emerging and developing worlds, helps to create a favorable atmosphere for SSI adoption. As societies adopt more creative solutions, there is a greater willingness to incorporate new technology, providing a favorable environment for the acceptance and implementation of self-sovereign identification on a worldwide scale.

Why SSI matters

In the digital age, when our lives are increasingly centered around online interactions and transactions, managing and safeguarding our digital identities has become critical. Traditional identification systems, which are frequently centralized and controlled by several groups, have created serious privacy and security problems. SSI emerges as a transformational answer to these concerns by empowering individuals to govern their digital identities. In this section, we will look at the fundamental notions of SSI, investigate its significance, and examine why it is important from a technical aspect. SSI signifies a paradigm change in the administration of digital identities. It overcomes the technological issues of centralized systems by providing better privacy, security, and user control. While SSI presents technological problems such as scalability, DID method fragmentation, regulatory compliance, and user experience, continuing research and development in the area is paving the way for a more secure and user-centric digital identity future. Greater interoperability, improved usability, a thriving decentralized identity ecosystem, privacy-preserving technologies, and a reaction to the quantum computing threat all promise to reshape how we maintain our digital identities in the future.

In today's digital environment, SSI is important because it tackles the basic concerns of control, privacy, and security in how we maintain our online identities. Consider the possibility of having your own digital *identity wallet* on your smartphone. This wallet stores vital information about you, such as your name, age, and educational credentials, in the same way that your actual wallet stores your driver's license or ID card. The key distinction is that with SSI, you have complete control over what is in your digital wallet, who sees it, and when they see it. It's like having a superpower that allows you to choose how much and with whom you share information. This approach permits you to preserve your privacy while also lowering the chance of your personal data being mistreated, as is common when you register for online services or social media platforms. SSI guarantees that your digital identity is safe and that you own it, rather than giant corporations or governments.

SSI is important for a variety of reasons, the most important of which is the increased security it provides. Traditional identification systems rely on numerous firms and organizations to keep your personal information secure. Unfortunately, these organizations are common targets for cyberattacks, and if they are penetrated, your data might wind up in the hands of hackers. SSI changes this by protecting your digital identity with powerful cryptography. Consider it like locking your personal information in a digital vault with only you as the key. This cryptographic technology protects your data from prying eyes and any compromises. Furthermore, SSI allows you to establish your identity without disclosing too much. As a result, if you need to confirm your age to visit a website or buy a product, you may do so without revealing your precise birthday or other sensitive information. In a sense, SSI acts as a personal fortress for your digital identity, keeping it safe and in your hands.

Cryptography in SSI

Cryptography is critical in ensuring that your digital identity stays safe, private, and under your control in the world of SSI. While the technical foundations of cryptography might be complicated, we'll cover the essentials in layman's terms. This section will go over essential cryptographic approaches, public and private key ideas, digital signatures, and the importance of verified credentials in the context of SSI.

Cryptography is the foundation of SSI security and trust. It's like the enchantment that keeps your digital identity secure and under your control. The tools that allow you to safeguard your personal information while safely sharing it with others include public and private keys, digital signatures, and verified credentials.

Remember that these cryptographic approaches are the locks and keys to your digital identity as you explore the world of SSI. You may use them to preserve your privacy and securely verify your credentials, all while benefiting from the advantages of a self-sovereign and secure digital identity.

Cryptographic techniques

Cryptography is a mystical language that computers employ to protect their secrets. It helps to keep your personal information safe from prying eyes while allowing you to safely communicate it when necessary. SSI uses a variety of cryptographic approaches to do this.

Encryption is one of the most important approaches. Consider encryption to be the process of converting your data into a secret code that only you and the intended receiver can decipher. This implies that even if someone intercepts your data, they won't be able to make sense of it unless they have the *key* to decode it.

Hashing is another important method. Hashing your data is similar to establishing a digital fingerprint of it. It converts your data into a unique string of characters, similar to a secret code that symbolizes your data. If even a little portion of your data changes, the entire fingerprint will alter. This is useful for maintaining data integrity and ensuring that it has not been tampered with.

Public and private keys

Public and private keys are like a pair of magical keys that only you have in the realm of SSI. Consider the public key to be a unique lock that anybody may see and use to deliver encrypted communications or verified credentials to you. Sharing your public key with anyone is secure since it simply allows them to send you information and not access your personal data.

The private key, on the other hand, is like a secret key that only you have. It's used to decrypt communications or credentials sent to you using your public key. Your private key is incredibly valuable since it allows anybody who has it access to your personal information. It is critical to keep your private key safe and secure.

Digital signatures

Assume you need to sign an essential document, such as a contract. This is possible in the SSI digital world with a digital signature. It's similar to signing your name but using an unforgeable method.

This is how it works: you generate a digital signature for a piece of data using your private key. This signature is one of a kind and ensures that the data is genuine and has not been changed. When you share this signed material with someone, they may verify the signature using your public key. If everything checks out, they'll know the data originated from you and hasn't been tampered with.

In SSI, digital signatures are equivalent to mystical seals that ensure the legitimacy of personal information. They give confidence and protection in an online world fraught with danger.

Verifiable credentials

In SSI, verifiable credentials are analogous to digital certificates and badges. In the same way that you may have a real driver's license or a diploma, you can have digital counterparts in SSI.

These credentials are unique in that they include a digital signature from the issuer, which might be a university or a government entity. This signature is safe and unforgeable since it is made with the issuer's private key.

When you acquire a verified credential, it's as if the issuer has stamped it with their official seal, verifying its authenticity. You may then give these credentials to others, such as employers or service providers, to demonstrate certain aspects of yourself without disclosing any more personal information than is necessary.

You can, for example, use a verified credential to confirm your age without disclosing your precise birthdate. You keep your privacy while still giving the required documentation.

Now that we've introduced the notion of SSI, we will look at how blockchain and DLT might help shape its implementation. In the following section, we will look at how blockchain and DLT may help to build safe, decentralized identity ecosystems that allow individuals to exercise control over their personal data and interactions.

Blockchain and DLT in SSI

By placing individuals in charge of their personal data, SSI is revolutionizing the way we handle digital identities. The utilization of sophisticated encryption algorithms and DLT is central to SSI. Blockchain and DLT both play critical roles in allowing SSI but for different reasons. In this technical examination, we will look at the roles of blockchain and DLT in SSI and emphasize the differences between these two technologies.

Role of blockchain in SSI

Blockchain is a decentralized and distributed ledger that records transactions over a network of computers. It is frequently connected to cryptocurrencies such as Bitcoin. Blockchain, in the context of SSI, acts as a secure and immutable database for recording identity-related data. Here's how blockchain helps with SSI.

Data immutability

Once data is stored in a block or a blockchain, it is very hard to change or erase it. This immutability protects identity-related data by prohibiting unauthorized alterations. Data immutability is a basic aspect of a blockchain that ensures that once information is put into a block, it becomes very resistant to any type of manipulation or deletion. The cryptographic techniques and consensus procedures that underpin blockchain technology are responsible for its immutability.

Here's why data in a blockchain is thought to be immutable:

1. **Cryptographic hashing**: A cryptographic hash is a unique *fingerprint* of the data included in each block of a blockchain. This hash is produced using a complicated mathematical algorithm that takes the contents of the block and generates a fixed-length string of characters. Even little changes in the input data will result in a significantly different hash. This makes altering the data within a block without changing its hash computationally impossible.
2. **Linkage of blocks**: A blockchain's blocks are linked in chronological sequence. Each block has a reference to the hash of the preceding block, resulting in a chain of blocks. Because of this interconnectedness, each change in one block impacts not only the hash of that block but also the succeeding blocks. As a result, tampering with a single block necessitates modifying every following block, which is a computationally expensive and nearly impossible process in a secure blockchain network.
3. **Consensus mechanisms**: Blockchain networks commonly use a consensus technique such as **proof of work** (**PoW**) or **proof of stake** (**PoS**). Multiple participants (nodes) validate and agree on the content of new blocks under these processes. The network would reject any unauthorized changes to the data since the consensus mechanism assures that all participants agree on the same version of the blockchain. This consensus technique is critical to sustaining data immutability.
4. **Decentralization**: Because blockchain networks are decentralized, there is no single central authority in charge. Instead, the network is made up of numerous nodes that keep copies of the blockchain. Because of this decentralization, there is no single point of control or failure, making it incredibly difficult for any entity to change data across all nodes at the same time.
5. **Timestamping**: Each block in a blockchain has a timestamp that indicates when the data was added. This timestamp adds to data immutability by providing a chronological sequence of transactions. Any effort to modify the sequence or content of blocks would be visible in the timestamps and rejected by the network.

In summary, data immutability in a blockchain is achieved by the use of cryptographic hashing, block linking, the consensus process, decentralization, and the use of timestamps. These aspects work together to guarantee that once data is put into the blockchain, it becomes extremely difficult, if not impossible, to change or delete without network consensus, protecting the integrity of identity-related data in SSI systems.

Decentralization

Blockchain is based on a network of nodes (computers), which eliminates the need for centralized authority. This decentralized character is consistent with the ideas of SSI, which distributes power over personal data. The decentralization of power and authority over a network of nodes (computers) rather than depending on a single centralized organization is a basic feature of blockchain technology. Here's a full description of how blockchain decentralization works and how it corresponds with SSI principles:

- **Elimination of a centralized authority**: In conventional systems, a central authority or middleman controls and maintains data, transactions, and information access. This centralization can result in a number of problems, including a single point of failure, a lack of transparency, and the possibility of power exploitation or abuse.
- **Peer-to-peer network**: Blockchain is a **peer-to-peer (P2P)** network, which means that every node in the network is equal and has its own copy of the complete blockchain. These nodes interact with one another, verify transactions, and collectively maintain the blockchain. The network is not controlled by a single node or entity.
- **Consensus mechanism**: To validate and agree on the state of the blockchain, decentralized blockchains rely on consensus procedures such as PoW or PoS. Nodes contribute to the consensus process by either solving challenging mathematical problems (PoW) or holding and staking Bitcoin (PoS). When a consensus is obtained, a new block is added to the blockchain and all nodes' copies are updated. This guarantees that the blockchain is safe and trustworthy in the absence of a centralized authority.
- **Data integrity**: Data on a blockchain is very resistant to manipulation and unauthorized alterations since it is spread across numerous nodes. To change a record in the blockchain, one must either control the majority of the network's computer power (in the case of PoW) or stake a sufficient quantity of money (in the case of PoS). In a decentralized network, achieving this degree of control is exceedingly difficult.
- **Transparency and trust**: Blockchain's decentralized nature ensures system transparency and trust. Because all transactions are public to all participants, anybody may independently check the data on the blockchain. Transparency and trust are especially important in SSI, where individuals want control over their personal data as well as confidence that it is not being exploited or manipulated.

SSI principles emphasize putting individuals in control of their digital identities. Decentralization fully matches with these objectives by giving individuals authority over their personal data. Blockchain technology's decentralization fully matches with the ideas of SSI. It gives people complete control over their digital identities, improves privacy and security, encourages interoperability, and avoids the dangers associated with centralized authority. This synergy between decentralization and SSI is a significant driver of digital identity management progress.

Transparency

Because blockchains are transparent, participants may check the data and transactions. This openness in the context of SSI can assist users in tracing the history of their identity-related transactions.

Security

Blockchain cryptographic security protocols provide strong protection for identification data. Users may govern their digital identities by using public and private keys for secure access.

Verifiable credentials

Blockchain may be used to create and verify verifiable credentials, which are digitally signed and tamper-proof attestations of identity-related data.

User-centric control

Individuals have complete control over their digital identities using blockchain-based SSI systems. Users own their identification data and may choose when and with whom they disclose it.

DLTs

While blockchain is a type of DLT, other DLTs exist and contribute to SSI as well. The term *decentralized ledger technology* refers to a larger spectrum of distributed ledger solutions. DLT assists SSI in the following ways:

- **Flexibility**: DLT enables a variety of consensus processes and data types, providing design freedom for SSI systems. This versatility is critical since various use cases may necessitate different DLT solutions.
- **Interoperability**: Interoperability across different SSI systems and networks is frequently provided via DLTs. This implies that users' digital identities may be used across several services and apps.
- **Scalability**: DLT can alleviate scalability issues that are frequently encountered in SSI, especially as decentralized identity acceptance rises. It makes it possible to maintain a large number of **decentralized identifiers** (**DIDs**) and verified credentials in an effective manner.
- **Ecosystem integration**: DLT can interact with current identity ecosystems and standards, making SSI adoption simpler for businesses. It is compatible with open standards and protocols, making adoption easier.
- **Privacy-preserving techniques**: Some DLT systems include extensive privacy protection mechanisms to ensure that individuals' sensitive identifying information stays private while remaining verifiable.

While blockchain and DLT are comparable ideas, there are significant differences in the context of SSI:

- **Data structure**: For data storage, blockchain employs a linear, chain-like topology. DLT covers a broader range of data structures, such as **directed acyclic graphs** (**DAGs**) and other distributed ledger formats.
- **Consensus mechanisms**: PoW or PoS consensus procedures are frequently used in blockchain. DLT supports many consensus techniques, providing more alternatives for meeting specific SSI needs.
- **Interoperability**: DLT solutions are typically more concerned with ensuring interoperability among multiple SSI networks and platforms, whereas blockchain may be connected to specialized networks or ecosystems.
- **Privacy features**: Some DLT systems prioritize advanced privacy-preserving mechanisms in order to address the demand for user privacy in SSI. The approach to privacy in blockchain might differ.
- **Use cases**: While blockchain is frequently connected with cryptocurrency and financial transactions, DLT spans a larger spectrum of applications, making it more flexible to varied SSI use cases.

Finally, blockchain and DLT are key technologies in SSI, providing safe, decentralized, and user-centric control over digital identities. While blockchain offers distinct benefits in terms of data immutability and security, DLT delivers flexibility, interoperability, and privacy-preserving features. The choice of these technologies in an SSI implementation is determined by the use case's unique needs and the desired balance of security, privacy, and flexibility. As the SSI ecosystem evolves, blockchain and DLT will both play critical roles in creating the future of digital identity.

Data storage and decentralization

Data storage and management have advanced substantially in the digital age. The notion of decentralization is one of the most transformational breakthroughs in this field. Traditional data storage frequently depended on centralized servers and databases managed by single companies, resulting in data vulnerability, single points of failure, and restricted user control. Decentralization has emerged as a potent alternative, changing data storage and management. This section digs into data storage and decentralization principles, highlighting their fundamental topics and their influence on the digital world.

Historically, centralized data storage was the norm, with certain organizations or service providers controlling data repositories. This strategy had various advantages, including simplicity, ease of management, and rapid data access. It did, however, have substantial limitations such as a single point of failure, data exposure, limited user control, and scalability issues.

Decentralization has emerged as a viable alternative to centralized data storage. It represents a fundamental leap in data storage and management. Decentralization has resulted in the development of novel data storage technologies that overcome the inadequacies of centralized systems. Decentralization has various technological features that make it an interesting data storage strategy:

- **Resilience**: Decentralized systems are more robust to failure due to their dispersed structure. If one node fails, data is still accessible via other nodes, guaranteeing high availability.
- **Security**: To safeguard data, decentralized data storage frequently employs complex cryptography algorithms. The immutability of blockchain technology, as well as data redundancy among nodes, improves data security.
- **Privacy**: Users have more control over their data and may pick what information to share and with whom. User privacy is prioritized in decentralized identification systems in particular.
- **Scalability**: Scalability is provided via decentralization by adding more nodes to the network as the demand for storage or data processing grows. This adaptability lowers infrastructure costs while ensuring effective growth.

Decentralization offers a paradigm shift in data storage and management, aligning with user control, security, and privacy principles. Decentralization improves resilience, security, and scalability by dispersing data across a network of nodes while lowering the risks associated with centralized systems. As decentralized data storage systems improve, they provide exciting alternatives to established ways, altering the digital environment and giving consumers more control over their data. However, resolving complexity, regulatory, interoperability, and user experience difficulties will be important for the broad adoption of decentralized data storage solutions in a variety of applications, including SSI systems.

Building on the study of blockchain and DLT in SSI, we now turn our attention to another critical component: DIDs. In the following section, we will look at the role of DIDs in providing decentralized, interoperable, and privacy-preserving identity solutions, allowing users to maintain and govern their digital identities securely.

DIDs

DIDs are virtual keys that protect your online identity. They let you construct and manage your identity without relying on large technology firms or governments. Consider DIDs to be the digital equivalent of your passport or driver's license. DIDs are your online IDs that you control, much like you have a physical ID that confirms who you are. This is how they work:

- **Unique and permanent**: Consider DIDs to be unique digital identification signatures. A DID is your online fingerprint, just as your name is unique in the real world. It's a code that you'll always have, and no one else on Earth will have the same one. Assume you have a secret code to your favorite hideout and that it is unique to you. DIDs function in a similar manner. When you establish a DID, you're creating your own unique key to the internet world. This key remains constant, and it serves as a reminder of your individual digital presence. The beauty of this distinction is that it distinguishes you in the broad digital environment. It prevents someone from impersonating or pretending to be you. Your DID, like your face and name in the physical world, becomes a reliable indicator of your identity. It's your digital identity and your virtual passport, and it never changes, letting you traverse the online world with confidence, knowing that your identity is safe and indisputably yours.

- **No central authority**: Consider the internet to be a large metropolis in which you may construct your own digital house without the requirement for clearance from any central government or authority. Your DID is the key to this hidden internet haven. You do not need permission or to obey someone else's rules to develop and use it. You have complete control over the situation. This is like having your own private, secure online castle where you establish the rules. You pick what information to disclose and to whom. No one can enter your hideout unless you give them permission. It's a digital area that respects your autonomy, just like your actual space does. DIDs are built on this freedom, letting you explore the internet world with ease.
- **Privacy and control**: DIDs provide you with a one-of-a-kind online superpower: the capacity to govern your personal information. You determine what facts you disclose and with whom, just like a superhero with the ability to keep their secrets hidden. It's like having a virtual barrier around your digital identity. You may connect with the online world with confidence, knowing that your personal information is within your control. DIDs provide you superhuman privacy protection, guaranteeing that your digital interactions are as safe and selective as you want – a degree of control that is frequently elusive in the broad expanse of the internet.
- **Secure and tamper-proof**: DIDs use cutting-edge technologies to strengthen their security. It's like arming your digital identity with an impenetrable digital lock. This lock, which was established using complex cryptographic procedures, guarantees that your online identity is safe from modification or unauthorized access. DIDs secure your digital presence in the same way that locks protect your physical assets. They create an impenetrable barrier that makes it virtually hard for anyone to change or undermine your online identity. This digital lock acts as an impenetrable wall, strengthening the integrity and security of your personal information and providing you with the assurance that your online identity is within your control and out of the reach of prying eyes.
- **Use anywhere**: DIDs, like master keys in the real world, are your adaptable digital keys. DIDs, like a master key, provide you access to many online services and platforms throughout the internet. In this enormous digital universe, it's as if you have a single global key that opens countless doors. Your DID is the continuous, dependable key that allows you access whether you're logging into your email, visiting a secure website, or validating your identity in various online transactions. This ease of use and consistency streamlines your online experiences, providing a smooth and secure method to explore the many digital environments, all with a single, trusted key.

DIDs empower you to control your online identity, making the internet a more secure and private environment. They're your digital keys to a world where you, as in real life, determine who gets to know you.

Usage of DIDs in the SSI space

Our lives are becoming increasingly entangled with the internet world in the digital era. We leave digital fingerprints wherever we go, from social media accounts to email addresses. These digital footprints frequently rely on businesses or institutions to confirm our identity and manage our online presence. But what if you could have your own digital passport, one that is not controlled by any single body and allows you to explore the internet world securely, anonymously, and independently? That is exactly what DIDs provide.

A DID is essentially your unique digital nameplate. Just like your name is unique to you, your DID is unique to you, and no one else on Earth can have the same one. It's your virtual business card, and it never changes. Consider it your digital self, a virtual representation of yourself in the vast internet realm. DIDs are extremely unique in that they are not regulated by any central authority. It's like having your own hidden refuge online, where no one can interfere with your identity. DIDs do not require authorization from large IT corporations or governments to generate or use. They're yours to control, and it's all about reclaiming control over your digital identity. You control who has access to your DID and when they do so, just as you control who has access to your hidden hideaway.

DIDs are all about security and privacy. They employ cutting-edge technology to ensure that your digital identity is secure. It's similar to putting a digital lock on your online identity. This digital lock protects your personal information, ensuring that only those you authorize have access to it. Consider utilizing your DID in multiple locations in the same way as you do your physical ID. It's like having a single master key that unlocks several doors. You may use the same trusted DID to log into websites, authenticate your identity, and access online services. It streamlines your online life and provides a consistent, secure manner of navigating the digital world.

So, what is the significance of DIDs in the digital landscape? It all comes down to control and trust. You control your digital identity using DIDs. You, as in the real world, determine who gets to meet you. It is a digital empowerment that makes the internet more secure and private. DIDs, on the other hand, aren't only for people; they're also for corporations and organizations. They enable secure and trustworthy interactions with consumers and clients. It's a win-win situation for everyone.

DIDs act as a digital barrier in a world where data breaches and privacy concerns are on the rise. They provide you with control, privacy, and security, putting your online identity firmly into your hands. It's a new age for digital identification, one in which you yourself may be securely and anonymously online. So, remember that DIDs are your digital passport to freedom, and they're here to make the online world a better place for all of us.

DID methods

DIDs are the cornerstone of a disruptive paradigm change in the field of SSI. Individuals have authority over their digital identities thanks to DIDs, which provide a safe, private, and self-governing approach to online interactions. DIDs, on the other hand, do not live in isolation; they are part of a larger ecosystem, and here is where DID techniques come into play.

Before delving into DID methodologies, it's critical to understand the foundations. DIDs are identifiers that are unique, globally resolvable, and cryptographically verifiable. DIDs, unlike traditional identifiers such as email addresses and usernames, are not bound to a centralized registry, making them genuinely autonomous. They are created using advanced cryptographic procedures, providing a degree of security that traditional forms of identity cannot match.

The blueprints that dictate how a DID is produced, handled, and resolved are known as DID methods. They establish the rules and methods that govern a certain type of DID. Consider DID techniques to be several portals going to the same SSI reality. Each door requires a different key, but once inside, you're in the universe of self-sovereign identification.

DID techniques are available in a variety of flavors, each geared to certain use cases and needs. Among the most prevalent DID approaches are the following:

- **The Sovrin method**: This approach is closely related to the Sovrin Network, which was a forerunner in the SSI domain. It employs a permissioned blockchain to provide a secure and compatible method of managing DIDs.
- **The Ethereum method**: This technology, which makes use of the Ethereum blockchain, provides a decentralized and widely used solution. It's ideal for individuals who like the Ethereum network's security and decentralization.
- **The Sidetree method**: Sidetree is a new technique that emphasizes scalability. It enables DIDs to be anchored across different blockchains, increasing flexibility and robustness.
- **The Hyperledger Indy method**: This technique, built on the Hyperledger Indy platform, prioritizes privacy and secrecy. It's a fantastic solution for situations where data security is critical.
- **The Bitcoin method**: This method roots DIDs in the Bitcoin blockchain, providing a robust and well-recognized way for individuals who trust the Bitcoin network's security and durability.

DID methods are important because they address the various demands and preferences of users and applications. Choosing the best DID approach is analogous to picking the best tool for the task. For example, if your SSI application requires great security and decentralization, you can choose a technique based on a strong blockchain, such as Ethereum or Bitcoin. If anonymity and secrecy are critical, a Hyperledger Indy-based solution may be preferable. DID techniques also allow for compatibility inside the SSI environment. They ensure that various entities may recognize and resolve DIDs, hence promoting a unified and integrated digital identity ecosystem.

The SSI world is continually growing, as are DID approaches. New approaches are always being created as additional use cases emerge. The goal is to give options and flexibility so that SSI solutions may be tailored to the specific needs of individuals, organizations, and industries.

In conclusion, DID techniques are the SSI world's architects. They provide the structural underpinning for self-sovereign identification, allowing individuals to create their digital identities to meet their own requirements. As the SSI environment evolves, DID approaches will play a critical part in constructing a more secure, private, and user-centric digital future in which individuals have control over their online identities and interactions.

DID resolution and resolution protocols

In a quickly changing digital ecosystem where our online presence is essential to our everyday lives, how we manage our identities in a safe, privacy-conscious, and self-sovereign manner has taken center stage. Traditional systems frequently demand us to submit our identities to third-party businesses, thus giving up control over our personal information and creating risks. However, the rise of SSI has ushered in a new era in which we reclaim sovereignty over our digital identities. The notion of DIDs and their resolution is central to this shift.

DIDs constitute a paradigm change in that they provide unique, globally resolvable, and cryptographically verified identifiers. DIDs, unlike traditional methods of identification such as email addresses and usernames, are not tied to a centralized registry or reliant on third-party intermediaries. They are created using modern cryptographic algorithms, enabling hitherto impossible levels of security and autonomy in the digital arena. However, the capacity to fully realize the promise of DIDs is dependent on a critical component – DID resolution.

Significance of DID resolution

Consider DIDs to be one-of-a-kind digital keys that open the door to your digital identity. These keys are safe and completely self-sufficient, but in order to properly use them, we need a mechanism to convert these digital keys into actionable data. This is referred to as DID resolution. DID resolution serves as a link between cryptographically secure DIDs and the real-world information and services they represent.

In layperson's words, DID resolution is the process of converting your digital identity into actions and interactions. It enables you to utilize your DID to access secure services, log onto websites, and authenticate your identity online. DID resolution is the magic that happens behind the scenes to make your digital identity effective, safe, and adaptable.

The components of DID resolution

DID resolution involves several key components:

- **The DID**: The digital key that represents your identity is the DID. It is one of a kind, cryptographically secure, and an essential component of the DID resolution process.
- **The DID document**: Each DID is linked to a DID document. This document provides crucial information, such as the public keys needed for cryptographic verification, service endpoints, and other information that allows interaction with the DID.

- **The DID resolver**: The resolver is the instrument or program that is in charge of DID resolution. It is the *translator* that converts a DID and its accompanying document into something that programs can understand and use.
- **The DID method**: DID methods are the blueprints that govern how a certain type of DID is formed, handled, and resolved. Each DID approach has its own set of principles and practices for dealing with DIDs.

To bridge the gap between the secure and self-sovereign character of DIDs and their practical deployment in the real world, DID resolution is required. Individuals and organizations can use their DIDs to log onto websites, access services, and confirm their identities online. DIDs would be isolated and useless without DID resolution. Resolution protocols, which standardize DID resolution methods, are critical to ensure interoperability throughout the SSI ecosystem. DID resolution is the method that allows people to take control of their digital identities, resulting in a more secure, private, and user-centric online environment.

The role of resolution protocols

Resolution mechanisms are critical in ensuring that DIDs are seamlessly translated into usable data. These protocols specify the standardized DID resolution techniques and formats, promoting interoperability throughout the SSI ecosystem. There are several resolution procedures, each customized to unique use cases and criteria. These protocols are critical in ensuring that diverse entities can recognize, resolve, and interact with DIDs consistently.

Navigating the DID resolution landscape

Individuals and organizations may recover control of their digital identities thanks to DID resolution and resolution protocols, which are at the forefront of the SSI revolution.

We shall investigate the complexities of these critical components as we go deeper into the world of DID resolution and resolution procedures. We will investigate the technological foundations of DID resolution, the many resolution protocols in use today, and the significance of DID approaches in determining the future of digital identity.

This section has provided you with a thorough grasp of how DID resolution works, the crucial function of resolution protocols, and how these parts work together to contribute to the growth of self-sovereign and secure digital identities. You will now be better equipped to navigate the ever-expanding digital identity environment, using the full potential of DIDs and resolution protocols to build a more secure, private, and user-centric online world with this information.

Now that we've discussed the importance of DIDs in allowing decentralized identity solutions, we will investigate the larger SSI ecosystem. In the following section, we will look at the components, stakeholders, and interactions of the SSI ecosystem, emphasizing its potential to transform how people manage and exert control over their digital identities.

The SSI ecosystem

In an increasingly digital environment, the concept of SSI has emerged as a game-changing concept. Individuals have total control over their digital identities using SSI, liberating them from the limits of traditional, centralized identity management systems. Several significant players are at the center of the SSI ecosystem, each playing a critical part in redefining how we maintain our digital identities. In this in-depth examination, we will dig into the complexities of the SSI ecosystem, with a particular emphasis on three key components: **identity wallets**, **identity hubs**, and **identity agents**.

The conventional digital identification paradigm has long been defined by reliance on third-party intermediaries, central authority, and the surrender of sovereignty over personal data. This paradigm frequently results in privacy violations, security flaws, and a lack of user autonomy. As a remedy to these difficulties, self-sovereign identification evolved, guaranteeing individuals ultimate control over their digital identities and personal data. Users become the only custodians of their identities in the SSI ecosystem, deciding what information is shared and with whom.

SSI stakeholders

The SSI ecosystem is made up of various stakeholders, each with its own set of tasks and responsibilities. These parties work together to establish a streamlined and safe environment for SSI management. Among the key stakeholders are the following:

- **Identity wallets**: These are user-friendly programs or software tools that allow people to create, administer, and govern their digital identities. Identity wallets are safe stores for personal data and credentials, allowing individuals to easily access their information and selectively share it with others.
- **Identity hubs**: Identity hubs act as decentralized storage and processing centers for people's personal information and credentials. They serve as intermediates, facilitating safe data exchange and exchanges between identity wallets and third-party entities. Identity hubs place a premium on data protection and user control.
- **Identity agents**: Identity agents are software agents that operate on behalf of SSI ecosystem users. They help with responsibilities connected to identification, such as credential issuing, verification, and communication. Identity agents guarantee that interactions are efficient, safe, and tailored to the user.

Let us now examine each of these SSI stakeholders in further depth, gaining a better grasp of their roles and significance, and how they jointly contribute to the realization of SSI.

Identity wallets

Identity wallets are user-centric programs that act as secure digital vaults for personal information, credentials, and verified claims. Their principal function is to give people control over their digital identities. Users may safely store and manage their data, selectively release information, and connect with a variety of services and companies while maintaining their privacy. User control is prioritized in identity wallets, which provide easy online authentication, safe data storage, and user-centric data management. They are critical components of the SSI ecosystem, allowing individuals to navigate the digital world with autonomy and security. These wallets serve as the user's portal to their SSI, offering a safe and simple method to manage and govern their digital persona.

The role of identity wallets is as follows:

- **Data storage**: Personal data, credentials, and verified claims are securely stored in identity wallets, guaranteeing that users have quick access to their information.
- **Selective disclosure**: Users can choose to expose their data as needed, without disclosing more information than is required for a specific transaction.
- **Interoperability**: Identity wallets should be interoperable with a variety of identity systems and standards, allowing for easy integration into the SSI ecosystem as a whole.
- **User control**: User control is a major aspect of identity wallets. Users have the last say over who has access to their data and under what circumstances.
- **Security**: Identity wallets prioritize security. They protect the data they keep using strong encryption and security procedures.

Identity hubs

Within the SSI ecosystem, identity hubs are decentralized data storage and processing centers. They are critical to the safe management of personal data, credentials, and verified claims. Identity hubs serve as middlemen, permitting the selective transfer of data between identity wallets and third-party stakeholders. They prioritize user control, allowing people to customize their data-sharing settings. Identity hubs play an important role in the issue and verification of verified credentials, as well as in promoting user-centric, privacy-conscious data management. These hubs enable users to preserve data ownership while providing smooth integration and safe data interchange throughout the SSI ecosystem. Personal data, credentials, and verified claims are stored and processed in decentralized identity hubs. They serve as middlemen, allowing safe data flow between identity wallets and other parties.

The role of identity hubs is as follows:

- **Data storage**: Identity hubs safeguard the storage of personal data and credentials, removing the need for third-party data repositories
- **Selective data sharing**: Users may design their identity hubs to exchange data with certain partners only when necessary, preserving their privacy and control
- **Interoperability**: Identity hubs are meant to connect with a variety of identity systems, standards, and identity wallets, allowing for smooth integration within the SSI ecosystem
- **User-centric data processing**: Identity hubs place a premium on user control and privacy, ensuring that data is processed in accordance with the user's wishes

Identity agents

Within the self-sovereign identification ecosystem, identity agents are software entities that represent persons. Their principal function is to act on behalf of users, carrying out a variety of identity-related duties. Identity agents provide and verify verified credentials, increasing the trustworthiness of shared data. They provide safe and confidential communication between users and third-party entities, ensuring that interactions are consistent with user wishes. Identity agents are crucial in protecting user privacy and managing data leakage during online transactions. These digital delegates enable users to navigate the digital world with confidence, security, and user-centricity, therefore contributing to SSI realization. Identity agents are software agents that operate on behalf of SSI ecosystem users. They handle a variety of identity-related responsibilities, such as credential issuing and verification, as well as contact with other parties.

The role of identity agents is as follows:

- **User representation**: Identity agents serve as users' digital representatives, carrying out identity-related actions on their behalf
- **Credential issuance**: Agents can create and sign verified credentials for users, increasing the trustworthiness of the information exchanged
- **Verification**: Agents validate credentials and data given by third parties, verifying the information's validity and dependability
- **Secure communication**: Identity agents enable users and other entities to communicate in a safe and confidential manner

The SSI ecosystem, which consists of identity wallets, identity hubs, and identity agents, works in tandem to provide individuals with self-sovereign digital identities. Users maintain their data in identity wallets that are secure and user-centric, while identity hubs protect and promote data interchange. Identity agents engage on users' behalf, boosting data trustworthiness and guaranteeing safe, user-centric interactions.

This ecosystem provides a variety of practical use cases, ranging from online authentication and secure data sharing to data privacy and ownership. It marks a paradigm shift in how we manage our digital identities, providing individuals with control, security, and autonomy over their online personas.

As the SSI environment evolves, it is critical to understand the roles that various stakeholders play and how they interact. They constitute the cornerstone of self-sovereign identification, paving the way for a more secure, private, and user-centric digital environment.

Now that we've reviewed the main components and players in the SSI ecosystem, we turn our attention to the critical element of SSI interoperability. In the following section, we will look at how interoperability facilitates the easy interaction and exchange of verified credentials across various SSI platforms, resulting in increased acceptance and scalability of decentralized identity solutions.

SSI interoperability

The idea of SSI envisions a digital future in which individuals have complete control over their personal data, allowing them to communicate in a secure and privacy-preserving manner. However, this digital utopia can only be completely realized if SSI systems and solutions can connect and collaborate smoothly. This is where SSI interoperability comes into play as the cornerstone that bridges the gap and unlocks the true potential of this game-changing paradigm.

Consider a scenario in which you have a digital wallet that stores your verified credentials, such as your driver's license, passport, and educational certifications. This wallet is your digital key, allowing you to authenticate your identity and access services without having to divulge sensitive personal information again. This is a world in which you, the user, have complete control, and data privacy is key. The problem is that there are several entities in the digital realm, each proposing their own SSI solutions. They're like jigsaw puzzle pieces, each one distinctive in its own way. These elements must come together flawlessly to form a full image, allowing for easy communication and digital credential verification. This is where SSI interoperability enters the picture.

As we progress through this part on SSI interoperability, we will look at the key components that make it all work. From standards and protocols to technical frameworks, we'll look at the tools and procedures that allow various SSI solutions to work together effortlessly. We'll talk about the challenges and complexities of SSI interoperability, as well as the great prospects it presents. You will learn about trust frameworks, the importance of DIDs and verified credentials, and how blockchain technology and distributed ledger systems contribute to interoperability. We will also go over the necessity of adhering to existing and new legislation in order to ensure that interoperable SSI solutions meet privacy and security standards.

Our objective as we explore the SSI interoperability environment is to offer you a thorough grasp of this critical part of the SSI ecosystem. By the conclusion of this section, you will understand and have the tools to appreciate how SSI interoperability unlocks the promise of a user-centric, privacy-respecting digital identity world, in which the pieces of the digital identity puzzle fit together to provide a holistic and trusted user experience.

Importance of interoperability

The digital identity ecosystem is vast and diversified, with new platforms, applications, and standards developing all the time. This variety is great for creativity and competitiveness, but it also poses a problem. We risk fragmenting the digital identity space without interoperability, making it difficult for users and dependent parties to navigate and trust the system. The interoperability of SSI is the vital glue that ties this ecosystem together. It guarantees that credentials given by one SSI solution are accepted and recognized by another. It's like having a universal translator for digital identity, allowing diverse platforms and standards to efficiently communicate.

Your digital wallet becomes a universal key with SSI interoperability. It may be used to open not just one door, but several – various services, applications, and platforms. It makes the user experience easier by eliminating the need for different wallets and credentials. You may go from one environment to another with ease, establishing your identity and enjoying services. The interoperability of SSI also promotes confidence. You may trust that your digital credentials will be acknowledged across diverse SSI ecosystems just as you believe that your actual driver's license obtained in one nation will be recognized in another. This confidence is required for the widespread acceptance of SSI and the full realization of its promise.

SSI in a multi-SSI network

The idea of interoperability has cleared the way for the construction of multi-SSI networks in the field of SSI. These networks function as dynamic ecosystems in which users, relying parties, and verifiers communicate in real time, even if they utilize different SSI solutions. Using SSI in a multi-SSI network is a crucial step toward realizing the full potential of user-controlled digital identities. Consider the following scenario: as an individual, you have an SSI wallet comprising your verified credentials provided by multiple entities such as government agencies, educational institutions, and financial services. These credentials, which are provided in a widely recognized format, let you confirm your identity and gain access to services across a variety of SSI ecosystems. The capacity to combine disparate SSI solutions under a single framework is at the heart of multi-SSI networks.

DIDs and verifiable credentials are at the heart of multi-SSI networks. DIDs give consumers a unique, worldwide identification, ensuring that their digital identity is consistent across several SSI systems. Verifiable credentials, on the other hand, are digital representations of real-world identification traits supplied by trustworthy third parties. They are not connected to any one SSI solution but are based on open standards, making them portable and widely recognized. DIDs and verifiable credentials serve as the common language of a multi-SSI network, allowing different SSI systems to communicate and interact seamlessly. When you offer your verifiable credential from one SSI solution to a relying party in another, it is recognized and accepted, resulting in a consistent user experience.

Blockchain and DLT are critical to the integrity and reliability of multi-SSI networks. These technologies lay the groundwork for constructing a tamper-resistant and auditable record of SSI interactions. The blockchain records every credential issuance, presentation, and verification, establishing an immutable audit trail. This audit trail improves the multi-SSI network's security, transparency, and trust. It guarantees that all players may independently check the integrity of provided credentials, removing the need for centralized middlemen and increasing the whole ecosystem's dependability.

Trust frameworks and governance models are critical in a multi-SSI network to ensure that various SSI solutions comply with agreed-upon standards and practices. These frameworks describe the network's rules and regulations, resulting in a shared understanding of how SSI interoperability is accomplished. Parties in the multi-SSI network agree to follow these trust frameworks, which provides confidence that user data will be managed with care and privacy will be protected. Compliance with these guidelines is required to preserve network trust and security.

Using SSI in a multi-SSI network is an important step toward realizing the future of digital identification. It gives people control over their digital personas, protects their privacy and data ownership, and streamlines interactions with online services. As we continue to investigate the complexities of SSI interoperability inside multi-SSI networks, it becomes clear that the future of digital identity is user-centric, open, and widely accepted. The potential for these networks to transform how individuals maintain their digital identities by allowing them to move fluidly across diverse SSI ecosystems offers hope for a world where digital identity is genuinely self-sovereign. The multi-SSI network emerges as the cornerstone of this disruptive digital identity environment with security, trust, and interoperability at its heart, establishing a future where individuals really own and manage their digital identities.

Following this overview of SSI interoperability, we now shift our focus to the importance of regulatory compliance in the SSI domain. In the subsequent section, we will look at the changing regulatory landscape surrounding SSI deployments, including compliance issues and techniques for negotiating legal frameworks in this emerging industry.

SSI and regulatory compliance

One of the most important issues in the quickly changing realm of SSI is how it fits with and conforms to the ever-expanding environment of data protection and privacy legislation. This section digs into the complicated relationship between SSI and regulatory compliance, utilizing the European Union's **General Data Protection Regulation** (**GDPR**) as a case study. We'll talk about the importance of data security, the rise of compliance standards, and the legal problems that SSI confronts in this setting.

The management of personal data is inextricably tied to digital identity. Because the underlying concept of SSI is user-controlled digital identity, the management of personal information becomes critical to its operation. In this setting, guaranteeing the greatest levels of data protection and privacy is a core concept, not a need. Individuals can use SSI to store and manage their verified credentials, which typically contain sensitive personal data. These credentials are digital replicas of physical identification qualities such as birth certificates, driver's licenses, and academic records. The management and preservation of this data are governed by a variety of privacy laws and regulations designed to protect the individual's rights and privacy.

The convergence of SSI and regulatory compliance, notably GDPR, demonstrates SSI's inventive potential in changing digital identity. It emphasizes the significance of SSI in adhering to legal norms that prioritize user privacy and data security.

We will go over these important legal factors in depth as we progress through this section. We will look at GDPR principles, compliance structures, and the legal problems that SSI faces in a more regulated environment. Understanding the connection between SSI and regulatory compliance is critical to developing a safe and legally compliant digital identity ecosystem, ensuring that users have the tools they need to defend their digital identities while following data protection and privacy rules.

GDPR and data protection

The GDPR, established by the European Union in 2018, is one of the most significant and far-reaching data privacy legislations in the world. The GDPR establishes severe requirements for the handling of personal data and places individuals firmly in control of their data. Its key ideas of openness, accountability, and data subject rights are strongly aligned with those of SSI.

Personal data can only be handled for particular, legitimate reasons under GDPR, and individuals must express explicit agreement to have their data processed. Data controllers and processors are required to safeguard personal information and notify authorities and people in the case of a data breach. Individuals have the right to see their data, have mistakes corrected, and have their data erased (the *right to be forgotten*). These concepts are echoed in the SSI founding principles, which emphasize user control, permission, and privacy.

The core concepts of SSI are strongly aligned with GDPR standards, particularly in terms of data protection. One of the technological consequences of this alignment is the requirement for strong data encryption and access restrictions. Verifiable credentials, which frequently contain sensitive personal data, must be safely saved and communicated in SSI. This mandates the use of robust encryption technologies to ensure that data remains secure, even if intercepted.

Furthermore, SSI's user consent processes are technically significant. The GDPR requires that data processing be based on explicit and informed permission. In the case of SSI, this translates to the requirement for procedures that enable users to effectively grant and manage permission. To guarantee that user permission is both verifiable and revocable, technical methods such as cryptographic consent receipts can be deployed.

Data minimization is one of the GDPR's fundamental principles and emphasizes that only required data should be gathered and handled. This notion is technically realized in SSI via selective disclosure. Users have the option of displaying only the characteristics necessary for a transaction. This reduces personal data exposure and ensures that only relevant information is provided.

Data reduction relies heavily on technical solutions such as **zero-knowledge proofs** (**ZKPs**). ZKPs enable a party to demonstrate possession of certain information without disclosing the information itself. This method is consistent with both data reduction and GDPR's privacy-by-design principles. SSI systems can potentially provide efficient and privacy-preserving transactions by utilizing ZKPs.

The technological obstacles to aligning SSI with GDPR are numerous. Data portability is one such issue. Individuals have the right under GDPR to collect and utilize their personal data for their own purposes. It is a fundamental problem in an SSI environment to ensure that users may technically export and import their verified credentials while maintaining data quality and security.

Furthermore, the cross-border nature of SSI creates difficult technological data transmission problems. The GDPR places stringent limits on personal data transfers outside of the **European Economic Area** (**EEA**). Because of the decentralized and cross-border nature of SSI, these data transfer problems must be addressed technically to maintain compliance.

Compliance frameworks

Because GDPR and SSI principles are so closely aligned, various compliance frameworks have evolved to assist SSI developers, operators, and users in attaining GDPR compliance within an SSI setting. These guidelines outline how SSI ecosystems may be developed and administered to guarantee that user-controlled digital identities comply with GDPR standards.

These compliance frameworks address concerns such as informed consent, data reduction, data portability, and encryption, which are all critical in an SSI environment. They also emphasize the need for user transparency, which allows users to fully understand how their data is used and to properly control it.

Several compliance frameworks have evolved to help negotiate these obstacles and achieve GDPR compliance in the technological arena. These frameworks give technical guidelines on how to construct and run compliant SSI environments. Specifications and best practices for secure data handling, consent management, and auditability are frequently included.

A compliance framework, for example, may specify the technical requirements for establishing cryptographic consent receipts. These receipts are intended to guarantee that user consent is received, documented, and readily verified. They offer a technical audit record of consent events, which is critical for showing GDPR compliance with consent-related regulations.

Legal challenges

While there is clear congruence between SSI and GDPR, there are legal problems arising from SSI's unique features. One such issue is the issue of liability and accountability. In conventional identification systems, centralized identity providers are frequently held liable for data breaches and misuse. However, in SSI, when power is delegated to users, accountability becomes more complicated.

Furthermore, the cross-border nature of SSI, as well as the GDPR's geographical reach, pose questions of jurisdiction and cross-border data transfers. SSI is designed to work effortlessly across international borders, making it critical to traverse the complex web of data protection requirements that differ from jurisdiction to jurisdiction.

Furthermore, ensuring that SSI systems comply with increasing data protection standards necessitates ongoing monitoring and adaptation. Because of the dynamic nature of SSI ecosystems, where new participants and standards arise, compliance frameworks and legal interpretations must be kept up-to-date.

The legal issues surrounding SSI and GDPR compliance are primarily technological in nature. Addressing these difficulties demands continual technical innovation as the SSI landscape evolves. One of the legal issues is that of liability and accountability. In traditional ID systems, centralized providers are frequently held responsible for breaches; however, under SSI, accountability becomes increasingly difficult as users acquire autonomy. Decentralized responsibility may be controlled using technological tools such as liability models and audit trails.

The dynamic nature of SSI ecosystems, where new participants and standards arise, is another key legal concern. Technical adaptation is required to maintain compliance with changing data protection requirements. This necessitates ongoing monitoring, updating, and modification of technological frameworks and implementations in order to keep up with the evolving legal context.

In conclusion, the technological implications of SSI and GDPR compliance highlight the need for data security and privacy in SSI ecosystems. GDPR compliance is not simply a legal necessity, but also a fundamental design consideration. The convergence of SSI's user-centric principles with GDPR privacy safeguards demonstrates SSI's ability to revolutionize digital identity while respecting individual rights and regulations. Building safe, privacy-respecting SSI systems that allow users to control their digital identities within a compliance framework requires balancing technical innovation with regulatory limitations.

Now that we've delved into the intricacies of regulatory compliance in the SSI arena, we turn our attention to forecasting future trends and developments. In the following section, we will look at developing trends and prospective developments in SSI technology, and how they may affect the future of decentralized identity systems.

Future trends in SSI

Several significant factors are impacting the future of digital identification as SSI evolves. This section delves into three basic characteristics that are at the forefront of SSI innovation and development. We will look at emerging technologies that are changing the way SSI works, as well as the critical need for scalability and performance improvements to accommodate a growing user base. We will also delve into the fascinating world of SSI in a post-quantum world, where the security landscape faces new challenges and opportunities.

Emerging technologies

Emerging technologies are propelling SSI evolution, offering new capabilities and advancements to the digital identity space. These innovations are transforming how verified credentials are issued, maintained, and displayed. Emerging technologies such as ZKPs have significantly influenced SSI. ZKPs allow users to demonstrate the veracity of a statement without disclosing any more information. This cryptographic approach is at the heart of SSI's privacy-preserving transactions. It enables users to selectively reveal information while keeping the rest of their data private. ZKPs are game-changers because they enable speedy and secure transactions while minimizing data exposure.

DID approaches are another developing technology. DID approaches broaden the possibilities for SSI by offering distinct and standardized procedures for creating and managing DIDs. These solutions facilitate interoperability among SSI ecosystems and provide consumers with more alternatives for maintaining their digital identities. They enable customers to select the technique that best meets their requirements while also providing interoperability with a wider range of SSI systems.

Aside from these foundational technologies, improvements in secure enclaves and hardware-based security are improving the security of users' private keys and credentials. These technologies offer a safe environment for storing and processing sensitive data, making it exceedingly difficult for bad actors to corrupt the information. This technological advancement improves SSI system security by protecting the secrecy and integrity of users' digital identities.

While blockchain is not a new technology anymore, it is evolving, with blockchain-based SSI systems becoming more common. These systems provide a secure, tamper-resistant ledger for recording verified credentials, assuring the data's integrity and auditability. The immutability and distributed nature of blockchain correlate nicely with the ideas of SSI, making it a viable future technology.

Scalability and performance improvements

Scalability and performance become critical problems as the user base of SSI systems expands. Maintaining efficiency and responsiveness while accommodating a large number of users and making transactions is a technological problem that necessitates creative solutions.

The usage of Layer-2 solutions is one of the technological developments addressing scalability. Layer-2 solutions, which are frequently developed on top of blockchain platforms, attempt to offload some of the primary network's transaction processing. This reduces congestion and improves the scalability of SSI systems, allowing them to handle a higher volume of transactions without sacrificing speed.

Furthermore, progress in consensus techniques is essential to SSI scalability. Traditional blockchain consensus algorithms, such as PoW and PoS, have speed and scalability limits. New consensus techniques, such as **delegated proof of stake** (**DPoS**) and **proof of authority** (**PoA**), enable quicker transaction processing while consuming less energy. These advancements improve the scalability and performance of SSI systems.

Techniques for caching and optimization are also important in enhancing SSI speed. SSI systems can enable quicker reaction times and lower computing costs by storing frequently used data and optimizing data retrieval and verification operations. These technological solutions are critical to providing a consistent user experience, especially in high-demand conditions.

Additionally, the creation of specialized SSI infrastructure and scalable protocols is on the horizon. These solutions are designed to meet the unique requirements of SSI systems, allowing them to handle an increasing number of users and verified credentials. These technical innovations ensure that SSI stays efficient and responsive in the face of rising use by concentrating on scalability from the bottom up.

SSI in a post-quantum world

With the advent of quantum computing, new problems and possibilities for SSI security emerge. To ensure the continuous safety of user data in a post-quantum world, SSI systems must evolve.

The development of quantum-resistant cryptography algorithms is an important technical trend in preparing for a post-quantum world. Existing encryption techniques employed in SSI may be broken by quantum computers. To reduce this danger, researchers are developing and implementing new encryption algorithms that are resistant to quantum assaults. These algorithms will replace insecure encryption approaches, assuring the security of verified credentials in the future.

Quantum key distribution (**QKD**) is yet another technological advancement for SSI in the post-quantum age. Even in the presence of quantum computers, QKD allows for the safe exchange of encryption keys. QKD provides a degree of security that is immune to quantum assaults by exploiting quantum physics concepts. This solution protects the keys used to safeguard verified credentials against the threat of quantum decryption.

Furthermore, the notion of quantum-resistant blockchains is gaining popularity. Quantum-resistant blockchains are built to withstand quantum computer assaults while remaining safe. These blockchains use quantum-resistant cryptographic methods to lay the groundwork for SSI systems to run in a quantum-secure environment.

In conclusion, the future of SSI is inextricably linked to developing technologies, scalability and performance enhancements, and the reaction to quantum computing concerns. As SSI continues to influence the digital identity environment, these technological developments will be critical to improving security, efficiency, and resilience, ensuring that SSI maintains a strong and user-centric approach to digital identity in the years ahead.

Now that we've explored future developments in SSI, we'll focus on a vital component of its implementation: scalability in relation to blockchain technology. In the following section, we will look at the difficulties and potential solutions for improving scalability within the SSI ecosystem, with a special emphasis on the interaction with blockchain technology.

SSI and blockchain scalability

Blockchain technology serves as the foundation for safe and decentralized data management in SSI systems. However, as SSI acceptance grows, blockchain scalability becomes a key issue. In this section, we look at the scalability issues that SSI systems confront, the role of Layer-2 solutions in addressing these issues, and future scaling possibilities that promise to maintain the smooth evolution of SSI ecosystems.

Scalability challenges

Scalability is a major challenge that occurs when SSI systems expand, and more users and transactions are introduced to the network. Understanding the scalability problem is critical to resolving the technological restrictions that may prevent SSI from being widely used.

The low throughput of blockchain networks is one of the key scalability difficulties in SSI. Traditional blockchains, such as Ethereum and Bitcoin, have limited capacity for transaction processing. The blockchain's capacity to conduct a high amount of transactions becomes a bottleneck when the number of SSI users and verified credentials grows. This can lead to longer transaction processing times and increased transaction costs, negatively impacting the customer experience.

The issue of data storage is another scaling barrier. Verifiable credentials and related data must be stored on the blockchain, resulting in an ever-expanding blockchain size. Larger blockchains are more difficult to maintain and might cause synchronization difficulties for new network players. As SSI ecosystems grow in size, the feasibility of keeping all data on-chain becomes an issue.

Furthermore, most blockchain consensus algorithms, such as PoW and PoS, have intrinsic restrictions. While these measures are useful for security, they can result in delayed transaction completion and poorer throughput. To improve the performance of blockchain networks, scalability frequently necessitates a re-evaluation of consensus techniques.

Layer-2 solutions

Layer-2 solutions provide a viable method for tackling blockchain scaling concerns while keeping the security and decentralization that are essential components of SSI systems.

Layer-2 solutions are intended to run on top of current blockchains, offloading part of the main chain's transaction processing. The Lightning Network, which was originally built for Bitcoin but has since been extended for other blockchains, is one of the most notable Layer-2 solutions. The Lightning Network provides low-fee, off-chain transactions, making it perfect for microtransactions and frequent contacts, both of which are important in SSI systems.

The use of state channels is another important Layer-2 solution. State channels allow users to engage directly off-chain while the main blockchain settles the final state on a regular basis. This method decreases congestion on the main chain and speeds up transaction processing. State channels are especially valuable in SSI systems where users often exchange verified credentials.

Furthermore, Plasma chains are a novel Layer-2 solution that enables the formation of child chains that are linked to the parent blockchain. These child chains handle transactions and offer their own security assurances. Plasma networks are extremely scalable, allowing SSI ecosystems to manage a huge volume of transactions without taxing the main blockchain.

Future scaling options

While Layer-2 solutions give immediate respite for SSI scalability concerns, the future of SSI scalability also includes the investigation of unique scaling approaches that have the potential to revolutionize the way SSI systems work.

The development of blockchain sharding is one of the most potential future scalability possibilities. Sharding separates the blockchain into smaller, more manageable shards, each of which may execute transactions independently. This method significantly boosts the capacity of the blockchain network, allowing SSI systems to expand without experiencing transaction processing bottlenecks.

Sidechains are also gaining popularity as a future scalability solution. Sidechains are independent blockchains that are linked to the main blockchain and may carry out transactions using their own consensus processes. Sidechains may be adapted to unique use cases, giving personalized solutions for SSI features such as credential issuance, verification, and revocation. SSI ecosystems can divide transaction processing across many chains utilizing sidechains, decreasing congestion on the main blockchain.

Furthermore, advances in consensus techniques such as DPoS and PoA offer intriguing future scaling alternatives. These mechanisms are useful for SSI systems because they allow quicker transaction processing and lower energy usage. Furthermore, the change to a more energy-efficient consensus process adds to the long-term viability of SSI ecosystems.

Finally, blockchain scalability is a critical factor in the long-term growth of SSI systems. Scalability issues are solved by using Layer-2 solutions, which give instant relief by offloading transaction processing from the main network. Looking ahead, innovative scaling alternatives, including sharding, sidechains, and sophisticated consensus techniques, promise to revolutionize the scalability environment, guaranteeing that SSI systems can support an increasing user population and a larger amount of verified credentials. The scaling technological breakthroughs will allow SSI to continue its path toward a user-centric, decentralized, and internationally adopted digital identity ecosystem.

Now that we've discussed the difficulties and solutions to blockchain scalability, we will look at another important part of SSI: the use of tokens. In the following section, we will look at how tokens are used in the SSI arena to enable safe and efficient transactions while keeping user control over identity data.

Use of tokens in SSI

Tokens are crucial in the field of SSI, serving as digital keys to enable safe, decentralized, and user-centric identity management. In this section, we'll look at the importance of token standards in ensuring interoperability and security, as well as dive into the confluence of SSI wallets and tokens, where the magic of digital identification actually happens.

Role of tokens

Tokens are digital assets that allow users in the SSI ecosystem to govern their identities. They are the keys that allow users to engage with confidence and privacy by unlocking the doors to verifiable credentials.

In SSI, tokens serve as digital representations of verified credentials. They form the foundation of a person's digital identity and provide the distinct benefit of user control. Each token represents a different type of credential, such as a driver's license, passport, or academic degree. These tokens are safely stored in the user's wallet and provide them with exclusive access to their digital identification data.

Tokens play a role in SSI by allowing for selective disclosure. Based on the transaction or engagement, users may pick which tokens to present and to whom. This selective disclosure option gives users control over their personal data, allowing them to avoid unwanted exposure while still establishing their identity or traits when needed. It's like having a wallet full of multiple identity cards and having to decide which one to reveal at different checkpoints.

Tokens are also important in the revocation of verified credentials. The relevant token is tagged as invalid, revoked, or expired when a credential becomes invalid, revoked, or expired. This prevents users from presenting expired or revoked credentials, ensuring the authenticity of digital identity interactions. It works similarly to how a credit card is canceled when it is lost or hacked, guaranteeing that it cannot be misused.

Tokens are also essential to facilitating confidence in SSI. They are issued by reputable organizations such as governments, educational institutions, and enterprises, and their authenticity can be independently confirmed. This trustworthiness is fundamental to SSI, since users rely on tokens to demonstrate their identities and traits in a variety of scenarios. It's analogous to depending on a respected source's key to access a secure facility.

Token standards

SSI token standards guarantee interoperability, security, and consistency across various SSI ecosystems. They establish the groundwork for tokens to function smoothly in the digital identification world.

The set of rules and specifications that govern the development, issuance verification, and maintenance of tokens in SSI is known as the token standards. These standards give a common foundation for diverse SSI ecosystems and players to uniformly interpret and interact with tokens. Some of the important token standards utilized in SSI are as follows:

- **Verifiable Credentials Data Model (VC Data Model)**: This is a W3C definition that describes the form and representation of verified credentials as tokens. It guarantees that tokens are understood and handled globally, independent of the SSI system in use.
- **DID auth tokens**: This standard specifies the usage of tokens for authentication in the context of DIDs. It uses tokens to enable safe and standardized authentication practices.
- **Verifiable Presentation Data Model**: This specification defines the creation and display of verified presentations as tokens. Verifiable presentations are an important component of selective token disclosure, allowing users to reveal their credentials in a regulated way.
- **JSON web tokens (JWTs)**: A JWT is a commonly used token standard that contains statements about the token's subject. While not unique to SSI, JWTs are frequently used to represent verified credentials in the SSI setting.

Token standards are critical to assuring the security and interoperability of tokens. They offer a common language via which SSI participants may communicate, issue, and validate tokens. These standards make SSI ecosystems run more smoothly and increase confidence among users and reliant parties.

SSI wallets and tokens

SSI wallets are digital vaults that store and handle tokens, letting users safely transport their digital identities. The interaction of SSI wallets and tokens gives consumers more control and mobility.

SSI wallets are user-friendly programs for storing and managing tokens. These wallets provide users with control over their digital identities, much like real wallets give people control over their identity cards, credit cards, and personal papers.

Tokens are organized, classified, and safely kept in SSI wallets. Users may access and manage their tokens, as well as select which ones to display for certain interactions. SSI wallets are built with user privacy and security in mind, and frequently include features such as biometric verification and PIN protection to keep tokens secure.

Tokens held in SSI wallets are used to get access to various services and to prove one's identity. During a traffic stop, for example, a user can submit a driver's license token from their wallet to a law enforcement official, just as they would a traditional driver's license. Tokens, unlike real credentials, are under the user's control and may be shared digitally, increasing ease.

SSI wallets also allow users to keep their digital identities consistent across devices. A user may access their SSI wallet by smartphone, tablet, or computer, making their digital identity extremely portable. This mobility is analogous to carrying one's money wherever they go.

To summarize, tokens are at the heart of SSI, allowing people to securely govern their digital identities. They are critical to selective disclosure, revocation, and trust. Token standards provide interoperability and consistency, while SSI wallets let users manage and govern their tokens. Individuals may navigate the digital world with confidence, privacy, and user-centric control over their digital identities thanks to this dynamic combination of tokens, standards, and wallets.

Moving on from the examination of tokens in the SSI arena, we now turn our attention to another exciting application: SSI in the **Internet of Things** (**IoT**). In the following section, we'll look at how SSI concepts are used in IoT ecosystems to provide secure and decentralized identity management for connected devices and systems.

SSI and identity in IoT

The IoT is changing how we interact with the world around us by linking gadgets, sensors, and systems in ways we never imagined. In this section, we will look at the dynamic link between the IoT and digital identity, the role of SSI in the IoT landscape, and the major security problems that must be solved to create a safe and trustworthy digital environment.

IoT and identity

Identity is critical in enabling devices and entities to communicate and interact safely in the broad and interconnected world of the IoT.

The IoT is all about devices, and gadgets have identities. An identity in the context of the IoT is similar to a digital fingerprint for devices, guaranteeing that they can be recognized and trusted by other devices, systems, and people. These identities are required for communication, access, and confirming the legitimacy of data and orders.

As an example, consider a smart home ecosystem. Each smart item in the home has its own identity, from thermostats and security cameras to light bulbs and voice assistants. This identification guarantees that only authorized people and devices have access to or control over them. For example, because both devices have established their identities and can check each other's legitimacy, your smartphone can safely connect to your smart thermostat.

Identity has a function in the IoT that goes beyond device-to-device interactions. It is also crucial to the data created and transferred by these devices. IoT devices generate data on a constant basis, which is frequently forwarded to central systems for processing and analysis. The identities of the devices are critical in ensuring the data's integrity and origin. In a healthcare IoT application, for example, medical sensor identity assures that the patient's vital signs data is trustworthy and has not been tampered with during transmission.

SSI in the IoT

SSI applies user-centric control and decentralized trust concepts to the IoT landscape, delivering a safe and user-friendly way to manage digital identities in a highly linked context.

Traditional identity management can be time-consuming and possibly risky in the IoT era. Centralized identification systems may pose security flaws and privacy problems. SSI takes a more elegant and user-friendly approach. Devices, people, and entities may all keep their digital identities separate via SSI. These identities are self-sovereign, which means they are controlled by the device or user and may be selectively shared.

The reduction of single points of failure is one of the key benefits of SSI in the IoT. Traditional identity management systems are frequently centralized, which means that a breach or compromise of the central authority might have disastrous repercussions. Devices and entities can keep their identities separate via SSI. This decentralized architecture ensures that compromising the identification of one device does not jeopardize the entire IoT ecosystem.

SSI also gives consumers access to their IoT devices. Users have the ability to give and revoke access to their devices as well as the data generated by these devices. In an IoT-controlled smart automobile, for example, the owner can provide access to a trusted technician, allowing them to perform diagnostics and repairs. Access may be removed after the maintenance is completed, ensuring the automobile owner's control over their vehicle.

The capacity to protect privacy is another key component of SSI in the IoT. IoT devices frequently capture sensitive data, such as personal health information or facts about home security. Users may share only the information they need with SSI, reducing data exposure. A fitness tracker, for example, may communicate your steps and heart rate with your healthcare provider without disclosing your personal identity or other irrelevant data.

Security challenges

As the IoT and digital identity merge, various security problems must be solved to maintain the ecosystem's integrity and trustworthiness.

The protection of device IDs is one of the most difficult security concerns in the IoT. Tampering, unauthorized access, and impersonation must all be prevented on devices. Inadequate security measures may result in unauthorized access or data theft. To protect device identities, effective device authentication and encryption must be implemented.

In the IoT, data security and privacy are critical. Devices frequently acquire sensitive information, and their transmission and storage must be safe. Encryption and access restrictions are critical to maintaining data confidentiality and integrity during transmission.

The sheer size and diversity of IoT devices make managing and preserving their identities difficult. To maintain the continued security of the IoT ecosystem, identity life cycle management, including device provisioning, revocation, and upgrades, must be addressed.

Furthermore, the resilience of IoT systems against assaults such as DDoS attacks must be assured. Devices must be resistant to malicious efforts to overload or disrupt them.

Finally, the merger of SSI and the IoT represents a dynamic shift in how we handle digital identities in a more linked society. SSI adds user-centric management, decentralization, and selective disclosure to the IoT environment, protecting the security and privacy of identities and data. Addressing security issues, on the other hand, is critical to realizing the potential of SSI in the IoT and guaranteeing a safe and trustworthy digital ecosystem.

Now that we've discussed the implementation of SSI in the IoT, we will turn our attention to the more general ethical and philosophical implications of this transformational technology. In the following section, we will look at the ethical and philosophical issues raised by SSI, including its influence on human autonomy, privacy, and societal institutions.

Ethical and philosophical implications of SSI

SSI not only changes how we maintain our digital personas, but it also generates a profound investigation of ethical and philosophical implications. This section delves into the ethical issues raised by SSI, the philosophical implications it brings to light, and how it allows individuals to take control of their digital lives.

Ethical considerations

The ideals of autonomy, privacy, and equal access to identity management guide SSI's ethical path.

The idea of individual autonomy is central to the ethical questions regarding SSI. Individuals get control over their digital identities via SSI. This transition from centralized control to user-centric administration is consistent with the ethical principle that individuals should have the ability to control who has access to their personal information and for what purposes. It enables people to declare their digital selves in ways that are congruent with their beliefs and interests.

Another ethical pillar of SSI develops as privacy. Personal data is a valuable commodity in the digital era. The emphasis on minimum disclosure and selective sharing of information in SSI is an ethical response to the pervasiveness of intrusive data-collecting practices. Ethical issues emphasize the significance of preventing unwanted exposure of personal information, fostering openness, and allowing individuals the freedom to understand and manage the data transmitted.

In the SSI context, equity in access to identity management is a critical ethical concern. The digital divide, which occurs when not all people have equal access to technology and services, emphasizes the significance of ensuring that SSI solutions are inclusive and accessible to all. To guarantee fair access to SSI benefits, ethical issues demand the removal of entrance barriers, whether financial, technological, or infrastructural.

Philosophical implications

The philosophical significance of SSI extends to digital conceptions of identity, trust, and self-determination.

SSI questions traditional concepts of identity. Identity is frequently defined and managed by centralized authority, whether governmental or corporate, in traditional identity management systems. SSI presents a more philosophical view of identity as a self-sovereign idea, in which individuals construct and maintain their notions of identity in a decentralized manner. This conceptual shift calls into question traditional notions of identity, posing issues about what it means to be a digital being in the modern world.

SSI's philosophical foundation is built on trust. In a world where trust is frequently assigned to intermediaries, SSI emphasizes individual trust reclaiming. SSI's concept proposes that trust may be developed through direct contacts, cryptographic verification, and transparency rather than depending on institutional trust. It demonstrates a philosophical movement away from trusting institutions and toward trusting technology and individual interactions.

In the philosophical ramifications of SSI, the notion of self-determination takes center stage. Philosophically, it is consistent with the idea that individuals should be able to control their digital destinies. SSI empowers individuals to establish their digital identities and choose which pieces of their identities to expose, embracing a digital self-determination concept.

Individual empowerment

SSI enables people to take control of their digital lives by improving their control and decision-making in the digital domain.

One of the key ideas of SSI is individual empowerment. Individuals get control over their digital identities via SSI. This empowerment is analogous to seizing control of one's digital life, allowing individuals to choose who gets access to their personal information and for what objectives. It moves authority away from centralized authorities and toward people, allowing them to traverse the digital world on their own terms.

Individuals can use SSI to preserve their privacy. The capacity to exchange information selectively and disclose only what is required is a valuable tool for protecting personal data. This empowerment is analogous to being able to select what to divulge in a discussion while still maintaining privacy.

Furthermore, SSI permits people to exercise their digital rights. It allows them to connect safely without being watched or tracked. This empowerment is analogous to having a digital shield that protects one's digital existence from unwelcome encroachment.

Finally, SSI adds significant ethical and philosophical elements to the digital identity ecosystem. It emphasizes autonomy, privacy, and equal access while questioning long-held beliefs about identity, trust, and self-determination. At its foundation, SSI allows individuals to take control of their digital lives while adhering to ethical norms and philosophical notions that are relevant in the changing digital world.

Now that we've explored the ethical and philosophical implications, we shift our focus to the practical problems and hazards connected with implementing SSI. In the following section, we will look at the many challenges and potential problems that businesses may face while implementing SSI solutions, as well as techniques for avoiding these risks.

Challenges and risks in SSI implementation

While SSI has the potential to alter digital identity management, its implementation is fraught with difficulties and threats. In this section, we'll look at the challenges that organizations and people may encounter while implementing SSI, as well as the regulatory risks that must be addressed and the technological risks that must be reduced in order to maintain a safe SSI environment.

Adoption challenges

Adoption issues in the field of SSI cover a wide range of challenges, from user awareness to interoperability difficulties. Let's look at these challenges:

- **User awareness**: User awareness is one of the most significant SSI adoption issues. Many people are still unaware of the notion of SSI and how it may help them. It is critical to raise awareness and educate users about the value and concepts of SSI in order to encourage its adoption.
- **Interoperability**: SSI is intended to be decentralized and user-focused. However, this decentralization might pose interoperability issues. It is critical to effective adoption to ensure that SSI systems, wallets, and identity providers can all operate together effortlessly.

- **Legacy systems integration**: When integrating SSI, organizations with current identity management systems may face issues. Transitioning from traditional systems to SSI can be difficult since a proper migration plan is required.
- **Resistance to change**: People and organizations are frequently reluctant to make changes, especially when it comes to something as basic as identity. It requires a big effort to overcome this reluctance and encourage acceptance.
- **Infrastructure and accessibility**: Access to critical infrastructure and digital devices, particularly in impoverished regions, can be a hindrance. It is a significant task to ensure that SSI is available to everybody.

Regulatory risks

The regulatory risks associated with SSI implementation are related to the changing legal and compliance landscape. Let's look at these risks:

- **Data protection and privacy regulations**: Compliance with data protection standards, such as the European Union's GDPR, is a major regulatory risk. SSI brings new dynamics in data processing that may necessitate strict adherence to current standards.
- **Legal recognition**: Obtaining legal recognition for digital identities, verified credentials, and SSI in various jurisdictions can be a difficult regulatory procedure.
- **Cross-border considerations**: As a worldwide solution, SSI may be required to overcome cross-border regulatory differences. Managing the legal complications of international adoption is a significant undertaking.
- **Liability and accountability**: Organizations must address the legal risk of determining liability and accountability in the case of identity-related difficulties. The changing environment of digital identification may need the development of new legal frameworks.
- **Identity attestation**: In the lack of centralized authority, ensuring the legality of identification attestation poses regulatory issues. It is critical to establish trust frameworks that meet with legal standards.

Technological threats

The security, resilience, and infrastructure of SSI ecosystems are all related to technological threats. Let's look at these threats:

- **Security vulnerabilities**: SSI, like any other digital system, is vulnerable to security flaws. Keeping SSI safe against data breaches, hacks, and fraud is a constant technical risk.
- **Scalability**: Ensuring that SSI ecosystems can grow to handle millions or perhaps billions of users is a big technological problem. Scalability is critical to enabling expansion while maintaining performance.

- **Key management**: Another technical risk is the secure administration of cryptographic keys. Key management is critical to the integrity and security of SSI, and errors can lead to vulnerabilities.
- **Data storage and encryption**: Safeguarding sensitive data storage and encryption is another technical problem. To protect personal information, strong encryption and secure data storage are required.
- **Decentralization challenges**: The decentralization of SSI poses technological complications. It is a technological problem to ensure that users have consistent access to their digital identities and credentials, even in a decentralized system.

Finally, implementing SSI poses a number of acceptance obstacles, regulatory issues, and technological challenges. Individuals, organizations, and regulatory authorities must all work together to navigate this terrain. While the path to SSI acceptance may be difficult, overcoming these difficulties is critical to realizing the promise of user-centric, privacy-respecting digital identities.

Summary

This chapter presented a full introduction to SSI, focusing on its disruptive potential in digital identity management. SSI promotes user autonomy over personal data, using decentralized and cryptographic concepts, blockchain technology, and distributed ledger systems to build and maintain digital identities without relying on central authority. It emphasizes the need for encryption, hashing, public and private keys, and digital signatures for assuring data security and integrity within the SSI framework. The chapter also emphasized the connection of SSI principles with GDPR regulations, highlighting the need for data protection, consent management, and safe data processing. It also highlighted the empowerment, privacy preservation, and digital rights provided to individuals by SSI, as well as its possible uses in industries such as banking, healthcare, and the IoT.

In the next chapter, we dive into the privacy-by-design paradigm and its necessity and usage in the identity management space.

Summary

10
Privacy by Design in the SSI Space

Privacy by design (**PbD**) is a basic approach that prioritizes privacy issues throughout the design and development process, guaranteeing that privacy is built into any system or technology from the beginning. This idea, coined by Dr. Ann Cavoukian, calls for proactive methods to integrate privacy features into the architecture of systems, applications, and processes, rather than treating privacy concerns as an afterthought. The primary objective is to minimize data breaches and abuse by emphasizing privacy issues throughout the development life cycle. This approach is especially important in the ever-changing digital world, where technology manages massive volumes of personal information. PbD focuses on user-centricity, openness, and user empowerment, with the goal of developing systems that not only comply with privacy legislation but also encourage a culture of trust and respect for individuals' privacy rights. As privacy concerns rise, embracing PbD principles becomes critical for companies devoted to developing ethical, safe, and privacy-conscious solutions.

In this chapter, we will cover the following topics:

- PbD in SSI
- The value of PbD
- PbD frameworks
- Best practices from a security perspective

PbD in SSI

In an era when our lives are becoming increasingly intertwined with the digital realm, preserving our online identity has never been more critical. Traditional methods of identity management typically involved handing over control of our personal information to central authorities and third-party organizations, leaving us with little control over how it was used and shared. This model not only posed security risks, but it also invaded our privacy. However, a new era of digital identity is on the horizon, and it centers on SSI. This section will explain how SSI reclaims control of your digital identity while ensuring privacy and security in a world where data is the new currency.

Consider a digital cosmos where you own your own identity empire. You become the ultimate monarch of your digital avatar in the SSI domain. If you have a self-sovereign identity, you own it entirely. You decide what information you want to share, when, and with whom. It's like having your own digital fortress, impervious to data-hungry companies and scammers. With SSI, your digital identity is truly your own, and privacy and security are of the highest priority.

To properly comprehend the significance of SSI's privacy and security features, first understand the drawbacks of traditional identity management. Previously, we relied on centralized agencies to validate our identities. Consider your usernames and passwords for various online accounts, as well as the personal information you supply when creating a new account or signing up for a service. All of this data eventually ends up in the hands of companies, leaving it exposed to data breaches and manipulation.

The concept of DIDs underpins SSI's privacy and security. These are one-of-a-kind digital IDs that are cryptographically secure and unique to you. They are not administered by a centralized authority or stored in an unsecured database waiting to be hacked. DIDs are your personal digital superpower, ensuring the safety of your online identity. They use cutting-edge technology to ensure that they cannot be tampered with, forming an impenetrable fortress around your important data.

You have complete control in the universe of SSI. You get to choose when and what to share. It's like having the ability to open several doors with a single key, but only if you want to use it. You may connect online with SSI while disclosing only the relevant facts, rather than your complete life narrative. This selective disclosure function protects your privacy and allows you to share information on your own terms.

SSI employs privacy-first resolution techniques behind the scenes to ensure that your data is handled in your best interests. These processes serve as your digital identity's guardians, working tirelessly to protect your privacy. They ensure that your digital identity is accessed or shared with care and consideration for your privacy.

As we go through this chapter, you will get a deeper understanding of the privacy and security concepts that underpin SSI. You'll discover how to utilize your digital identity as a privacy fortress, controlling who comes in and who stays out. You'll learn how SSI employs cutting-edge technology to safeguard your identity and how selective disclosure enables you to share only what's necessary. In the world of SSI, your digital identity is not a weakness waiting to be exploited; it is a precious asset built on the foundations of privacy and security. Accept the SSI revolution and join the march to a digital future where your privacy and security are protected.

The value of PbD

The value of privacy cannot be stressed enough in our fast-paced digital world. We leave traces of our personal information every time we engage with technology, from accessing social media to purchasing online. In this digital universe, the notion of *PbD* stands out as a beacon of hope, giving a solution to the all-too-common privacy worries. In this section, we'll delve into the world of PbD, looking at how it's altering the digital sphere and guaranteeing that privacy is at the center of every technological advancement.

When you engage with the digital world, privacy is about securing your personal information and ensuring that it is respected and safeguarded. You deserve the same degree of privacy when you're online as you would when you close the curtains at home. Unfortunately, the digital age has witnessed a significant increase in data breaches, privacy violations, and a general lack of control over our personal information.

PbD is a philosophy that prioritizes your privacy. It entails designing systems, procedures, and technologies with privacy in mind from the start. Consider whether every product, website, or app was built with your privacy as a primary focus. PbD is a guarantee that your data will be handled with the highest level of care, regardless of where you are online.

The following diagram depicts the seven pillars of PbD:

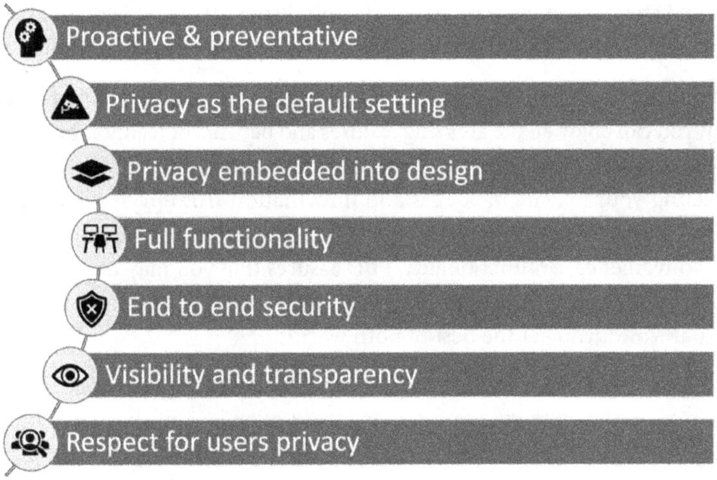

Figure 10.1 – Pillars of the PbD model

At its core, PbD follows a set of key principles:

- **Proactive not reactive**: The notion of PbD is based on proactive precautions rather than reactive ones. Privacy issues and data breaches are frequently afterthoughts in the digital realm, with organizations trying to control the damage once it has occurred. By taking a proactive approach, PbD reverses the script. Assume you're building a house and want to make your backyard a secluded retreat. Instead of waiting for neighbors or passersby to peep into your yard and then hurriedly installing barriers, you are required by PbD to construct a solid fence before anybody ever dares to breach your privacy. This fence is built as a well-thought-out, important component of your property's design, not as a hasty reaction. PbD in the digital era implies that privacy precautions are woven into the fundamental fabric of any product, service, or system from the start. Privacy issues are built into the heart of each technology breakthrough, just as the fence is an intrinsic feature of your property's layout. This strategy guarantees that possible privacy threats are detected and managed before they become problems. It's similar to having a strong, proactive security system that keeps attackers out rather than waiting for a break-in to occur before acting. PbD strives to establish a digital environment in which data protection and user privacy are the default settings by anticipating and resolving privacy problems early in the development process. It represents a significant departure from the reactive and frequently insufficient privacy controls of the past to a proactive and comprehensive strategy that prioritizes the protection of your personal information.

- **Privacy by default**: In the world of PbD, privacy is not a choice to be sought out in perplexing contexts; it is the fundamental and default condition of affairs. It's like going into your house and seeing that the doors have automatically locked, ensuring your safety without you having to rummage for keys. This method simplifies and increases data protection, making it an essential component of all digital interactions. From the minute you interact with a product or service, your personal information is protected by default. Privacy is no longer an afterthought; it is now integrated into the fabric of the digital experience, providing peace of mind from the start.
- **Full functionality**: You may have your cake and eat it too in the world of PbD, or, to put it another way, you can enjoy all the amazing features and benefits of technology without sacrificing your personal privacy. It's like having a super-smart phone that performs everything you want while protecting your private messages and information from prying eyes. Privacy is not a bargain; it is a given. When using digital products and services, you don't have to give up on enjoyment, convenience, or functionality. PbD assures that you may have it all without fear of your personal information being compromised. It's like eating your favorite dessert guilt-free – in the digital world, you get the best of both worlds.
- **End-to-end security**: When we talk about keeping your data safe, from collection to storage and beyond, we're comparing it to guarding a prized treasure on an amazing trip. Consider your data to be a valuable jewel, and PbD guarantees it is protected at all times. It's not just putting it in a safe at home; it's like having a squad of guards keeping an eye on it while it goes, ensuring nothing bad happens to it. PbD protects your data from the moment it is collected, such as when you submit your information online until it is kept in databases. It's the assurance that your treasure will not be stolen along the road, as well as a commitment to protecting your data during its entire journey.
- **Visibility to transparency**: When it comes to your personal data, you're in control in the world of PbD. It's like having a clear map and a tour guide for the journey of your data. You have the right to know where your data is going, what it is being used for, and who has access to it. PbD ensures that everything is as plain as a clean window, allowing you to see precisely how your data is handled. Assume you're giving your favorite toy to a pal. PbD guarantees that you both understand the rules before you hand it over. You know when your friend can play with it, how long they can play with it, and what they can do with it. PbD is like having a trustworthy guide that walks you through every step of your data's journey, from the time you share it until it reaches its final destination. It's all about ensuring that you have control and can select what's best for your data. Transparency and clarity are the watchwords, and you get to direct your data's journey. So, you're not just a passenger; you're the captain, and PbD is your compass, guiding you confidently across the data environment.

As we progress through this section, we will look at how PbD is changing the digital environment. You'll learn how it gives you control over your data and guarantees your privacy, no matter where your digital travels take you. PbD is no longer a pipe dream; it is a reality that is reshaping our digital world, one invention at a time. In this section, you'll learn how privacy is becoming an essential component of every digital encounter. In this environment, privacy is not an afterthought, but rather the basic basis upon which our digital future is based. So, let us begin this journey into PbD, where your digital privacy is always prioritized.

PbD frameworks

At its foundation, PbD is an ethical and user-centered strategy that emphasizes privacy concerns throughout a technology or system's lifespan. The idea is to avoid privacy-invasive events before they occur, rather than treating privacy as an add-on or afterthought. In today's digital era, data security has become a top priority for enterprises across sectors. With the growing amount and complexity of data collection, storage, and processing, as well as shifting privacy legislation and higher consumer expectations, organizations must prioritize privacy by creating frameworks to secure sensitive information and preserve stakeholder confidence.

Safeguarding data in the digital age

In today's fast-paced digital environment, where personal data has become a valuable commodity, the notion of PbD has evolved as a critical paradigm for protecting individuals' privacy rights. PbD is more than just a theoretical notion; it is a real and proactive method to integrate privacy concerns into the fundamental fabric of technologies, systems, and processes. In this inquiry, we will look at a few PbD frameworks, learning about their principles, their implementation methodologies, and the role they play in promoting a privacy-centric digital environment. Numerous frameworks have been created to help corporations efficiently implement PbD. These frameworks provide an organized method to guarantee that privacy is a top priority in technological development.

PbD is a proactive strategy that incorporates privacy issues into the design and development of systems, processes, and technologies from the beginning. Organizations may reduce privacy risks, improve data protection, and assure regulatory compliance by incorporating privacy principles and practices at every step of the data life cycle. Let's look at how using PbD frameworks might assist in the protection of data in the digital age. Some of the main principles are as follows:

- **Data minimization and purpose limitation**: PbD frameworks stress the concepts of data reduction and purpose limitation, advocating for the collecting and processing of only the personal data required to achieve defined goals. By restricting the extent and length of data collection and usage, companies may lessen the risk of sensitive information being accessed, misused, and exposed. Implementing data reduction strategies such as pseudonymization, anonymization, and aggregation can improve privacy while allowing for important insights from data analytics.

- **Transparency and user control**: Transparency and user control are essential elements of PbD frameworks, allowing people to understand and govern how their personal information is gathered, used, and shared. Organizations should offer clear and accessible privacy disclosures, consent processes, and user-friendly privacy settings to educate users about their data rights and choices. Organizations may establish trust and responsibility with their customers, workers, and other stakeholders by promoting transparency and giving them more control over their data.
- **Privacy-enhancing technologies (PETs)**: PbD frameworks encourage the use of PETs to reduce privacy risks and safeguard data against illegal access, disclosure, and modification. PETs include a wide range of technologies, approaches, and procedures that aim to improve privacy safeguards while maintaining data usefulness and functioning. PETs include encryption, tokenization, differential privacy, secure multi-party computing, and homomorphic encryption. PETs let enterprises incorporate strong security mechanisms and cryptographic methods to protect sensitive data throughout its lifespan, from collection and storage to transmission and processing.
- **Privacy Impact Assessments (PIAs)**: PIAs are an important technique advocated by PbD frameworks for identifying, assessing, and mitigating possible privacy risks connected to new initiatives, goods, and services. PIAs are the systematic evaluation of the privacy implications of data processing operations, taking into account elements such as data sensitivity, data sharing procedures, data retention durations, and the possible influence on individuals' privacy rights. Conducting PIAs early in the development process allows firms to address privacy issues, implement appropriate protections, and show compliance with privacy rules and best practices.
- **Accountability and governance**: PbD frameworks highlight the need for responsibility and governance in ensuring the successful implementation and continual enhancement of privacy safeguards. Organizations should develop clear policies, methods, and controls for handling personal data, delegating responsibility for data protection, and monitoring compliance with privacy regulations. Organizations may improve their data governance policies, reduce privacy concerns, and boost trust and confidence in their data handling methods by cultivating a culture of privacy and accountability.

To summarize, preserving data in the digital age necessitates a proactive and comprehensive approach to privacy protection, and utilizing PbD frameworks is critical to attaining this aim. Organizations that incorporate privacy principles and practices into their operations can reduce privacy risks, improve data protection, and build trust with their stakeholders, ensuring compliance with privacy regulations while also maintaining the integrity and confidentiality of sensitive information in today's rapidly evolving digital landscape.

User-centric privacy controls

In the context of SSI, user-centric privacy controls are critical to giving individuals authority over their personal data. At the heart of SSI is the notion of putting people in control of their digital identities, granting them the ability to manage and distribute their information as they see appropriate. Privacy and design frameworks in the SSI field strive to accomplish this by incorporating strong mechanisms for user-centric privacy controls.

Consent management is an important part of SSI's user-centric privacy measures. Users should be able to give informed consent for the collection, processing, and sharing of their data. This includes creating user-friendly interfaces that clearly express the goals for which data is gathered and allow users to select their consent options at a granular level. Furthermore, privacy frameworks should include means of acquiring and managing consent in an open and auditable way, giving consumers insight and control over how their data is used.

Furthermore, privacy measures in the SSI domain should include options for data reduction and selective dissemination. These technologies allow users to reveal the least amount of information required to complete a certain transaction or engagement. By implementing principles such as zero-knowledge proofs and selective disclosure protocols, SSI frameworks can enable users to assert fine-grained control over the characteristics they expose to various parties, lowering the risk of overexposure and unwanted data access. Overall, user-centric privacy controls are critical in protecting individuals' privacy rights and building confidence in the SSI ecosystem.

Consent management

Consent management is an essential component of privacy and data protection frameworks, especially when it comes to digital interactions and the collecting, processing, and exchange of personal data. The need for gaining informed permission from consumers has grown in recent years as internet services, social media platforms, and digital transactions have proliferated. This section delves deeply into the subject of consent management, including its relevance, problems, best practices, and new trends.

Consent management is the process of getting individuals' explicit, informed, and voluntary consent before collecting, processing, or disclosing their personal data. In the field of data privacy rules, such as the GDPR in the European Union, enterprises managing individuals' personal data are required by law to get valid consent. Beyond regulatory compliance, consent management is critical to establishing trust between organizations and individuals, promoting openness, and respecting individuals' autonomy and rights to their personal information.

Transparency is a crucial component of consent management. Organizations must offer individuals clear and understandable information about the reasons for which their data will be used, the categories of data gathered, the organizations engaged in data processing, and any possible dangers or repercussions of data sharing. Transparent communication allows individuals to make educated decisions about whether to grant or withhold consent, giving them more control over their personal information.

Furthermore, consent management entails acquiring affirmative agreement from individuals, which requires that consent be provided freely, actively, and explicitly. Organizations must develop user interfaces and consent procedures that encourage meaningful interactions while reducing the possibility of coercion or manipulation. Consent forms, for example, should be prepared in straightforward English, without jargon or technical vocabulary, and presented in an easy-to-understand style. Furthermore, enterprises should provide consumers with granular control over their consent preferences, allowing them to opt in or out of certain data processing activities or categories of data.

However, acquiring valid consent is a continuous process that necessitates continual communication with individuals throughout the data lifespan. Organizations must guarantee that individuals can withdraw or alter their permission at any time, and that changes to data processing activities or privacy policies are disclosed clearly and quickly. Furthermore, enterprises should put in place effective methods for recording and documenting consent transactions, such as timestamping consent events, preserving audit trails, and keeping track of individuals' consent preferences.

In addition to gaining express consent from individuals, organizations must guarantee that consent is received only from those who are capable of providing it. This requires organizations to develop age verification measures and get parental or guardian approval before processing children's personal data. Furthermore, enterprises should examine the validity of permission received in various contexts, such as online transactions, mobile applications, or **Internet of Things** (**IoT**) devices, and adjust their consent management processes appropriately.

Furthermore, consent management includes the notion of informed consent, which requires enterprises to offer consumers enough information to fully comprehend the consequences of their consent decisions. This involves notifying people about any potential dangers, implications, or secondary uses of their data, as well as their data protection rights, such as the ability to access, correct, or delete their personal information. Organizations should also give individuals ways to obtain further information or clarify data processing policies, such as through privacy statements, FAQs, or customer service channels.

In addition, as data flows become more complex and linked, companies must develop systems to manage consent across numerous data controllers and processors. This entails creating data-sharing agreements, contracts, or protocols that control the flow of personal data between different companies while also ensuring that individuals' permission choices are followed along the data supply chain. Organizations could also consider using technological solutions, such as consent management platforms or APIs, to provide smooth integration and interoperability between diverse systems and data sources.

To summarize, consent management is a comprehensive process that includes transparency, affirmative consent, continuous engagement, age verification, informed consent, and interoperability. Organizations that use strong consent management procedures may improve trust, accountability, and compliance with data protection rules while also empowering individuals to exert control over their personal information. As the digital landscape evolves, companies must prioritize consent management as a core component of their privacy and data protection plans, ensuring that individuals' rights and choices are honored in an increasingly data-driven society.

Data reduction

Data reduction is a core idea in data management and privacy that aims to reduce the quantity of data gathered, processed, kept, and shared by businesses while still achieving the desired results. In the context of privacy and data protection regimes, data reduction methods are critical to reducing privacy concerns, improving data security, and encouraging openness and accountability in data processing operations. This section delves further into the subject of data reduction, including its relevance, methodologies, problems, and best practices.

At its core, data reduction is the planned and systematic reduction of data quantities and granularity to meet particular goals such as reducing privacy threats, increasing data quality, optimizing storage and processing resources, and improving data governance and compliance. Organizations can decrease their susceptibility to privacy breaches, data breaches, and regulatory fines by restricting the extent and size of data collecting and processing operations, as well as reducing the impact on people's privacy and autonomy.

Data minimization is one of the most common strategies of data reduction and entails gathering and maintaining only the data required to fulfill a certain goal or target. This principle is enshrined in a number of privacy and data protection regulations, including the **General Data Protection Regulation** (**GDPR**) in the European Union, which requires organizations to collect data for specific, explicit, and legitimate purposes and not process it in ways that are incompatible with those purposes. Data reduction allows companies to restrict the amount of personal data they gather and manage, lowering the risk of unwanted access, abuse, and exploitation.

Another approach to data reduction is data anonymization, which includes deleting or encrypting **personally identifiable information** (**PII**) in datasets in order to make it anonymous or pseudonymous. Masking, hashing, tokenization, and generalization are examples of anonymization techniques that convert sensitive data into non-sensitive or less sensitive forms while keeping it useful for analysis, research, or other reasons. By anonymizing data, companies may preserve individuals' privacy and confidentiality while still extracting insights and value for lawful objectives.

Furthermore, data minimization includes data retention and deletion strategies, which entail developing rules and processes for managing data throughout its lifespan, including storage, retention, and destruction. Data retention rules specify how long data should be maintained based on legal, regulatory, contractual, or operational needs, whereas data deletion procedures guarantee that data is safely and permanently wiped once it is no longer required for its original purpose. Organizations that employ strong data retention and deletion processes can reduce the risk of unauthorized access, data breaches, and compliance violations while also lowering storage costs and complexity.

Furthermore, data reduction tactics use data aggregation and summarization techniques, which entail merging or summarizing individual data points or records to produce higher-level aggregates or summaries. Aggregation techniques involve grouping, clustering, and summarizing data based on shared qualities or characteristics, such as demographics, geographic location, and transactional activity. Aggregating and summarizing data allows companies to minimize the granularity and complexity of individual data points while retaining the broader patterns, trends, and insights included in the data.

However, companies face various obstacles and concerns when implementing data minimization methods, such as balancing privacy with utility, assuring data quality and integrity, and adhering to legal and regulatory requirements. For example, while data reduction and anonymization techniques can improve privacy and security, they may reduce the data's usability and value for analysis, research, and other uses. Similarly, data retention and deletion strategies must find a balance between meeting legal and regulatory obligations and maintaining access to historical data for business continuity, audit trails, or other valid objectives.

Furthermore, organizations must examine the unintended effects of data minimization tactics, such as the possibility of data loss, bias, and distortion, as well as the influence on people's rights and freedoms. For example, indiscriminate or excessive data reduction might result in insufficient or erroneous representations of reality, leading to poor judgments, policies, or consequences. Furthermore, businesses must ensure that data reduction procedures do not disproportionately affect specific groups or persons, such as disadvantaged and vulnerable populations, and that data processing operations adhere to the principles of justice, equity, and non-discrimination.

To summarize, data reduction is a diverse and nuanced term that includes many approaches, practices, and considerations for reducing privacy threats, improving data security, and encouraging openness and accountability in data processing operations. By implementing strong data reduction policies, companies may strike a balance between privacy and utility, reduce the dangers of unwanted access and abuse, and create trust and confidence among individuals and stakeholders. As the digital world evolves, companies must prioritize data minimization as a core component of their privacy and data protection plans, ensuring that they respect individuals' private rights and adhere to ethical and legal standards in their data processing operations.

Selective dissemination

Selective dissemination is the targeted and controlled release of information to select persons, groups, or institutions based on predetermined criteria or preferences. This technique enables organizations to adjust information transmission based on recipients' requirements, interests, and permissions, improving the relevance, efficacy, and efficiency of communication operations. Selective distribution is important in many situations, including marketing, education, research, and information management, where a timely and focused supply of relevant information is critical to attaining desired results.

In marketing and advertising, selective distribution allows firms to tailor their promotional efforts to certain audience groups or demographics, thus increasing the impact and efficacy of their campaigns. Marketers may use data analytics, consumer insights, and segmentation strategies to discover and target people who are most likely to be interested in their products or services, improving the possibility of conversion and engagement. Selective distribution enables marketers to provide individualized and relevant material to customers, improving the entire customer experience and promoting loyalty and brand advocacy.

In the realm of education and learning, selective dispersion enables individualized and adaptable learning experiences that are adapted to each student's requirements, preferences, and learning styles. Data-driven insights, assessment data, and learner profiles may be used by educational institutions and e-learning platforms to tailor course material, learning resources, and instructional tactics to the specific learning objectives and problems of each student. Selective distribution encourages student engagement, motivation, and academic performance by offering tailored learning pathways and targeted assistance, resulting in enhanced learning outcomes and student satisfaction.

In the sphere of research and knowledge management, selective distribution allows academics, researchers, and professionals to keep up-to-date on the newest discoveries, trends, and insights in their particular fields of interest. Academic journals, research repositories, and digital libraries use selective distribution techniques including email alerts, RSS feeds, and personalized recommendation algorithms to inform subscribers about relevant articles, conferences, and events. By selecting and providing relevant content to users based on their research interests, skills, and citation networks, selective distribution improves the efficiency and efficacy of information retrieval and knowledge exchange, hastening discovery and innovation.

Furthermore, selective distribution is critical to information management and organizational communication, allowing corporations, government agencies, and non-profit organizations to expedite internal communication procedures while improving cooperation and decision-making. Enterprise content management systems, collaboration platforms, and intranet portals use selective dissemination features such as user permissions, access restrictions, and content tagging to ensure that workers get the correct information at the right time. Organizations may improve data security, compliance, and governance by allowing selective access to sensitive or secret information and supporting targeted communication channels, while also fostering openness, accountability, and knowledge exchange across stakeholders.

However, companies face various obstacles and concerns when implementing selective dissemination, such as data privacy, information overload, and algorithmic bias. Organizations must comply with applicable data protection requirements, such as the GDPR in the European Union, by gaining users' express agreement before collecting and utilizing their personal data for selective release. Furthermore, enterprises must address concerns about information overload and filter bubbles by offering consumers tools and controls to manage their information choices and personalize their content consumption experiences.

Additionally, companies must be wary of algorithmic bias and discrimination in selective distribution systems, especially when employing machine learning and artificial intelligence to create content suggestions and predictions. Biases in data collecting, algorithm design, and decision-making procedures can result in unequal treatment, exclusion, and injury to people or groups, weakening the integrity and credibility of selective distribution systems. Organizations must thus employ open and responsible approaches to algorithmic decision-making, such as bias detection and mitigation techniques, algorithmic audits, and user feedback channels, to ensure justice, equity, and ethical behavior in information distribution.

To summarize, selective distribution is an effective and adaptable system for delivering tailored and relevant information to individuals, groups, and organizations depending on their requirements, preferences, and permissions. Organizations may improve the efficacy, efficiency, and impact of their communication and information management operations by using data analytics, customization approaches, and user-centric design concepts. Organizations must, however, address privacy, transparency, and fairness concerns to guarantee that selective distribution methods foster trust, inclusion, and responsible information sharing in the digital era.

Security best practices

Protecting our digital identities is a significant responsibility in our increasingly digital environment. SSI promises to return control to people, yet with great power comes tremendous responsibility. In this section, we will look at the security best practices, dos, and don'ts to ensure that your SSI infrastructure or solution stays a fortress of privacy and data protection.

The dos of securing your SSI infrastructure are as follows:

- **Choose a trustworthy identity wallet**: Choosing a reliable and safe identity wallet is a critical decision regarding where to keep your most important things. Just as you would put your important possessions in a strong and dependable vault, selecting the correct identity wallet guarantees that your digital assets, such as personal information and credentials, are protected from potential attacks. It serves as your digital fortress, protecting your information from unauthorized access and cyberattacks. In the realm of SSI, your identity wallet is the first line of defense, setting the tone for the security of your digital self.
- **Implement strong authentication**: Enabling strong **multi-factor authentication** (**MFA**) is akin to strengthening the security of your online presence. It's like adding another layer of security to your internet *front door*. Just as you wouldn't rely exclusively on a single key to secure your actual house, MFA adds an extra layer of security. Consider it like having a standard key plus a secret password. To obtain access, you must have both, something you have (the key) and something you know (the password). MFA adds this second layer, making it far more difficult for unauthorized persons to infiltrate your digital fortress and ensuring your online identity stays safe and well-protected.
- **Regularly update your software stack**: Updating your identity wallet and accompanying software on a regular basis is analogous to bolstering your digital defenses. It's the same as swiftly locking a susceptible window in your home to keep burglars out. Updates act as a reinforcement of that window, closing off possible entrance sites for cyberattacks. These upgrades, like safeguarding your physical surroundings, provide vital security improvements that repair vulnerabilities and improve the durability of your digital fortress. Neglecting updates may expose your digital identity to new dangers, but remaining current guarantees that your SSI infrastructure stays resilient and protected, preventing cyber invaders and providing a solid barrier against future attacks.
- **Encrypt your data**: Encryption is like using an unbreakable code to protect your digital conversations. It guarantees that your data stays safe and secure even if it is intercepted during transit. Consider it like putting your most treasured secrets in a closed box. To unlock and access the contents, only the intended receiver has the key. Encryption provides an unbreakable layer of security, ensuring that your information remains private even in the unpredictable and sometimes perilous digital realm.

- **Practice selective disclosure**: Accepting the SSI feature of selective disclosure is similar to having discretion over what information you provide about yourself, exactly as you would only divulge your age when purchasing age-restricted things. It allows you to reveal only the information required for a certain interaction while keeping the rest of your personal information secure. This practice guarantees that you keep a level of privacy and control in the digital arena, similar to providing only the exact amount of information necessary without disclosing your complete life story, hence improving your security when engaging in online transactions.
- **Back up your data periodically**: Backing up your data on a regular basis is analogous to making duplicates of important papers and keeping them safely. Unforeseen disasters may occur in the digital realm, just as physical papers can be lost or destroyed. Backups provide a safety net, guaranteeing that even if your digital identity is compromised, your data remains intact and can be readily recovered. It's a preventative approach to protecting your digital assets, similar to keeping spare copies of important documents in a secure location for peace of mind.
- **Educate yourself**: Understanding how your SSI works and the hazards linked to it is your most powerful defense in the digital age. Understanding your SSI empowers you to defend your digital identity in the same way that learning how to recognize and avoid typical scams protects you from fraud. Knowledge enables you to make educated decisions, recognize possible hazards, and take protective steps. It's like having a strong awareness armor that allows you to confidently traverse the digital terrain while keeping your digital identity private and under your control.
- **Use reputable sources**: Relying on reliable sources and services for verifiable credentials is analogous to obtaining recommendations from individuals you can trust. Just as you respect your friends' counsel for its dependability, picking credible sources assures the legitimacy and integrity of the credentials you exchange. It's about connecting with a company who has a demonstrated track record in securing your data and respecting your privacy, just the way you rely on recommendations from friends who have your best interests at heart. This practice increases the security and credibility of your SSI contacts.
- **Establish a recovery plan**: Neglecting security settings is like leaving your digital fortress's entrance wide open. Just as you wouldn't leave your physical door unlocked, failing to pay attention to security settings exposes your digital area to prospective dangers. These settings serve as digital locks and alarms and disobeying them might lead to difficulties. It is critical to ensure that they are correctly configured in order to protect your digital privacy and personal information, as well as to prevent unauthorized access to your online identity.

The don'ts of securing your SSI infrastructure are as follows:

- **Don't share everything**: Choosing a reliable and secure identity wallet is like selecting a strong vault for your digital assets. An identity wallet is your virtual protection, assuring the security of your personal data, credentials, and digital identity in an online environment full of possible threats, much as you would trust a strong vault to secure your goods.

- **Don't use weak passwords**: Using simple passwords is like having a weak, easily picked lock on your front door. Simple passwords can make your online accounts vulnerable to unauthorized access, just as a weak lock provides a security risk to your home. They're essentially an open invitation to anyone who wants to break in. To safeguard your digital identity, use strong, unique passwords that act as sturdy locks on your digital *front door*, making it far more difficult for outsiders to get access.
- **Don't click on suspicious links**: It's like opening a mystery parcel when you click on links or download files from unknown sources. These activities, like receiving an unexpected present from a stranger, might result in injury or security problems. To prevent any internet hazards, stick with reliable sources.
- **Don't forget to log out**: Leaving your identity wallet or accounts logged in on shared computers is equivalent to leaving an open safe unattended. It's as if you've left your personal information and possessions out in the open for everyone to find. Logging out is critical to preserving your digital valuables, just as you wouldn't leave your safe unlocked in a public place. It protects your sensitive data by keeping it private and safe, preventing breaches and unauthorized access. Always handle your digital identity with the same care as you would your actual things, and remember to lock the *safe* while not in use.
- **Don't ignore software updates**: Neglecting software updates is analogous to leaving your house's windows open to possible burglars. These upgrades frequently include critical security changes, such as fortifying your windows to prevent break-ins. Ignoring them can expose your digital area to threats, just as open windows welcome intruders. Keeping your software up-to-date is a critical step in protecting your digital identity and fortifying it against potential threats.
- **Don't trust unverified sources**: Refusing verifiable credentials from unverified or untrustworthy sources is analogous to turning down offers from strangers. Just as you would be wary of accepting stuff from strangers, only trust and accept credentials from recognized, trusted sources in the digital realm. This protects your online identity from threats and ensures that your data is kept safe, comparable to only accepting favors from people you trust in the real world.
- **Don't neglect periodic backups**: Backing up your data on a regular basis is analogous to having a spare key in case you lose your primary one. Data backups, such as that spare key, guarantee you have a copy of your digital information in case of unforeseen loss or disasters. Without these backups, you risk losing access to your precious digital assets, making recovery difficult and possibly locking you out of your own digital existence.
- **Don't share recovery information**: Sharing your recovery keys or phrases is akin to giving someone your digital lifeline, much like giving someone your ATM PIN. These keys serve as critical access points to your digital identity, and disclosing them to outsiders puts your security at risk. Protect your recovery keys and phrases, just as you would your ATM PIN, to keep control over your online persona, preventing unauthorized access and breaches.

- **Don't overlook security settings**: Neglecting SSI security settings is similar to leaving the entrance to your digital castle exposed to possible intruders. Just as you wouldn't leave your front door open, failing to pay attention to these settings might expose your digital identity. Security settings are your virtual locks and alarms, and failing to use them exposes you to security threats. It is critical to ensure that they are appropriately set in order to protect your SSI infrastructure, defend your digital identity, and prevent unauthorized access to your personal information.

The obligation of securing your SSI comes with the terrain of digital sovereignty. By adhering to established practices and being aware of potential hazards, you can guarantee that your SSI stays a haven of privacy and data security. Remember, you own your digital destiny, and with the proper measures, your digital identity may be as secure as your physical one.

Threats and mitigations

The necessity for robust security measures is critical in the ever-changing context of SSI, where individuals have increasing authority over their digital selves. SSI software and infrastructure, like any other digital system, are vulnerable to a variety of risks that, if ignored, might jeopardize the integrity of personal data and user privacy. Understanding these vulnerabilities and the associated mitigation ideas is critical to preserving the trustworthiness of SSI ecosystems.

Cybercriminals may try to fool people into exposing their SSI credentials by acting as reputable businesses via false emails or web pages. Malicious actors might obtain login passwords, recovery phrases, or encryption keys, allowing them to gain unauthorized access to an individual's SSI wallet. Intercepting communication between users and SSI programs allows attackers to possibly manipulate data transmitted between the parties. Malicious software can infect devices, collecting keystrokes or undermining the SSI program's integrity. Impersonation of trustworthy entities might result in the issuing or verification of digital credentials being false.

By demanding multiple forms of authentication for user access, MFA adds an extra layer of protection and reduces the chance of credential theft. Raising knowledge about phishing dangers and encouraging good online hygiene assist people in recognizing and avoiding phishing efforts. Protecting credentials within the SSI wallet in well-encrypted, tamper-resistant storage reduces the danger of theft. Using strong encryption for data in transit and at rest ensures that the data remains unreadable to unauthorized individuals, even if intercepted. Using reputable sources and services for verification, as well as following verification process standards, helps protect against identity spoofing. Updating SSI software is critical since updates frequently include security fixes that resolve vulnerabilities. The SSI feature of selective disclosure guarantees that only information required for a specific transaction is exposed, reducing vulnerability in the case of an attack. A significant privacy protection method is the implementation of protocols that allow the sharing of verified credentials without disclosing the underlying personal data.

Identity wallet security is ensured by rigorous access restrictions, biometrics, and PIN protection. The SSI infrastructure is decentralized, which reduces the possibility of single points of failure and centralized data repositories. Zero-knowledge proof approaches enable data verification without disclosing the data itself, hence increasing privacy. Because blockchain is immutable, it offers a tamper-proof audit trail for SSI transactions, increasing confidence and transparency. Adhering to data protection standards and privacy legislation helps to guarantee that SSI solutions are designed to be compliant and secure. The interoperability of SSI components enables a uniform and secure experience across diverse SSI environments. Continuous education and user awareness programs assist users in understanding security dangers and taking proactive steps to secure their digital identities.

SSI represents a fundamental shift in how people engage with their digital identities, providing greater control and privacy. This enhanced power, however, implies a heightened responsibility for security. As digital risks increase, integrating these mitigation principles becomes increasingly important in ensuring that SSI maintains a safe and trustworthy basis for digital identity management. The key to a successful and safe SSI environment is to balance user empowerment with rigorous security measures.

Summary

This chapter explored SSI as a solution to privacy and security problems in digital identity management. It underlined the need for proactive privacy protections, default privacy settings, and end-to-end security to safeguard digital assets. The chapter outlined security best practices, such as the need to use strong, unique passwords, perform frequent data backups, and obtain verified credentials from reliable sources. It also emphasized the significance of educating oneself about SSI and its related hazards, as well as the use of selective disclosure to retain privacy and control in the digital realm.

In the next chapter, we shall look at the relationships between decentralized identifiers and self-sovereign identities.

11

Relationship between DIDs and SSI

Distributed identifiers (**DIDs**), also known as decentralized identifiers, and **self-sovereign identity** (**SSI**) are closely related concepts in the field of decentralized identity.

Identity management (**IdM**) has become a significant topic in an increasingly digital society. Traditional identification systems have frequently been centralized, resulting in privacy problems, security flaws, and a lack of user autonomy. However, DIDs and SSI are changing the way we think about digital identification. In this chapter, we'll go deep into DIDs and SSI, looking at their origins, applications, problems, and the tremendous influence they have on the connections that characterize our digital lives.

In this chapter, we will cover the following topics:

- DIDs as the backbone of SSI
- DIDs and SSI relationship basics
- Emerging DID methods and innovations in the space
- DID issuers and verifiers
- Benefits of DID methods and enhancing privacy and security approaches
- Technological challenges and future directions

DIDs as the backbone of SSI

DIDs are the linchpin that binds the entire ecosystem together in the realm of SSI. DIDs are more than simply cryptographic keys; they are the underlying component that allows individuals to exercise control over their digital identities while also assuring security, privacy, and interoperability. This section will go into DIDs' important position in SSI, examining their relevance, structure, and transformational potential in the digital identity ecosystem.

To comprehend the essential significance of DIDs in SSI, we must first comprehend the essence of DIDs. DIDs are identifiers that are unique, durable, and globally resolvable and are not attached to any centralized registry, **certificate authority** (**CA**), or intermediate. They are a revolutionary idea aimed to enable self-sovereign digital identities and their importance is based on the following factors:

- **User-centric control**: DIDs enable people to freely own and manage their digital identities, decreasing reliance on centralized authorities and institutions
- **Decentralization**: DIDs adhere to decentralization principles, ensuring that identity ownership is diffused among the user community rather than centralized in the hands of a few companies
- **Interoperability**: DIDs are intended for worldwide interoperability, letting users utilize their identities in several SSI systems and applications

Grasping the role of DIDs in SSI requires a thorough grasp of their anatomy. DIDs are made up of different components that define their structure and functioning. They are as follows:

- **Method**: The method defines the DID scheme and how DIDs are produced, resolved, and managed.
- **Identifier**: The unique string of characters that differentiates one DID from another is known as the identifier.
- **DID document**: The DID document provides critical metadata as well as the DID's public keys. It is crucial in creating trust and facilitating secure interactions.

DIDs are more than simply identifiers; they are cryptographic keys that improve SSI security. The following are the components of security in implementing a DID-based SSI ecosystem:

- **Cryptography**: DIDs are based on public-key cryptography, which ensures data confidentiality, integrity, and validity
- **Decentralized trust**: DIDs help to build decentralized trust networks, lowering the danger of **single points of failure** (**SPOFs**) and improving the security of digital interactions
- **Privacy protection**: DIDs provide selective disclosure, enabling users to transmit only the information they need in a safe and private manner

DIDs have an important function in SSI that extends to interoperability:

- **Cross-platform compatibility**: DIDs are platform-independent, allowing users to utilize their identities across several SSI systems, applications, and services
- **Cross-domain compatibility**: DIDs let users communicate in secure and trustworthy ways across several domains, building a global identity ecosystem
- **Standards**: The importance of standardization and best practices in ensuring that DIDs work seamlessly within the SSI context

This chapter will present real-world examples and case studies of organizations, projects, and individuals using DIDs to redefine digital identities to emphasize the fundamental role of DIDs in SSI. The case studies will highlight how DIDs are used in a variety of industries, ranging from healthcare to banking, and how they enable users to exercise control over their online identities. Finally, the chapter will emphasize the dynamic character of DIDs in the SSI environment. DIDs are not static; they change, adapt, and respond to changing user, organizational, and technological demands. We recognize DIDs as a catalyst for the digital identity revolution, changing the future of SSIs in an interconnected society, by understanding their important significance.

To begin with, DIDs decentralize identity ownership. Users are subject to central authorities or intermediaries who govern how their digital identities are handled in traditional identification systems. DIDs challenge this paradigm by allowing users to autonomously generate, own, and manage their identifiers. This decentralization shifts the focus of power away from centralized bodies and toward the users themselves, which aligns with SSI basic concepts.

DIDs also foster user-centricity in the digital arena. They put people at the center of their digital interactions, giving them control over when, how, and with whom they disclose personal information. DIDs allow users to choose just the information needed for a certain engagement, increasing their privacy and liberty. This user-centric strategy helps people to establish their digital selves, minimizing their reliance on third parties and cultivating a feeling of self-sovereignty.

DIDs, in essence, are the mechanisms that allow SSI to be fully decentralized and user-centric. They provide the foundation for the transition away from institutional control of digital identities and toward people having the last word. DIDs empower individuals to decide who they are in the digital world and how they want to participate, paving the way for a future in which users are at the center of their digital identities, a concept key to the very spirit of SSI.

Interoperability between DIDs and the various components of SSI is analogous to ensuring that different smartphone manufacturers may all exchange text messages with each other. It is essential for a smooth and user-friendly digital identity experience.

Consider having a DID that you use to authenticate your identity while accessing various online services or platforms, similar to how you use your phone to send messages to friends, family, and coworkers. If your DID cannot *talk* to the SSI components used by others, such as verified credentials or identity wallets, it is analogous to your phone being unable to send messages to someone using a different brand of phone. Interoperability guarantees that your DID is compatible with a variety of SSI tools and services. That is, you can use your digital identity wherever you choose, just as you can message anybody, regardless of phone brand. This adaptability is critical for a flourishing SSI ecosystem because it enables safe and trustworthy connections across platforms, services, and even nations.

Interoperability assures that your digital identity is as adaptable and generally accepted as your smartphone in the digital sphere, in a future where everyone uses their DIDs with various SSI components, just like everyone uses their phones to communicate. It is essential for a user-centric, self-sovereign, and trouble-free digital identification experience.

DIDs and SSI relationship basics

DIDs are a new type of identifier that is created and managed by individuals themselves rather than by a centralized authority or organization. DIDs are based on blockchain technology and use public-private key cryptography to provide a secure and decentralized way to manage digital identity. DIDs can be used to authenticate identity, establish trust relationships, and control access to personal information.

SSI is a concept that is closely related to DIDs. SSI refers to the idea that individuals should have complete control over their personal data and how it is used. In an SSI system, individuals create and manage their own digital identity, which is stored on a decentralized network. This gives them complete control over their personal information and allows them to share it only with trusted parties on a need-to-know basis.

Traditional IdM systems are based on centralized institutions such as CAs, corporate directory services, and domain name registries. In terms of cryptographic trust verification, these centralized authorities are the foundation of trust in and of themselves. Federated IdM must be created to guarantee that IdM runs smoothly across these disparate platforms.

The introduction of **distributed ledger technology** (**DLT**), often known as blockchain technology, provides an unprecedented chance to build completely decentralized IdM. Entities in a decentralized identification system, ranging from organizations to persons and diverse objects, are free to use any shared basis of trust. Globally distributed **peer-to-peer** (**P2P**) networks, distributed ledgers, and other similarly capable systems provide opportunities to monitor the foundation of trust without creating a SPOF or central authority. When combined, DLTs and decentralized IdM systems enable entities to develop and maintain their own IDs across many distributed platforms, encouraging sovereign roots of trust and reducing reliance on a centralized authority. This not only improves security but also provides resilience and autonomy in IdM.

The DID communication model components are represented by the following diagram:

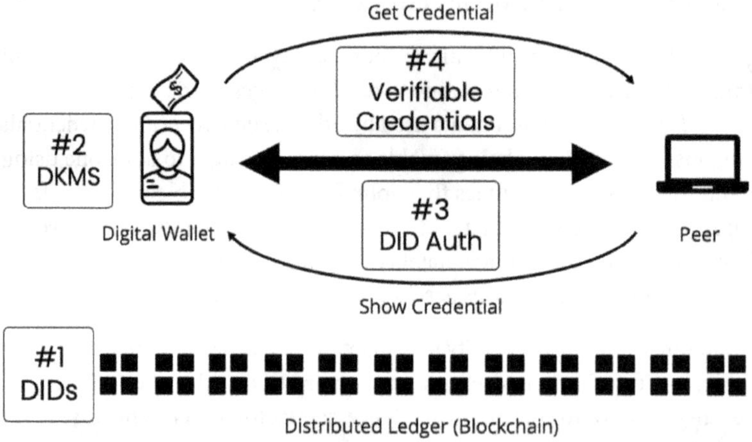

Figure 11.1 – DID communication model components

DIDs are a novel type of identification for decentralized, verifiable digital identities. These new identifiers are intended to empower the bearer of a DID by allowing them to exert control over it autonomously and function independently of any centralized registry, **identity provider** (**IdP**), or CA.

DIDs are used to identify things and provide authentication using proofs such as privacy-preserving biometric techniques and digital signatures. These DIDs link to the appropriate DID papers. A DID document contains a set of service endpoints that allow users to communicate with the entity identified by the DID, also known as the DID subject. Following the principles of Privacy by Design, each entity can have several DIDs, as well as their associated DID documents and service endpoints. This feature enables entities to accomplish their chosen segregation of identities, contexts, and personas in accordance with the common understanding of these concepts.

DID procedures are techniques for generating, accessing, modifying, and deactivating a DID and its related DID document inside a given distributed ledger or network. The setup and functioning of DID methods are described in separate specifications for each method.

SSI is a remarkable innovation, with implications well beyond traditional concepts of identity. It is poised to transform digital interactions and connections between individuals, organizations, and entities. SSI, with its influential and complex character, is both daring and familiar, promising to break away from the long-standing contradiction of having to choose between security and user experience or vice versa. It is possible to achieve significantly higher levels in both areas. Importantly, SSI offers life-changing opportunities for underprivileged persons globally, promising positive changes in their situations.

Emerging DID methods and innovations

This section delves into cutting-edge techniques and technologies that are altering how we manage and safeguard digital identities: an important investigation of the ever-changing world of DID, a subject that is always evolving to meet the difficulties of IdM, privacy, and security in our increasingly digital and linked society.

The concept of decentralized identification, which differs from traditional, centralized identity systems, is one of the core principles. Rather than depending on single, monolithic institutions such as governments or companies to identify and maintain our identities, DID techniques are paving the way for a more user-centric and self-sovereign paradigm.

We introduce various innovative DID methodologies and technologies that are transforming the scene. The incorporation of blockchain technology into DID systems is one of the most fascinating advancements. Blockchain's DLT ensures trust and transparency by recording identity-related transactions in a secure and tamper-proof manner. These solutions provide users with entire ownership over their digital identifiers, a departure from traditional identity suppliers that have traditionally held the keys to our digital identities.

Zero-knowledge proofs (**ZKPs**) are mathematical structures that enable users to demonstrate a fact about themselves (for example, their age) without exposing the underlying knowledge. ZKPs provide a new degree of privacy and selective disclosure, allowing users to reveal only the information needed for a given transaction or encounter while protecting more sensitive data.

As biometric identifiers such as fingerprints and face recognition become more widely used, DID techniques are investigating how to safely combine these distinct physical characteristics into digital IdM. This gives a more easy and safer method of verifying identification while maintaining privacy and security. Users may carry their digital identity across several services and platforms using SSIs, making it portable and interoperable. This notion has the potential to transform how people interact with digital services by reducing the need for them to continually register and verify their identity on numerous sites.

To attain their full potential, developing DID techniques and technologies must follow common standards, providing easy integration and communication between diverse systems and platforms. Standards such as the **Decentralized Identity Foundation's** (**DIF's**) Universal Resolver and Verifiable Credentials are assisting in the creation of a single ecosystem for DID technology. The move toward user-centric, decentralized identity models, the incorporation of blockchain and cryptographic techniques, the growth of self-sovereign identification, and the drive for standards and interoperability are all factors contributing to the continued evolution of digital identity. As we go deeper into the digital era, it is critical to stay aware and adapt to these advancements, which promise a more secure, private, and user-centric digital environment.

Development of new DID methods

The ongoing development of new DID approaches and technological advances offers a dynamic frontier in the expanding digital identity landscape. As our dependence on digital interactions and online services grows, so does the demand for safe, user-centric, and privacy-preserving identification solutions. To meet these needs, inventors and technologists are pushing the frontiers of what is possible.

> **Important note**
>
> *Decentralized identifiers* and *distributed identifiers* are frequently used interchangeably to describe the same notion. Both phrases refer to a sort of identity that is generated, controlled, and validated using decentralized technology or DLT, such as blockchain. These IDs are intended to allow safe and privacy-preserving digital IdM without reliance on centralized authorities. Individuals can have more control over their digital identities by employing DIDs, which prevent personal information from being kept or maintained by a single centralized institution.

One notable tendency in this continuous journey is the development of innovative DID techniques to address specific use cases and sectors. DID techniques optimized for healthcare, for example, can allow for the safe exchange of medical information, while those optimized for financial services can enable smooth and trustworthy identity verification in the realm of online banking. These approaches are designed to fulfill specific needs and legal requirements of their respective fields, demonstrating the adaptability and variety of decentralized identification systems.

Technological developments are at the heart of these advances. Blockchain, with its decentralized, immutable ledger, continues to play an important role in enabling DID systems and creating trust and transparency. ZKPs offer selective disclosure and privacy preservation, allowing users to communicate identity-related information while protecting sensitive data. The use of biometrics such as face recognition and fingerprint recognition not only improves security but also simplifies user authentication. Furthermore, SSI solutions are gaining pace, allowing users to maintain their digital identities across platforms, services, and devices without the need for recurrent registration and verification.

As the technology environment evolves, the ongoing development of new DID techniques and breakthroughs demonstrates a dedication to tackling the ever-changing concerns of IdM, privacy, and security in our increasingly digital society. This constant quest promises a future in which individuals have greater control over their digital personas and safe, smooth, and user-centric interactions throughout the digital sphere are the norm.

Relevance of new DID methods

Emerging DID approaches are critical to improving the capabilities of the SSI ecosystem. SSI is at the center of this dramatic change in IdM, with its basic idea of giving individuals control over their digital identities. These new DID approaches will help to make SSI more robust, adaptable, and broadly applicable. One of these approaches' most noteworthy contributions is their applicability to numerous use cases and industries. SSI may be smoothly incorporated into a broad range of applications by adapting DID approaches to specific sectors such as healthcare, banking, or education. This makes identity verification more secure and user-centric. This versatility guarantees that SSI can fulfill particular requirements and regulatory norms of diverse industries, allowing it to be used more widely.

Blockchain integration and ZKPs are two technological breakthroughs that improve the security and privacy of SSI. The use of blockchain technology creates an immutable, tamper-proof record that secures the integrity of identity-related transactions. Individuals can exchange information with ZKPs without disclosing the underlying data, safeguarding privacy while validating identities. These advancements are critical to increasing trust and confidence in the SSI ecosystem. Furthermore, the incorporation of biometric authentication methods and self-sovereign identification solutions brings SSI in line with current trends and user expectations. Biometric authentication provides a more seamless and safe method of identity verification. In contrast, SSI technologies enable users to retain consistent, transferable, and interoperable digital identities across many services and platforms.

DIDs are rapidly being used in a variety of real-world contexts, proving their ability to improve security, privacy, and user control over personal data. The following are some significant instances of DID techniques in real-world applications, notably in healthcare data exchange:

- **Healthcare data exchange**: MedCreds, developed by ConsenSys Health, uses DIDs to enable safe and confidential health data sharing. Patients may manage their health records and share them with healthcare practitioners as appropriate, ensuring that sensitive information is transferred in a safe and verified way. MedCreds uses DIDs to ensure that data stays decentralized, lowering the risk of breaches and illegal access.

- **Vaccination Credential Initiative (VCI)**: The VCI is a group of public and commercial entities that has created a standard for digital vaccination certificates. Using DIDs, the program enables individuals to keep and share verifiable proof of immunization status. This strategy assures that data is safely preserved and can be independently checked without relying on a central authority, hence increasing privacy and confidence.
- **ID2020 and MyData**: ID2020 is an alliance promoting digital identities for all. In healthcare, it argues for the use of DIDs to grant patients an SSI. Similarly, MyData, a global movement, promotes human-centered ownership over personal data, including health information. These efforts employ DIDs to provide users with more control over their data, improving privacy and security.
- **CareChain**: CareChain uses DIDs to build a decentralized healthcare data exchange network. This technology enables healthcare practitioners and patients to safely and effectively exchange and access health records. DIDs make it easier to verify identities and guarantee that data exchange follows privacy standards, increasing confidence and collaboration among healthcare stakeholders.

These examples demonstrate the expanding use of DIDs in healthcare, which offer safe, verifiable, and privacy-preserving solutions for health data communication. DIDs, which decentralize authority over personal information, present a potential method for tackling concerns of data security and privacy in healthcare.

In conclusion, these developing DID approaches are not only keeping up with the expanding digital identity landscape but are also actively moving the SSI ecosystem ahead. Because of their versatility, security upgrades, and conformity with current technology developments, they are critical components on the road toward more secure, user-centric, and privacy-respecting digital identities. As the SSI ecosystem evolves, these techniques help to shape a future in which individuals have greater control over their digital identities, resulting in a more dependable and simplified digital landscape.

Need for the standardization of DID methods

The significance of DID technique standardization and best practices cannot be more strongly emphasized. The necessity for shared standards and best practices becomes increasingly important as the digital identity environment changes and decentralized identification systems gain significance. These criteria provide interoperability, security, and dependability, clearing the path for DID technologies to be widely adopted.

DID method standardization is analogous to the basic architecture that enables disparate systems and platforms to connect effortlessly. DID approaches help ensure that digital identities stay consistent and widely recognized across multiple services and applications by conforming to established standards. This degree of consistency is critical for developing a more user-centric and adaptable identity ecosystem.

Furthermore, standardized DID procedures help to improve security. They provide a standardized set of security features and protocols that have been carefully tested and updated, eliminating vulnerabilities and guaranteeing that digital identities are resistant to unauthorized access and fraudulent activity. In a world where data breaches and identity theft are common, standardized DID procedures are an important step toward protecting people's privacy and security.

Best practices, on the other hand, are the result of years of knowledge and skill condensed into a set of principles that aid in the design, implementation, and management of DID techniques. These best practices assure the dependability of identification solutions while also streamlining development and deployment procedures. They provide a road map for DID developers, lowering the likelihood of mistakes and complexities that may stymie the deployment of these systems.

Furthermore, best practices aid in dealing with ethical and regulatory elements of digital identity. As digital identity becomes more important in our lives, it is critical to address concerns such as permission, data ownership, and user rights. Best practices can provide insights on how to preserve user privacy while also ensuring that DID approaches are in line with increasing regulatory frameworks and compliance needs. Another key benefit of DID technique standardization and best practices is that they stimulate innovation and trust within the ecosystem. They let developers focus on unique ideas and breakthroughs by providing a solid base, removing the need to reinvent the wheel. This leads to a faster rate of innovation and the creation of new features and functions that benefit both consumers and organizations.

In conclusion, the foundations of a trustworthy, secure, and widely used decentralized identity ecosystem are standardization and best practices for DID procedures. They improve security, expedite development processes, handle ethical and regulatory problems, and foster innovation. As the digital world evolves, these standardized procedures will be critical in creating an environment that values user-centricity and privacy. This option will provide individuals more autonomy over their digital identities while also giving enterprises the confidence to use decentralized identity solutions.

Distributed identity issuers and verifiers

When it comes to using blockchain technology for IdM, it's important to recognize that three different parties are involved:

- Individuals who own the identity
- Identity issuers
- Identity verifiers

The entity functioning as the identity issuer, which is often a trusted authority such as local government, has the capacity to offer or confer personal credentials to the identity owner, also known as the user. Through the issuing of a credential, the identity issuer certifies the legitimacy of the personal information contained in that credential, such as the date of birth and last name. Subsequently, the identity owner can save these credentials in their personal identity wallet and use them later to back up assertions about their identity to a third party, known as the verifier.

An issuer holder and verifier relationship can be visualized as follows:

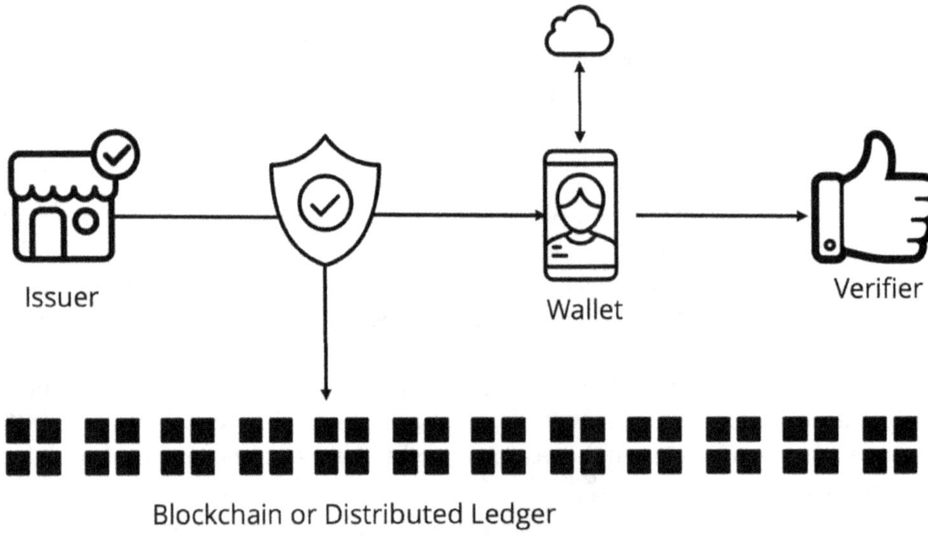

Figure 11.2 – Verifiable credentials relationship

A credential is made up of numerous identification attributes, each of which reflects a specific piece of data about an individual (such as their name, age, or date of birth). Secondary parties supply these credentials, confirming the veracity of the data contained inside. The reliability and value of a credential are totally dependent on the issuer's trustworthiness or reputation.

Using a blockchain architecture eliminates the need for individuals in charge of verification to directly check the veracity of the data in the proof presented. Instead, they can use the blockchain to ensure the authenticity of the attestation and the attesting party (such as the government). Based on this information, they can decide whether to validate the proof.

Assume someone wishes to establish their age by presenting a paper. Instead of verifying the actual date of birth on the certificate, the individual reviewing it would first ensure that the government's signature on the paper is legitimate. Then, they would decide whether to believe the government's judgment of the information's veracity. So, whether or not the proof is accepted is determined by whether the person reviewing it believes the person who confirmed it is trustworthy.

DIDs are a novel class of different identifiers used to validate digital identities that are wholly controlled by the identity owner. DIDs function independently of centralized registries, IdPs, and authorities, allowing a new degree of control over digital identification.

DIDs, unlike IP addresses or email addresses, should not be reassigned, as this poses security and privacy problems. Decentralized identities must be both resolvable and cryptographically verifiable to ensure a decentralized structure. Using DLT creates confidence by establishing a **single point of truth** (**SPOT**) for data included inside credentials.

Basics of verifiable credentials and digital identity

Verifiable credentials are similar to digital badges in that they allow you to demonstrate who you are and what you can accomplish online. These credentials are extremely crucial in today's digital age, when we disclose a lot of our personal information online. They provide us with more control over our data, improve our privacy, and make our online lives safer.

Consider your digital identity to be a collection of information about you that exists in the internet world. It may include your name, birthday, schooling, or even immunization information. This digital identity is used whether we join up for services, apply for employment, or seek healthcare. The issue is that in the past, we had to give far too much personal information in order to perform these things. Credible credentials alter this.

You may now obtain a verified credential, which is similar to a digital certificate that states something factual about you. For example, if you receive a university graduation certificate, you may convert it into a verified credential that verifies you graduated without disclosing all of your school records. This allows you to choose who sees your diploma, keeping your online life more private.

Verifiable credentials make the internet safer as well. They safeguard your data with robust encryption and technology. When you show someone your verified credential, they may verify its authenticity without viewing all of your personal information. It's the same as displaying your ID without giving your home address, phone number, or other personal information.

Verifiable credentials are, thus, a game changer. They let us own and govern our digital identities while keeping our personal information safe and secure. In a world where privacy and security are paramount, these credentials empower us to disclose just what is necessary, making online interactions more safe and trustworthy.

Understanding verifiable credentials

A digital identity is a basic idea in today's digital world that plays a critical part in our online interactions. Consider it a digital mirror of who you are: a collection of data that characterizes you, your online presence, and your actions. Personal information such as your name, email address, social media profiles, and other characteristics are frequently included in your digital identity. It is impossible to overestimate the importance of digital identity in online interactions. It acts as your virtual passport, allowing you to shop, bank, connect, and even work remotely. It is the key to accessing the huge diversity of possibilities and services available in the digital world.

Traditional identification systems, on the other hand, have serious flaws. When using services or executing transactions in these systems, we are frequently compelled to provide excessive personal information. For example, when you establish an account on a website or sign up for a service, you are usually asked for your complete name, address, date of birth, and, in some cases, your social security number. The more personal information you disclose, the more likely it will come into the hands of the wrong people due to data breaches or abuse. Furthermore, conventional systems are often centralized, which means that your data is managed and stored by a single individual or organization. Because of this concentration, there is a SPOF. If the company in charge of your data is hacked, significant identity theft and privacy breaches might occur.

Verifiable credentials provide a game-changing answer to the problems of existing identification systems. These digital credentials function similarly to online badges in that they allow you to demonstrate something about yourself without disclosing all of your personal information. Assume you have a verified credential stating that you are over a particular age limit, which may be used for age verification when purchasing age-restricted things. Instead of exposing your entire birthdate and identity, merely display this certificate to demonstrate that you are of legal drinking age. You reclaim control of your data by utilizing verifiable credentials. You pick who and what information to disclose. Your personal information is kept hidden, and you only expose what is required for a given transaction or contact, greatly improving your privacy and security.

The decentralized nature of verified credentials is also important in overcoming traditional identification system flaws. Verifiable credentials, as opposed to centralized methods, employ distributed technologies such as blockchain, in which data is held across a network of computers. Because of this decentralization, the danger of a SPOF is reduced, making it far more difficult for bad actors to corrupt the whole system. Furthermore, to maintain data security, these credentials use advanced cryptographic techniques such as ZKPs. You may use ZKPs to verify anything about yourself without providing the real facts. For example, you can demonstrate your degree without displaying your whole academic transcript. This cryptographic wizardry provides another layer of anonymity and security, protecting your digital identity even more.

Verifiable credentials are similar to digital certificates that state the truth about you. Issuers, which might be educational institutions, companies, or even government bodies, establish these credentials. A university, for example, can provide you with a valid diploma that certifies your graduation. The wonderful aspect is that verified credentials are all about privacy and security. They are kept in your digital wallet, and when you need to verify something about yourself, you just share the credentials without disclosing any personal information. Assume you want to show a prospective employer that you graduated from that university. Rather than giving your entire academic record, you only show them the verified certification. Without knowing your whole educational history, the employer can check to determine whether it is genuine. It's the same as displaying your ID without giving your home address, phone number, or other personal information.

In practice, verified credentials may be used for a variety of purposes. Consider renting a flat or a home. You may use a verified credential that shows you have a good credit score instead of handing up your credit score, bank statements, and other sensitive data. The landlord may verify it without having access to your whole financial history, making the leasing process safer and more confidential.

Verifiable credentials, in essence, are changing the way we handle identity and personal data in the digital world. They put you in charge, protect your privacy, and make online interactions safer, all while allowing you to communicate just what is necessary. It's a win-win situation for both individuals and businesses, and it's influencing the future of digital identity and online trust.

Key components of verifiable credentials

Verifiable credentials are a novel and game-changing approach to digital identification. These components work together to build trust in digital interactions and provide people with more control over their personal information. Let's break this down:

- **Issuers**: The entities that develop and distribute verified credentials are known as issuers. These entities might range from organizations to institutions to people. A university, for example, can provide a verified credential to a graduate, while a government agency can provide a driver's license credential or an employer can provide an employment verification credential. The role of the issuer is to ensure the correctness and validity of the information included in a credential. They are authorized sources who attest to the credential's claims. For example, an institution that issues a diploma certification must confirm that the recipient truly graduated.
- **Holders**: Individuals or businesses that obtain and keep verified credentials in a digital wallet are referred to as holders. This wallet might be a computer or mobile device software program. Holders have full control over their credentials and may decide when and with whom they disclose them. Holders are responsible for managing their digital identities. They may acquire and gather credentials from different issuers, storing them safely in their digital wallet. Holders choose which credentials to share and provide permission for their usage, giving them more control over their personal information.
- **Verifiers**: Verifiers are entities that must validate the validity and correctness of information given by the bearer of a verifiable credential. Employers, colleges, and internet platforms are examples of **service providers** (**SPs**). Verifiers play an important part in the trust equation. When someone offers a verifiable credential, the verifier verifies its authenticity as well as the issuer's credibility. If the verifier believes the issuer and the digital signature on the credential are legitimate, they can accept the credential without requiring the possessor to provide their whole personal information. When compared to standard identity verification methods, this procedure is more safe, efficient, and privacy-preserving.

The relationship between issuers, holders, and verifiers is critical for building confidence in digital transactions. Here's how these components interact:

- **Credential issuance**: A verified credential is created by an issuer, such as a university. The issuer validates that the information contained in the credential, such as the recipient's graduation, is true and authentic during this procedure. When the issuer is happy with the correctness of the information, they digitally sign the credential using cryptographic techniques to ensure its integrity.
- **Credential storage**: After that, the credential is provided to the holder, who saves it in their digital wallet. The possessor has complete control over their wallet, and the credential is encrypted within it. The holder may monitor, maintain, and selectively distribute the credential depending on their requirements and consent, guaranteeing their data's privacy and security.

- **Verification request**: When the holder wants to prove a certain claim, the verifiable credential can be presented to a verifier. For example, while seeking a job, the holder may provide a potential employer with a work history certificate.
- **Credential verification**: The verifier examines the digital signature of the credential to ensure that it was issued by the authoritative source (the issuer) and that it has not been tampered with. If the digital signature is legitimate, the verifier can trust the information included in the credential without needing to access the entire personal information of the bearer. This authentication procedure lowers the likelihood of identity theft and unauthorized access to critical data.
- **Trust establishment**: Trust is developed at several levels as a result of this interaction:
  - Verifiers believe that the issuer has properly validated and authenticated the information in the credential. If the issuer has a reputation for accuracy and honesty, the credential will be trusted as well.
  - Holders have control over their credentials and only present those that are relevant to the engagement, which increases confidence between holders and verifiers.
  - Verifiers are critical in maintaining the trust structure. They develop confidence with holders and issuers by appropriately confirming credentials and preserving privacy.
  - The system of verified credentials is intended to boost confidence in digital interactions. It improves security, privacy, and data control, resulting in a more dependable and secure digital identity ecosystem.

Finally, verified credentials are comprised of issuers, holders, and verifiers, and their interactions serve as the foundation for safe and privacy-preserving digital IdM. This novel technique helps individuals to govern their digital identities, selectively reveal personal information, and build trust in online interactions. The digital world is evolving toward a more secure, efficient, and user-centric identity ecosystem with verified credentials, lowering the dangers associated with existing identity verification techniques and data exchange.

Privacy and security considerations

In the digital era, privacy and security are vital, and verified credentials are at the forefront of resolving these issues. They improve privacy by allowing users to take control of their personal information and expose just what is required in digital interactions.

Users can share particular qualities or assertions from a credential without releasing their complete information with verifiable credentials. When visiting an age-restricted website, for example, users can show they are above the age of 18 without providing their actual birthday or other sensitive details. This feature decreases the possible surface area for data breaches. Verifiable credentials frequently make use of cryptographic techniques such as ZKPs. Individuals can use ZKPs to show the veracity of a statement without exposing the real facts. It's the same as displaying your diploma without disclosing your whole academic record. This improves privacy by allowing verification without revealing personal information.

Verifiable credentials store data using blockchain technology and decentralized networks. Decentralization lowers the danger of centralized data breaches while also giving individuals more control over their digital identities. The data reduction concept is essential for verified credentials. Users disclose just the information required for a specific interaction, exposing the least amount of data possible. This method complies with privacy standards and protects against needless data collecting. Before revealing their verified credentials, users provide express approval. This informed consent guarantees that personal data is only disclosed when individuals freely choose to do so, providing another degree of protection to the process. Credential revocation is possible with verifiable credentials. Users can remove access to a credential if necessary, which improves security.

Security features of verifiable credentials

Verifiable credentials combine numerous security elements to safeguard personal information and assure credential authenticity:

- **Blockchain technology**: Blockchain technology is used in many verified credential systems. Blockchain is a decentralized and tamper-resistant ledger that improves credential security. The immutability of information kept on a blockchain reduces the possibility of unauthorized adjustments or data breaches.
- **Cryptography**: To protect verifiable credentials, cryptographic techniques such as digital signatures are utilized. Digital signatures ensure that the origin of the credential is validated and that its information remains intact. The addition of cryptography to the credential verification procedure offers an extra degree of protection.
- **Decentralized trust anchors**: Decentralized trust networks are used to verify credentials. Rather than depending on a single centralized authority, trust is spread across numerous institutions that collectively vouch for a credential's legitimacy. This decentralized strategy improves security and resilience.

Verifiable credentials use a variety of security measures, including blockchain technology for tamper-resistant ledgers, cryptography for origin validation and data integrity, and decentralized trust anchors, which improve security and resilience by spreading trust across multiple entities rather than relying on a single authority.

Verifiable credentials applications and use cases

Verifiable credentials have numerous uses in a variety of industries, considerably boosting the efficiency and security of digital transactions. Here are some instances of their uses:

- **Online services**: For access to online services, verifiable credentials are used, allowing users to authenticate their identity or claims without disclosing unneeded personal information. Users, for example, can verify their age to gain access to age-restricted material or services, boosting privacy and security.

- **Employment and human resources**: Verifiable credentials make background checks and employment verification easier in the job market. Job seekers may safely and effectively communicate their career history, degrees, and qualifications, saving time and decreasing the danger of fraudulent claims.
- **Education**: Verifiable credentials are revolutionizing the education industry. Students earn digital diplomas and certificates that educational institutions and companies may easily verify. This avoids the need for comprehensive document verification and assures that academic results are genuine.
- **Healthcare**: Verifiable credentials enable the secure transmission of patient information in healthcare. A patient, for example, can present a valid credential to a healthcare professional to prove their vaccination status without disclosing their whole medical history. This procedure is crucial for protecting patients' privacy.
- **Government services**: Verifiable credentials are used to gain secure access to government services. Citizens can demonstrate their eligibility for benefits or services without revealing personal information. A citizen, for example, can exchange a verified certification to verify residency for tax purposes.
- **Travel and immigration**: Verifiable credentials are making the travel and immigration processes more efficient. Without exposing their whole trip history, travelers can safely communicate their vaccination status or visa information with authorities. This speeds up border inspections while also improving data privacy.
- **Financial services**: Verifiable credentials are used in financial services to verify identification, lowering the risk of fraud and identity theft. Users can authenticate themselves to banks or financial organizations while keeping their personal information private.
- **Supply chain**: Verifiable credentials improve the tracking of commodities and products in **supply chain management** (**SCM**). These credentials offer a safe method of verifying the validity and origin of products, lowering the possibility of counterfeit goods entering the market.

Verifiable credentials provide considerable benefits in each of these use cases by boosting privacy, security, and efficiency. They provide people control over their personal information, allow businesses to expedite procedures, and lower the risk of identity-related fraud and data breaches. As the use of verified credentials grows, their beneficial influence on a variety of industries is poised to transform the digital world, providing better safety for individuals and organizations alike.

Potential benefits and concerns

Adopting verified credentials marks a paradigm change in the field of digital identification, with several potential benefits such as simplicity and enhanced security. Let's take a closer look at this:

- **Convenience**: Verifiable credentials make identity verification easier. Individuals may easily transfer digital credentials online, saving time and decreasing friction in many encounters rather than depending on cumbersome paperwork or long in-person verifications.

- **Reducing data storage**: One of the most noticeable benefits of verified credentials is the option to transmit only information that is required. Users can share particular qualities or claims without providing their whole personal information, reducing the danger of identity theft and oversharing.
- **Improved security**: To improve security, verifiable credentials use sophisticated cryptographic algorithms, digital signatures, and decentralized technologies such as blockchain. This makes it far more difficult for hostile actors to tamper with or counterfeit credentials.
- **Privacy enhancement**: Individuals get ownership of their data when their credentials are verified. Users have control over when, with whom, and what information they disclose. This ability to regulate personal information protects user privacy, decreases the danger of data breaches, and guarantees that data is only shared with consent.
- **Minimized data exposure**: The data reduction concept is essential for verified credentials. Users only disclose data that is directly relevant to the transaction, limiting personal information exposure and decreasing the attack surface for prospective attackers.
- **Selective disclosure**: Selective disclosure is enabled via verifiable credentials, which allow individuals to pick which qualities to divulge. For example, while demonstrating age, individuals can demonstrate that they are over the age of 18 without disclosing their specific birthday. This feature protects user privacy while increasing user control.
- **Revocation**: The ability to revoke credentials is a useful feature. A user can prevent a verifier from accessing certain data, giving them even more control and protection over their digital identity.
- **Efficient verification**: Verifiable credentials help SPs and verifiers speed up the verification process. Verifiers can swiftly and correctly check the veracity of claims made in a credential by relying on digitally signed credentials and blockchain technology.
- **Decentralization**: Decentralized networks and blockchain technologies lessen reliance on a single, possibly vulnerable institution. The security and robustness of verified credentials are enhanced by decentralization.

While there are several advantages to using verified credentials, there are also problems and misunderstandings, notably with privacy and data ownership:

- **Privacy concerns**: Some people are concerned that verified credentials would still violate their privacy. It is critical to emphasize that verified credentials improve privacy by limiting the exposure of personal data and allowing users control over their data. Strong data protection legislation and user consent methods can help to alleviate privacy issues.
- **Data ownership**: One prevalent misperception is that while utilizing verified credentials, individuals may lose control of their data. Users, on the other hand, maintain ownership and control over their data. They have control over when and how they share it, allowing them to retain ownership while enjoying the ease and security of verifiable credentials.

- **Complexity**: Verifiable credentials may be perceived as technically challenging or difficult to utilize by certain people. It's critical to emphasize that user-friendly interfaces and applications are being created to make the experience easier. Verifiable credentials do not require users to be cryptography specialists.
- **Security myths**: Some people are concerned about the security of digital systems and blockchain technology. While no system is completely secure, verified credentials use strong security techniques, including encryption, to safeguard data. Furthermore, decentralization reduces the danger of SPOFs.
- **Regulatory compliance**: Concerns regarding regulatory compliance are prevalent, particularly in businesses subject to stringent data protection regulations. Verifiable credentials can be constructed to meet the requirements of various regulatory systems. Adoption of these qualifications may be in accordance with legal regulations.
- **User adoption challenges**: Individuals used to old identification systems may be resistant to widespread acceptance of verified credentials. Education and awareness efforts can assist people to comprehend the benefits and alleviate any reservations they may have.

While it is critical to address common concerns and misconceptions, it is also critical to recognize that verified credentials constitute a big step forward in digital IdM. As this technology becomes more widely used, it has the potential to change the digital world, making online interactions safer, efficient, and privacy-protecting.

The road ahead

Consider a future in which your digital identity is as safe, private, and adaptable as you want it to be. Because of a new concept known as verified credentials, this future may not be too far away. Verifiable credentials are similar to digital badges or certifications that indicate something accurate about you. These digital badges are generated and distributed by reputable organizations or institutions, such as your university or bank. What's intriguing is that you can keep and manage them in your digital wallet, just like you would a real wallet.

Let's look ahead to see how verified credentials could continue to revolutionize the digital identity landscape:

- **Streamlined access to services**: Verifiable credentials have the potential to make our lives easier. Consider applying for a job without having to fill out long application forms or provide a plethora of documentation. With your digital wallet full of verified credentials, you may safely and easily share what's needed, such as your university degree or employment history. This simplified service access will save you time and decrease paperwork difficulties.
- **Enhanced online safety**: We are all concerned about our data and privacy in the digital age. Verifiable credentials have the potential to improve online safety dramatically. You become the custodian of your data by disclosing just the required information and keeping the rest under digital lock and key. This reduces the chances of identity theft or oversharing of personal information.

- **Revolutionizing education and work**: Verifiable credentials are already having an influence on education and the labor market. Your school accomplishments and employment experience may be saved as digital credentials in your wallet in the future, making it much easier for employers to verify your qualifications. It's the equivalent of handing them a digital diploma without divulging your whole academic record. This type of transformation might save both businesses and job seekers time.
- **Healthcare in your hands**: You desire control and privacy when it comes to your health. Verifiable credentials can assist you in sharing your medical information safely and only when necessary. For example, if you need to show your vaccination status, you don't have to provide your whole medical history. Your privacy is preserved, and you may continue to use the services you require.
- **Travel made easy**: With all of the requisite documentation and verifications, traveling might be a bit of a pain. Credential verification might make this procedure easier. Consider securely presenting your immunization record, visa, or other travel-related papers from your digital wallet. It has the potential to speed up border inspections and lessen the danger of document fraud.
- **Easy access to government services**: Verifiable credentials have the potential to change the way we engage with government services. You might be able to demonstrate your eligibility for benefits or services without sharing all of your personal information. This protects your privacy while also making government transactions more efficient and safe.
- **Building trust in commerce**: Credential verification helps increase confidence in online transactions. When shopping online, you may prove your age, address, or other data without disclosing too much information. This enhanced trust may result in fewer fraudulent transactions and safer e-commerce.
- **Staying in the know**: As time passes, you'll discover that your digital wallet keeps you informed and protected. You will be notified of updates, notifications, and alerts regarding your verified credentials. It's like your wallet watching out for you, ensuring that your data stays under your control and is only used with your consent.

The tech industry is buzzing right now about ongoing innovations and initiatives in the realm of verified credentials. Many organizations, standards agencies, and open source initiatives are attempting to make verified credentials a reality. They're working together to make sure the technology is safe, accessible, and easy to use. The **World Wide Web Consortium** (**W3C**), in particular, has played a critical role in establishing worldwide standards for verified credentials. Their efforts are directed toward ensuring that verified credentials operate seamlessly across several platforms and systems. This worldwide standardization secures the technology's interoperability, which is critical for wider adoption.

Furthermore, several digital businesses and start-ups are building user-friendly tools and platforms to assist users in easily managing their verified credentials. These platforms strive to improve the user experience, whether you're sharing your educational accomplishments with a potential employer or verifying your identity to gain access to a service.

Finally, the future of verified credentials seems truly promising. This new technology has the ability to transform the way we maintain our digital identities as we continue to traverse the digital era, making our lives more convenient, safe, and private. Verifiable credentials are set to revolutionize the way we engage with services, organizations, and institutions in a more secure and privacy-respecting manner as they continue to shape the digital identity environment. So, keep your digital wallet at the ready because the future of verified credentials looks promising.

Enhancing privacy and security

SSI is a paradigm change in digital IdM that has the potential to greatly improve the privacy and security of DIDs. SSI tackles many of the long-standing difficulties associated with existing identification systems by providing individuals greater control over their digital identities, opening the path for a more secure and privacy-respecting digital environment.

Individuals are put at the center of IdM with SSI, giving them complete control over their digital selves. Personal information is often maintained and handled by third-party businesses in traditional identification systems, creating risks of data breaches and unauthorized access. SSI, on the other hand, allows users to securely store and maintain their identification traits in user-controlled digital wallets, lowering the danger of data exposure and privacy breaches.

The principle of selective disclosure is a major aspect of SSI. Individuals can use SSI to provide only the information needed for a specific transaction or interaction while keeping the rest of their identifying traits private. This reduces the risk of identity theft and information abuse by limiting the exposure of personal data.

To verify identification without disclosing the underlying data, SSI systems frequently employ advanced cryptographic techniques such as ZKPs. Individuals can use ZKPs to confirm their identification traits (such as age or credentials) without providing real information, boosting privacy while maintaining safe verification. This technology not only protects personal data secrecy but also adds a degree of confidence and security to the verification process.

In conventional identification systems, centralized IdPs have enormous authority and responsibility for user data management. This concentration of power provides a SPOF and makes the target appealing to attackers. By design, SSI eliminates the need for centralized IdPs by spreading identity attribute maintenance throughout a decentralized network. This decreases the possibility of large-scale data breaches and unauthorized access to massive volumes of personal data.

For secure IdM, SSI depends on cryptographic keys. Typically, these keys are produced and stored in a user's digital wallet. Key management in SSI, when properly done, may be incredibly secure, making it extremely difficult for unauthorized entities to get access to a user's digital identity. Encryption, safe key storage, and robust authentication techniques increase the system's security.

Users can cancel access to their identity attributes at any moment with SSI. Individuals may use this feature to regulate and limit who can use their data and for what purposes. If a user's personal information has been shared with an SP, for example, and that user decides to withdraw consent, the information may be revoked promptly, safeguarding their privacy and lowering the danger of continued data exploitation.

To record identity-related transactions, several SSI systems use blockchain technology. The usage of blockchain creates a secure and tamper-proof record from which users may validate the integrity of their identity interactions. Blockchain improves DID security by assuring the immutability of transaction records and removing the possibility of fraudulent alterations to identification data.

In SSI, trust is spread via a network of entities called trust anchors, which collectively swear for a DID's legitimacy and validity. This decentralized method lowers reliance on a single, possibly corrupted institution, making it harder for hostile actors to erode system confidence. This method improves the overall security of DIDs and SSI.

The notion of data reduction is promoted by SSI. SSI systems simply record and communicate the basic features required for a single transaction or contact rather than gathering and keeping large volumes of personal data. This decreases the attack surface for prospective attackers and restricts personal information exposure.

In SSI, the usage of blockchain creates an immutable audit trail of identity-related interactions. This implies that every DID-related transaction is forever logged and may be independently confirmed. This openness and auditability improve system security and give a method for dispute settlement.

In conclusion, SSI provides an intriguing option for improving the privacy and security of DIDs. SSI solves many of the long-standing problems of traditional identification systems by transferring ownership of digital identities to people, employing sophisticated encryption techniques, minimizing data sharing, and utilizing decentralized networks and blockchain technology. These developments not only allow users to keep their privacy, but they also improve the security of digital identities, lowering dangers associated with data breaches, identity theft, and unauthorized access. As SSI evolves and gains popularity, it has the potential to create a more secure and private digital environment in which individuals have complete control over their digital identities.

Technological challenges and future directions

Complex and multidimensional technological issues may affect the scalability and performance of DID systems and SSI connections. As these technologies gain pace, it is critical to solve these issues to ensure widespread acceptance and smooth functioning. This section will go into some major technical challenges that may impact the scalability and performance of DIDs and SSI connections. The issues that DID systems and SSI partnerships face in terms of scalability and performance have sparked a wave of innovation and collaboration in the IT industry. While these issues are complicated, there are various potential answers and breakthroughs on the horizon to solve them:

- **Blockchain scalability**: Many DID approaches rely on blockchain technology for its immutability and security; however, blockchain scalability remains a big barrier. The volume of transactions on blockchains rises as more people and organizations embrace decentralized identities, potentially contributing to network congestion, higher transaction fees, and slower confirmation times. Scaling solutions, such as layer 2 protocols and sharding, are being developed to address this problem, although full implementation and adoption remain hurdles. Various ways are being investigated to address the constraints of blockchain scalability. Layer 2 solutions, including sidechains and state channels, try to reduce congestion by processing transactions off-chain and settling them on the main blockchain. Furthermore, blockchain platforms are focusing on adopting sharding, which is a means of dividing the blockchain into smaller, more manageable portions, hence boosting transaction throughput. These methods promise to improve DID system scalability while keeping the security and decentralization that blockchains enable.
- **Privacy and confidentiality**: The heart of SSI is maintaining privacy while ensuring selective dissemination of personal information. However, striking this equilibrium is technically difficult. Although ZKPs are a promising strategy for enabling selective disclosure, their effective implementation at scale remains a technological difficulty. SSI's success depends on striking the proper balance between privacy, security, and performance. Innovations in privacy-enhancing technology are assisting in achieving the proper balance between privacy and performance. The efficient implementation of ZKPs is a current research topic. To improve data privacy and selective disclosure capabilities, advanced cryptographic techniques such as homomorphic encryption, secure multi-party computation, and confidential computing are being integrated with DID approaches. These innovations will help to create a more efficient and private SSI ecosystem.
- **Interoperability**: DIDs and SSI solutions must smoothly interoperate with current identity systems and standards in order to achieve widespread acceptance. This difficulty necessitates the creation of uniform standards and conventions, which may be a time-consuming process involving several players. A technological problem that must be solved for SSI to realize its full potential is ensuring that different DID methods can communicate efficiently and that SSI systems can interoperate with varied SPs. Common standards and protocols are being developed to overcome interoperability issues. Organizations such as the DIF are attempting to develop interoperable standards that may be used by various DID approaches. These standards will provide smooth communication and data interchange across the decentralized identity ecosystem, allowing multiple SSI systems to coexist peacefully.

- **User experience**: Achieving a streamlined and user-friendly experience is critical for SSI acceptance. Users should be able to simply maintain their decentralized identities and engage with services. This includes creating user-friendly interfaces, effective **key management systems** (**KMSs**), and reducing the complexity of cryptographic procedures while retaining security. Striking this equilibrium is a never-ending technological challenge that necessitates constant progress. Adoption of SSI requires a user-friendly experience. Individuals are finding it simpler to maintain their decentralized identities as user-centric design advances. To simplify the user experience, user-friendly wallet apps, biometric authentication, and user-friendly key management solutions are being created. These enhancements are intended to eliminate friction and make SSI more accessible to a larger user audience.
- **DID method standardization**: The variety of DID approaches provides a standardization difficulty. While flexibility is necessary for diverse use cases and sectors, a lack of standardization can restrict interoperability and split the ecosystem. Developing uniform standards for DID approaches and securing their widespread adoption is a technological problem that the community must face. The necessity for DID method standardization is recognized, and attempts are being made to develop common standards for many use cases and businesses. These standards will allow for the consistent implementation of DID methodologies, assuring best practices and compatibility with other systems.
- **Key management and recovery**: Securing and maintaining cryptographic keys is critical to DID and SSI security. Key management and recovery processes must be user-friendly, robust, and attack-resistant. A crucial technological problem is developing systems that assure the secure storage and backup of keys, as well as their retrieval in the case of loss or compromise. Users may now safely maintain and back up their cryptographic keys thanks to advancements in key management and recovery procedures. Solutions such as decentralized key recovery services and hardware-based secure key storage are becoming more widely available, assuring key security while allowing for key recovery in the event of loss or compromise.
- **Performance and scalability**: DID generation and administration may be computationally demanding, especially as the number of DIDs in the system rises. It is a constant struggle to keep the process of DID formation, resolution, and updating efficient and scalable. It is critical to optimize the performance of DID methodologies and their underlying technology. Optimizations in the formation, resolution, and updating of DIDs are being investigated to improve the performance of DIDs and DID techniques. Caching methods, efficient data structures, and load-balancing algorithms can all contribute to lowering the processing cost and latency associated with DID operations.

- **Consent and revocation:** It is technically difficult to allow users to grant and remove authorization for disclosing their identification characteristics. Implementing a system that gives users granular control over their data, with the ability to revoke access at any moment while keeping the process safe and efficient, is a technological difficulty in SSI systems. User consent and revocation management solutions are being developed. On blockchain networks, smart contracts can enable fine-grained control over data sharing, allowing users to give and withdraw access to their identification attributes as needed. These technologies will provide consumers with greater control over their data while also ensuring security and efficiency.
- **Cross-platform compatibility:** DIDs and SSI systems should work in unison across platforms, devices, and operating systems. Cross-platform compatibility is a technological problem that demands careful thought and adaptation, especially in situations with varying levels of support for decentralized identification systems. Developers are focused on building standardized libraries and SDKs for implementing DID techniques in order to ensure cross-platform interoperability. These technologies will make it easy to integrate DIDs into different platforms, devices, and operating systems, guaranteeing a uniform user experience independent of technology.
- **Trust anchors:** Trust anchors are used to check the validity of DIDs to assure the trustworthiness of DID techniques. It is difficult to scale trust anchors while retaining trust and security. Scaling alternatives, such as verified credentials and decentralized trust networks, are in the works, but widespread acceptance remains a technological issue. Scaling trust anchors is a huge difficulty, but innovative solutions such as decentralized trust networks, in which numerous entities jointly attest to DID trustworthiness, are being developed. These networks spread trust more widely, decreasing reliance on a single entity and encouraging scalability while ensuring security.

In summary, the technological issues affecting the scalability and performance of DIDs and SSI interactions are complex and numerous. To address these problems, the decentralized identification community must work together to develop novel solutions, establish shared standards, and constantly upgrade the technology. As the need for safe and user-centric digital identities develops, it is critical to overcome these technological barriers in order to realize the full potential of decentralized identity systems and facilitate their widespread acceptance in the digital world.

While the technological hurdles that DID systems and SSI interactions face are daunting, the community is continually seeking answers and advances to overcome these limitations. Collaboration between developers, organizations, and stakeholders is propelling advancements in blockchain scalability, privacy-enhancing technologies, interoperability standards, user-centric design, DID method standardization, key management and recovery, performance optimization, consent and revocation mechanisms, cross-platform compatibility, and trust anchor scalability. These advancements have the potential to pave the way for a more scalable, secure, and user-centric decentralized identity ecosystem, allowing individuals to have greater control over their digital identities while retaining privacy and confidence in the digital environment.

Summary

This chapter examined the usage of verified credentials and decentralized IdM, focusing on the advantages of privacy, security, and user control. It emphasized the significance of blockchain technology in allowing confidence and validity in digital credentials, as well as the possibility for quicker service access and increased online safety. The chapter also discussed user acceptance issues and current initiatives to standardize and secure verified credentials for wider usage. Overall, it provided a thorough overview of the revolutionary power of verified credentials in creating a more secure and privacy-conscious digital identity ecosystem.

In the next chapter, we will take a look at DID protocols and standards.

12

Protocols and Standards – DID Standards

In the ever-changing world of digital identification, **decentralized identification** (**DID**) protocols and standards serve as catalysts for a significant shift in how individuals and entities interact online. DID is a shift from centralized methods, giving consumers choice over their digital identities. This chapter dives into the fundamental processes behind this paradigm change. DID is fundamentally based on privacy, security, and user-centricity. This voyage delves into the complex environment of DID protocols, stressing their significance in promoting interoperability and seamless integration across several platforms. These protocols provide the foundation for safe and verifiable digital identities by leveraging cryptographic techniques such as digital signatures and zero-knowledge proofs.

Interoperability is a significant subject, bridging walled identity systems and establishing a unified digital environment. Standards outlining how DIDs are produced, managed, and confirmed are critical in this networked ecosystem. Beyond technicalities, the investigation delves into social and ethical dimensions, specifically how DID protocols align with principles of inclusivity, consent, and user control. The path toward DID protocols and standards also includes standardization initiatives, which try to develop a common foundation for widespread usage. This collaborative approach involves diverse stakeholders contributing to the shaping of digital identity's future. The investigation also delves into regulatory elements, asking how politicians handle innovation while protecting user rights.

In this chapter, we will cover the following topics:

- The need for protocols and standards
- W3C DID standards, DID documents, and DID resolvers
- Privacy by design
- W3C verifiable credentials

The need for standards

Protocols and standards play an important role in defining the digital world by tackling critical issues such as interoperability, security, privacy, efficiency, and innovation. They provide seamless communication across disparate systems, create security best practices, maintain privacy when managing personal data, simplify procedures for efficiency, and serve as a common basis for creativity, encouraging the creation of new ideas and applications. These essential parts work together to strengthen, rely on, and advance the digital ecosystem.

Overall, protocols and standards are essential for creating a digital society that is secure and efficient and respects individuals' privacy rights. Without them, digital systems would be more fragmented, less secure, and less interoperable, which would limit their potential to improve our lives and solve important societal challenges.

Decentralized identifiers (**DIDs**) are set to revolutionize digital identification by giving users more autonomy, privacy, and security. However, in order to fully utilize the promise of DIDs, standardized procedures must be established. These standards provide consistency and dependability in DID management by ensuring interoperability across different systems and services. They also place a premium on security, user trust, and regulatory compliance, lowering the danger of data breaches and privacy violations. Scalability, decreased friction, and long-term sustainability are all ensured by standardized protocols, promoting innovation and wider use of DIDs. These standards and protocols are critical to building a stable, secure, and user-friendly environment for DID solutions to survive in a quickly expanding digital identity world.

What do standards and protocols entail?

DID standards and protocols, such as specifications and requirements, establish what a project's goals are and the techniques it will employ to achieve them. Think of standards and protocols as unambiguous directions, similar to a baking recipe. While criteria may not expressly indicate that a project must conform to a given standard, it is commonly believed that adherence to such standards is implied. Standards are similar to the rules of a game or profession, whereas protocols are tools and step-by-step instructions described in simple language.

The **International Organization for Standardization** (**ISO**) model, often known as the **Open Systems Interconnection** (**OSI**) model, provides a conceptual foundation for standardized network communication protocols. The ISO architecture, which has seven layers ranging from physical to application, promotes interoperability and adherence to standards by providing a methodical and widely recognized structure for creating and implementing networking protocols. This approach ensures that disparate systems may communicate effortlessly by specifying unambiguous interfaces and functionality at each tier, therefore creating a standardized and interoperable basis for global network communication. The following diagram depicts the seven layers of the OSI model:

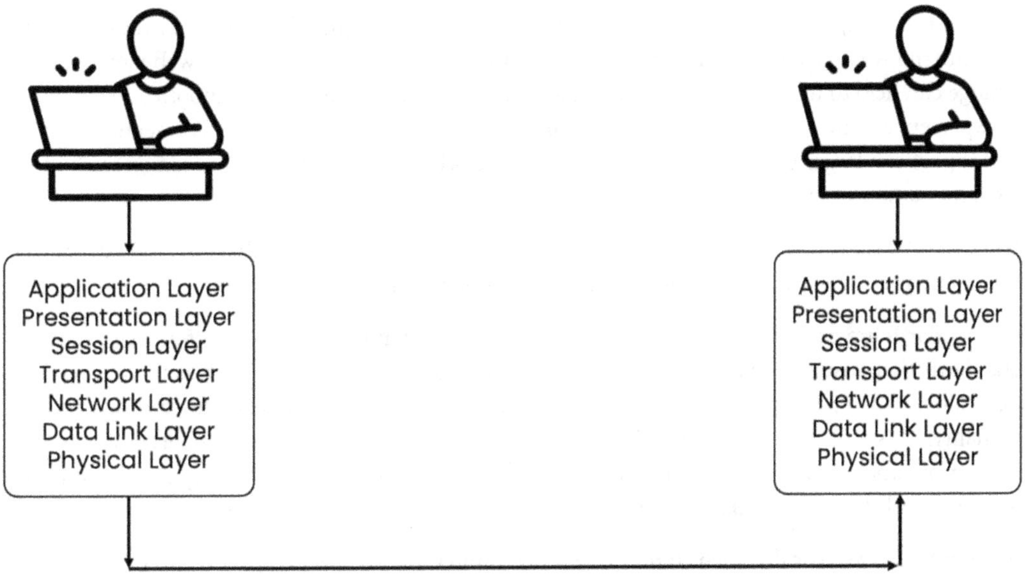

Figure 12.1 – The OSI model

In the realm of technology and digital communication, standards and protocols are fundamental. They specify the rules, formats, and standards that devices, applications, and systems must adhere to in order to be compatible, interoperable, and dependable. Standards and protocols, in essence, serve as a common language that allows various organizations to interact and collaborate seamlessly.

What do standards address?

Standards are a collection of agreed-upon specifications that describe how a given technology or system should operate. These standards are created through a consensus-building process that includes experts, organizations, and stakeholders from several fields. They can discuss everything from data formats and network interactions to security and accessibility. In the context of the internet, for example, the **Hypertext Transfer Protocol** (**HTTP**) provides a standard for how web browsers and web servers transmit information, allowing users to visit websites worldwide. For instance, a quality assurance department should work from a pre-defined standard of quality suitable to their projects and work. Such a quality standard can point out that there must be no bad or broken links, or that individual page weights should be falling between 10 KB and 50 KB.

In web development, the term *page weight* refers to the overall size of a web page, including all of its resources such as graphics, scripts, style sheets, and other files. It is commonly expressed in kilobytes (KB) or megabytes (MB). Page weight is an important measure since higher page weights can cause longer loading times, reducing user experience and potentially lowering search engine results.

The reference to page weights ranging from 10 KB to 50 KB is frequently related to best practices for optimizing web page performance. It shows that limiting the overall size of a web page within this range can help to improve loading speeds, which are critical for ensuring a smooth and efficient user experience, especially on mobile devices and in areas with slower internet connections. This advice is part of a larger push in web development to prioritize performance optimization for better accessibility and user experience.

In terms of DID standards, users are advised to use interoperable standards so that DID infrastructure is able to use present tools and software libraries developed for interoperability.

Some standards are extensively followed, like industry-wide standards. The ISO offers a huge array of standards that ensure — when the companies that adhere to these standards pass the numerous accreditations and implementation programs — consistent, foreseeable, and verifiable levels of agreement and quality. This means that one contributing company can be sure that a certain part *X* will work with their own part *Y*. If you have ever encountered a compatibility failure, you'll comprehend how truly wonderful standards compliance is.

Some standards are basically open standards, which implies that they were defined and made by the industry consensus. Other standards are proprietary and closed, which means they were shaped by a company and then came to control the market due to the success of that business's products. One well-known example of an open standard is HTTP, which is used for web communication. HTTP specifies how messages are structured and conveyed, allowing various systems to comprehend and communicate with one another via the internet. This open standard allows for the smooth interchange of information between web browsers and servers, providing the foundation of the World Wide Web. Other kinds of standards include style guides and benchmarks and are typically limited in scope to the DID projects.

What do protocols address?

Protocols, on the other hand, are a collection of rules or processes that govern how data is sent and received inside a network or system. Protocols regulate the sequence and structure of data packets, error handling, and connection setup. The **Internet Protocol (IP)** is an example of a set of rules for routing and addressing data packets across the internet to ensure they reach their intended destinations. The following table highlights the key differences between protocols and standards, providing a clear comparison of their roles and characteristics:

Aspect	Protocols	Standards
Definition	Rules for data exchange and communication between devices	Established guidelines and criteria for consistency and quality
Purpose	Ensure reliable communication and data transfer	Ensure interoperability, compatibility, and quality
Examples	HTTP, TCP/IP, FTP	ISO, ANSI, IEEE

Aspect	Protocols	Standards
Scope	Technical details of communication processes	Broad guidelines covering various aspects and industries
Flexibility	Typically more flexible and can evolve over time	Usually rigid and require formal revision processes
Enforcement	Often enforced by specific systems or devices	Enforced by regulatory bodies or industry associations
Usage	Primarily used in technology and networking	Applied across multiple sectors including manufacturing, healthcare, and so on
Development	Developed by technical working groups or committees	Developed by recognized organizations or standard bodies
Adoption	Based on need and compatibility requirements	Often required for compliance and certification

Table 12.1: Differences between protocols and standards

Devices must communicate. For instance, a printer must send messages to the computer commanding it that it has no paper or that it is ready to print the document, while a computer must send the data it desires to print to the printer. Computers must send data between themselves so that, for instance, email can be exchanged, and the internet can work. When two devices wish to successfully communicate, they must follow some rules about the manner in which they do it. These are called protocols.

When two software systems need to exchange data, they must use the terminology and protocol that these systems understand. The context property makes sure that two systems running on the same DID document are applying the mutually agreed terminology.

If a DID document publishes the service endpoint developed for the authentication or authorization of the DID subject, it is the accountability of the service endpoint provider, subject, and/or relying party to obey the necessities of the authentication protocol(s) supported at that specific service endpoint.

Communications protocol is the expression used to label a set of rules that message equipment adheres to when they transmit data to each other. If two devices send and receive data but use different rules, then the recipient device will not understand what was sent! Additionally, a communications protocol must outline various specifications before effective communication can take place.

Standards and protocols work together to provide flawless communication and collaboration in the technological world. They lay the groundwork for the creation of hardware and software that can co-exist happily, guaranteeing that diverse devices, applications, and systems can efficiently comprehend and communicate with one another. As a result, innovation is facilitated, new technologies are adopted more quickly, and the reliability, security, and efficiency of digital systems are improved. Standards and protocols are at the heart of contemporary technology, determining how we engage with the digital world, whether it's ensuring smartphone and charger compatibility, protecting online transactions, or making the internet a worldwide platform for information sharing.

DID standards and protocols

The framework of standards and protocols is a vital feature that enables the dependability, interoperability, and security of DID solutions in the ever-changing environment of digital identity. DIDs have emerged as a viable technique for empowering individuals and businesses to take control of their digital identities while also improving privacy and security. The actual promise of DIDs, however, can only be realized via the development of stable, widely agreed standards and protocols. This section goes into the complex realm of DID standards and protocol specifications, throwing light on their importance, development process, and influence on the DID ecosystem.

The digital keys to the kingdom of user-centric identity management are rooted in decentralized, interoperable, and secure protocols and standards, ensuring seamless communication, robust security, and individual control over personal information. DIDs let individuals own and control their unique identities while offering a safe and privacy-focused infrastructure. To fully realize this technology's promise and enable its seamless integration into the larger digital identity ecosystem, generally approved standards, specifications, and norms are required. DID standards and norms are very important in this context.

The development of DID standards and protocols is a complicated, collaborative process including experts, organizations, and stakeholders from the digital identity ecosystem. This procedure is often divided into many stages:

- **Identification of needs**: The procedure starts with determining the exact needs and requirements for DIDs within a given environment or application. These requirements might range from secure authentication to privacy protection, depending on the use case.
- **Consensus building**: Experts and stakeholders work together to reach an agreement on the best practices and standards to use. This method of reaching an agreement guarantees that the final standards and protocols are generally recognized and applicable.
- **Technical development**: Following the definition of the criteria, technical specialists work on the actual construction of the standards and protocols. This entails developing standards, norms, and documentation outlining how DIDs should be produced, maintained, and utilized.
- **Testing and validation**: To ensure functionality and security, the created standards and protocols are rigorously tested and validated. This step may include real-world testing in a variety of applications to discover possible problems or flaws.
- **Iterative development**: Standards and protocols do not remain static; they develop over time to accommodate changing technological and security contexts. Continuous improvement is critical to ensuring the relevance and efficacy of these requirements.
- **Publication and adoption**: The standards and protocols are published and made accessible for adoption by the larger community after extensive development and testing. They become the standard for implementing DIDs in a variety of applications and systems.

Finally, the creation and implementation of DID standards and protocols are critical steps in the advancement of user-centric identity management. These standards not only guarantee interoperability, security, and privacy, but also pave the way for a more inclusive and safe digital identity ecosystem.

The impact on the DID ecosystem

The DID ecosystem is built around DID standards and protocols. Their effect is felt in several dimensions:

- **Enhanced trust**: Stakeholders in the DID ecosystem may have confidence that DIDs are produced, managed, and utilized in a reliable and secure manner by adhering to standardized practices and processes.
- **Interoperability**: DIDs may interface with many platforms, systems, and services thanks to universal standards. This increases the versatility and accessibility of DIDs in a variety of applications.
- **Privacy protection**: Standardized protocols frequently include privacy-enhancing features, giving users better control over their personal data. This results in a more privacy-focused approach to digital identification.
- **Security assurance**: In the digital identity world, security is a top priority. Standards and regulations establish a solid foundation for safeguarding DIDs, lowering the chance of vulnerabilities and data breaches.
- **Innovation catalyst**: With well-defined standards in place, developers are encouraged to innovate and create new DID-based applications and services. This advancement broadens the utility and acceptance of DID systems.

Decentralized identity systems, anchored on DIDs and backed by complete standards and protocols, are poised to transform the way we maintain and safeguard our digital identities as we go deeper into the digital age.

W3C DID standards

The **World Wide Web Consortium** (**W3C**) has played an important role in establishing DIDs, setting the framework for a more secure, private, and user-centric digital identity ecosystem. A DID is similar to having a unique key for your online identity that you control. It's like having a digital passport; however, instead of being issued by a central authority, you build and administer it yourself.

The official W3C DID standard lays forth the principles and recommendations for how DIDs should be formatted, generated, and used across the internet. Let's break down the specification's essential components.

Anatomy of a DID

DIDs are intended to be globally distinct identifiers. They are divided into three major sections in the W3C draft. The first thing to notice is the `did:` prefix, which indicates that what follows is a DID. The procedure follows, which describes how the DID was formed and handled. It's similar to the procedure or system that created your digital identity. Finally, there is the particular identification, which is the one-of-a-kind portion issued to you.

When all of this is put together, a DID can look like this: `did:method:identifier`.

DID methods

The idea of a *method* in a DID is critical. It denotes the system or framework that is used to create and maintain DIDs. Consider it the technology or infrastructure that underpins your digital identity. The W3C DID standard draft allows for a variety of DID techniques to meet a variety of purposes and scenarios. Methods based on blockchains, distributed ledgers, or other decentralized technologies, for example, might exist. Each technique has its own set of rules and processes for developing and dealing with DIDs.

DIDs are the digital keys to your online identity, and knowing DID procedures is critical to realizing the full potential of this new approach. A technique in the field of DIDs is more like a unique recipe or framework that dictates how your digital identity is produced, managed, and guarded, rather than a step-by-step procedure. Consider it the technology underlying your digital passport, and each approach is a unique means of creating that passport, catering to different requirements and interests.

DID method examples

The idea of DID techniques offers a varied array of approaches to building and managing digital identities in the area of DIDs. Consider these strategies to be diverse recipes, each with its own set of components tailored to different tastes. Blockchain-based systems, such as Bitcoin and Ethereum, encrypt identities and store them in immutable ledgers. Privacy and security are prioritized in specialized toolkits such as Hyperledger Indy. The Sovrin Network develops a sense of shared identity. Through common qualities such as online handles, name-based systems facilitate identification. Finally, **self-sovereign identity (SSI)** allows people to govern their digital identities, similar to owning a virtual wallet.

The following examples of DID methods demonstrate the adaptability of DID approaches, allowing users to select a framework that meets their own needs and preferences:

- **Blockchain-based methods**: Consider a blockchain to be a digital ledger, a secure and immutable record of transactions. This technology is used by blockchain-based DID techniques to establish and manage IDs. It's similar to having a digital identity stamped on a blockchain, which ensures that once it's there, it's there for good. The Bitcoin blockchain, for example, may be a method, and your DID might be `did:bitcoin:identifier`. This solution connects your digital identity to the Bitcoin blockchain's security and immutability.

- **Decentralized ledger methods**: These extend the concept beyond Bitcoin. Ethereum is yet another blockchain, but it is more than simply money; it also enables smart contracts. Consider your DID to be intelligent and capable of carrying out agreements and transactions. DID approaches based on Ethereum enable this type of smart identity. As a result, `did:ethr:identifier` might be your Ethereum-based DID, tying your identity to the flexibility of smart contracts and decentralized apps.
- **Hyperledger Indy**: Hyperledger Indy acts like a specialized toolkit for creating decentralized identities. It is particularly intended for handling digital identities in a secure and privacy-preserving manner. If you use Hyperledger Indy, your DID may look like `did:indy:identifier.` This solution ensures that your digital identity is handled in accordance with the Hyperledger Indy privacy and security rules.
- **Sovrin Network**: The Sovrin Network functions similarly to a community-driven identity platform. When you use the Sovrin Network, your digital identity becomes part of a larger network, generating a sense of belonging and collaboration. In this scenario, your DID may look like `did:sov:identifier`, emphasizing your affiliation with the Sovrin identity community.
- **Name-based methods**: Let's take a break from complicated technology for a minute. Consider getting a DID based on your name. Your digital identity might be as easy as `did:web:john_doe`, where your web presence is linked to your digital identity. This name-based strategy is straightforward to use, remember, and access, making it an excellent choice for individuals who appreciate simplicity.
- **Self-sovereign methods**: SSI is a concept that empowers you to manage your digital identity. It's like keeping your identifiers in a virtual wallet and deciding when and where to share them. Using SSI as your technique indicates that your DID is all about self-sovereignty. Your DID may look like `did:ssi:identifier`, indicating your independence in controlling and displaying your digital identity.

In essence, DID approaches provide a menu of options for creating and managing your digital identity. It's like selecting the technology or system that best suits your interests and demands. Whether it's the security of blockchain, the flexibility of smart contracts, the community-driven approach of the Sovrin Network, or the simplicity of a name-based system, each choice adds a distinct flavor to how your digital identity is managed. As the world of DIDs evolves, these methods enable individuals to construct their online presence in the way that best suits them – simple, safe, and in complete control.

DID documents

DIDs do not exist in isolation; they are accompanied by something known as a DID document. This document serves as your identity's digital home, providing information about how your DID was formed, what it may be used for, and even public keys for safe communication. It's similar to your online profile, except instead of being held by a centralized company, you control it. DIDs are the digital keys to our online identities, and grasping the notion of DID documents is analogous to peeling back the layers of a digital identity. Consider a DID to be your unique identity, like a passport for the internet world, and that this digital passport comes with a complete document that has all of your personal information, including your preferences, contact information, and even your virtual signature. This extensive paper is known as the DID document.

The DID document is essentially your digital identity's home. Just as your real home has all of your personal information - your address, contact information, and maybe even a list of your favorite things - your DID document provides critical information about your digital existence.

Consider the DID paper to be the blueprint for your online self. It's the thorough blueprint outlining how your digital identity was formed, what it may be used for, and the keys that keep it safe. This digital blueprint guarantees that people connect with your online presence in a safe and trustworthy environment.

Security is critical in the digital environment, and the DID document is critical in guaranteeing the security of your interactions. It contains cryptographic keys, both public and private. The public key functions as a lock that others may use to authenticate your identification, but the private key is your unique key that allows you to open and govern your digital identity. It's like having a key to your house that only you have.

DIDs and the documentation that goes with them are adaptable. Depending on the context or application, your digital identity can be portrayed in a variety of ways. It's similar to having a business card with several parts, such as your professional information for employment, your hobbies for social encounters, and so on. The DID document guarantees that your identity adjusts to changing circumstances while retaining its integrity.

Consider the DID document to be the public record of your online presence. It contains the information required for others to communicate with you safely. If you're a member of a decentralized community or network, your DID document serves as your digital phonebook entry, allowing people to discover you and converse securely.

The beauty of DID papers is that they are universally accessible. They are intended to be globally accessible, making your digital identity recognizable and interactable across several platforms and systems. It's like having a common language for your online presence, guaranteeing that your digital identity isn't restricted to certain applications or services.

DID document examples

The DID document emerges as the digital handbook in the dynamic arena of DIDs, carefully defining the components of our online identities. It functions like a Bitcoin wallet's user manual, providing a blueprint for safe transactions, storing cryptographic keys, and ensuring complete control over digital assets. The DID document, similar to an email signature, becomes a personalized seal, certifying messages by containing critical facts. It allows users to organize and share particular information, similar to a social network profile, providing a varied depiction of identity. The following examples demonstrate the adaptability and user-centric character of DID papers, which contribute to the development of a secure and personalized digital identity ecosystem:

1. **Bitcoin wallet example**: Consider your DID document to be the user guide for your Bitcoin wallet. It includes information on how your wallet was formed, the public address that others may use to safely transfer Bitcoin to you, and your private key – the secret code that only you know to access and operate your wallet. This agreement protects the security of your transactions and ensures that you have complete control over your digital assets.
2. **Email signature example**: Consider your DID document to be your email signature. It includes your name, contact information, and maybe a remark about yourself. This signature, which is attached to your emails, guarantees that recipients can confirm they are indeed from you. You have complete control over what information is presented, just like you do with the contents of your email signature.
3. **Social media profile example**: Consider your DID paper to be your online profile. It includes the information you choose to publish, such as your name, profile picture, and sometimes a brief bio. The DID document, like customizing your social media profile, allows you to choose which elements of your identity you wish to broadcast to the digital world.

The DID document is essentially your digital identity guidebook, containing the core of who you are online. It's the blueprint, the key to secure communication, and the public record that assures your digital life is adaptable and under your control. Understanding and controlling your DID document is becoming increasingly important as the digital world evolves, guaranteeing a safe and user-centric online identity.

DID universal resolver

DIDs are intended to function smoothly across several systems and services. To that end, the W3C definition encourages interoperability. It also introduces the notion of a *universal resolver*, which functions as a worldwide directory for DIDs. This resolver aids in the translation of DIDs into the real information required to engage with that identity, making DIDs more widely available.

Assume you have your digital passport (DID) but are in a large digital metropolis. How do you utilize and navigate your digital identity across platforms and services? This is where DID resolvers may help. In our digital city, a DID resolver is like a useful guide. It's a specialized technology that understands DID language and can help you translate it into real-world actions. A DID resolver, similar to how you might approach a guide to discover a certain address in a city, assists systems and apps in locating and interacting with your digital identity based on your DID.

How do DID resolvers work?

When a system or program sees a DID, it does not know where to go for the related data. The DID resolver enters the picture and decodes the DID, recognizing its distinct structure and approach. After decoding the DID, the resolver functions as a navigator, leading the system to the correct location for information. It understands where your digital identification information is kept and how to obtain it. The resolver, like a digital detective, retrieves the pertinent data linked with your DID. Depending on the circumstances, this might include your public keys, authentication methods, or other information. Most importantly, DID resolvers are safe to use. They utilize encryption and cryptography techniques to protect your data while it is being retrieved. It's like a protective shield around your digital identity.

Implementing DID resolvers

DID resolvers may be deployed in a variety of methods to meet the varying demands of the digital world. Here are some typical approaches:

- **Blockchain-based resolvers**: Consider a blockchain to be a safe, decentralized database. Blockchain-based resolvers traverse this digital metropolis by referencing the blockchain, which stores your identification data. This provides an immutable and trustworthy source for your digital identification.

1. **Decentralized ledger resolvers**: Decentralized ledger resolvers, like blockchain, interact with distributed ledgers to ensure that your digital identity is securely handled across a network of interconnected services.

- **Hub resolvers**: Hub resolvers serve as central hubs for storing your digital identification information. They simplify the retrieval process, allowing programs to retrieve your data without having to engage directly with the underlying technology.

Finally, DID resolver implementation incorporates a variety of solutions geared to the digital landscape's unique demands. Whether through blockchains, decentralized ledgers, or hub resolvers, these techniques provide secure administration of your digital identity in a decentralized, dependable, and user-friendly manner, leading to a safer and more efficient digital environment.

Practical use of DID resolvers

Let's put this into context with a simple analogy: your email address. Consider your email address to be your DID, and your email server to be the DID resolver. Someone uses your email address (DID) to send you an email (interact with your digital identity). The email server (resolver) knows where to look for your inbox and guarantees that the email is sent securely to the intended recipient.

Similarly, when you log in to a website or use a service, the DID resolver assists the system in locating and verifying your digital identity. It handles the technical parts of getting your identity information, ensuring a smooth and secure connection.

Benefits of DID resolvers

The DID resolver is a critical facilitator at the heart of this transformational paradigm. This digital handbook converts the complex terminology of DIDs into real activities, guaranteeing that our digital passports are not only unique but also widely accessible and safe.

The advantages of DID resolvers are numerous. They usher in a new era of interoperability, allowing various systems and applications to connect with a plethora of DIDs with ease. As individuals maintain strict control over their digital identities, they may dictate what information is shared and with whom. DID resolvers' cryptographic expertise creates a strong firewall over sensitive data, ensuring the integrity and secrecy of our online personas. DID resolvers, which embrace user-centricity, provide individuals unparalleled control over their digital presence, signaling a major shift in the dynamics of identity management in the digital arena.

DID resolvers deliver several benefits to the digital identity world, including improved interoperability, privacy management, security, and a user-centered approach:

- **Interoperability**: DID resolvers improve interoperability by allowing various systems and applications to comprehend and interact with a diverse set of DIDs
- **Privacy control**: Users retain control over their digital identities, determining what information is shared and with whom because the resolver only retrieves data when it is required
- **Security**: DID resolvers use cryptographic methods to ensure the secrecy and integrity of your digital identification data
- **User centricity**: With DIDs and resolvers, the emphasis turns to user-centric identity management, giving people more control over their online presence

Finally, DID resolvers are critical in the developing context of digital identification. They provide a link between the abstract realm of DIDs and the concrete requirements of applications and systems. The user retains control, security is enhanced, and a new era of decentralized, user-centric identity management begins.

Decentralized trust

One of the most innovative parts of the W3C DID definition is its emphasis on decentralized trust. We have always relied on centralized institutions to verify our identification. This model is turned on its head by DIDs. Trust is dispersed over different systems and techniques rather than a single entity, making your digital identity more robust and less reliant on a single point of control.

In summary, the W3C DID standard process is ushering in a new age of digital identification, one in which individuals have greater control, privacy is prioritized, and interoperability is essential. It's like having a digital identity toolkit with explicit instructions on how to safely and universally build, administer, and use your online identity. The promise for a more user-centric and secure digital identity ecosystem becomes more evident as this standard matures and achieves greater use.

Privacy by design

In an age where digital traces are ubiquitous, the notion of **privacy by design** (**PbD**) has evolved as a guiding principle, directing the development of systems and technologies toward robust privacy and security. PbD, championed by Dr. Ann Cavoukian, is a proactive strategy to ensure privacy is incorporated into the fundamental fabric of technology from its development. We'll delve into the seven foundational principles of PbD, deciphering each one and revealing tangible implementation strategies with real-world examples, all presented in layperson's terms to make this critical aspect of our digital landscape accessible and understandable:

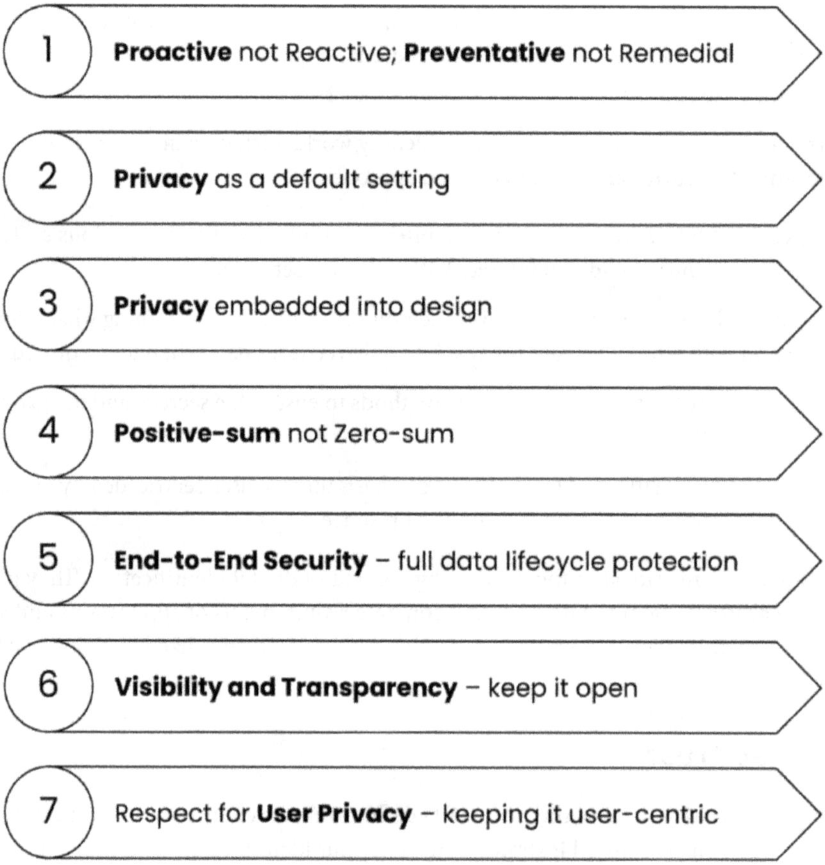
1
Proactive not Reactive; Preventative not Remedial
2
Privacy as a default setting
3
Privacy embedded into design
4
Positive-sum not Zero-sum
5
End-to-End Security – full data lifecycle protection
6
Visibility and Transparency – keep it open
7
Respect for User Privacy – keeping it user-centric

Figure 12.2 – The seven pillars of the PbD model

PbD is a fundamental notion that pervades the creation and implementation of digital systems, ensuring that privacy concerns are incorporated at all stages. This comprehensive approach promotes proactive efforts to preserve individuals' privacy rights, as opposed to the reactive bolt-on solutions of the past. Organizations may create trust, improve user control, and negotiate the complex environment of privacy in the digital age by seamlessly incorporating privacy principles into system design and operation.

Proactive not reactive; preventative not remedial

The first principle serves as a proactive compass in the field of PbD, directing the route of technology progress. This approach argues for anticipated privacy concerns before they arise, asking developers to incorporate preventative measures into the fundamental structure of their systems. It's like building a fortress against future privacy breaches, ensuring that defense is in place long before any threat arises.

- **Principle**: This principle emphasizes the need to anticipate and prevent privacy-invading incidents.
- **Implementation strategy**: This strategy integrates privacy concerns early in the system development process to ensure that privacy is a core element of the design process.
- **Example**: When developing a new mobile app, the development team considers privacy elements such as strong encryption and user-friendly permission methods from the beginning. By doing so, they avoid possible privacy violations rather than dealing with them after the software is released.

Privacy as the default setting

The second principle of PbD advocates for simplicity and user-centricity. Consider a world in which privacy is not a choice but an inherent default – a scenario in which users do not need to negotiate complicated settings to protect their digital identities. This concept announces that goal, urging developers to create systems in which privacy is the starting position, requiring users to consciously pick less private choices if they so choose. It changes the default option from neutral to a safe haven, where privacy is not just a choice but a guarantee.

- **Principle**: Privacy should be the default option, which means that users should not have to take extra actions to preserve their privacy – it should be automatically protected.
- **Implementation strategy**: Ensure that privacy settings are set to the most secure alternatives by default, with users needing to deliberately pick less private options if desired.
- **Example**: By default, a social media platform makes user profiles private, limiting access to personal information. Users can then select whether or not to make certain details public.

Privacy embedded into design

As we get further into the heart of PbD, the third principle invites us to consider privacy not as an afterthought, but as an essential component of the fundamental DNA of technology creations. It necessitates a shift in perspective – privacy as a cornerstone, seamlessly integrated into the fabric of a system's architecture. This principle alters the traditional approach to privacy by putting it at the forefront of design decisions, making it an inherent and non-negotiable characteristic.

- **Principle**: Privacy should be included in the system architecture rather than being handled as an afterthought

- **Implementation strategy**: Instruct developers and designers to think about privacy implications at every level of the design process, and to include privacy features directly into the system's structure
- **Example**: The architecture of a new data management system is meant to automatically encrypt critical information and offer extensive user controls over data access

Full functionality – positive-sum, not zero-sum

The fourth principle in the PbD symphony provides a harmonic idea in which privacy and utility co-exist, one improving the other. Rather than viewing privacy as a trade-off for functioning, this notion envisions a win-win situation in which both components thrive. It calls into question the concept of compromise, asking developers to consider methods in which privacy not only survives but also contributes to the total richness and functionality of the digital experience.

- **Principle**: The goal of privacy should not be at the price of utility; the two may live and complement one other
- **Implementation strategy**: Seek solutions where privacy features improve overall functioning and user experience, resulting in a win-win situation
- **Example**: A healthcare app uses differential privacy strategies to secure individuals' health data while still providing useful information for medical research

End-to-end security – full life cycle protection

The fifth principle emerges as a protector throughout the data life cycle as we go deeper into the privacy-conscious cosmos. It emphasizes the need to protect data not just during transmission but also from conception to disposal. It functions like a digital life cycle guardian, imposing security controls at every stage of data gathering, storage, and final disposal.

- **Principle**: Privacy must be protected throughout the data's lifetime, from collection to storage and final destruction
- **Implementation strategy**: Implement strong security measures at every level, such as encryption during data transfer, secure storage methods, and data disposal practices that are safe
- **Example**: End-to-end security is provided by an e-commerce platform, encrypting customer payment information during transactions, safeguarding it in storage, and safely deleting old transaction records in accordance with privacy requirements

Visibility and transparency

The sixth principle lifts the veil of transparency, allowing people to see within data ecosystems. The emphasis in this component of PbD is on demystifying privacy practices and providing consumers with a clear picture of how their data is managed. It argues for transparency as a basic right rather than a duty, ensuring that people have the tools and knowledge to make educated decisions about their digital footprint.

- **Principle**: Users should be aware of privacy practices and have complete transparency over how their data is gathered, utilized, and shared
- **Implementation strategy**: Provide clear and easy-to-understand privacy rules, utilize simple language, and provide tools that allow consumers to see and control their data
- **Example**: A cloud service provider presents a streamlined, user-friendly privacy dashboard, allowing consumers to see who has accessed their data and offering choices for managing permissions

Respect for user privacy

The seventh principle, a dedication to honoring user choices and managing data responsibly, marks the ultimate frontier in PbD. This concept provides a contract between systems and users, emphasizing the importance of explicit agreement, effective data control methods, and a genuine regard for individuals' privacy rights. It elevates user privacy from a notion to a lived experience, emphasizing organizations' and systems' ethical responsibilities to behave as guardians of user trust.

- **Principle**: Systems and organizations must respect user privacy by honoring user preferences and handling data properly
- **Implementation strategy**: Create systems for getting informed permission, giving people choice over how their data is used, and offering data deletion alternatives
- **Example**: Before obtaining location data, a mobile app obtains explicit user agreement, offering clear explanations of how this data will be used and allowing users to reject access at any time

In conclusion, PbD is a theory that develops a digital world in which privacy is not an afterthought but a fundamental concern at every level of creation. Developers, designers, and organizations may create systems that prioritize user privacy, encourage trust, and serve as beacons of responsible technology in an interconnected and data-driven environment by understanding and adopting these seven principles. In this day and age of digital transformation, pursuing PbD is more than an option; it is a need for creating a safe, ethical, and user-centric digital future.

Verifiable credentials

Verifiable credentials (**VCs**) emerge as important tools in the changing world of digital interactions, redefining how we express and authenticate information online. VCs are fundamentally digital credentials provided by trusted bodies such as institutions or organizations attesting to the validity of specified information about an individual. What elevates VCs to a transformative level is the establishment of robust standards that govern their creation, issuance, and verification. These guidelines serve as the foundation for a safe, interoperable, and user-centric digital credential ecosystem.

Consider your driver's license, a physical document that certifies your identity and privileges. Similarly, we can consider a digital equivalent, a VC; this digital representation contains information about you that has been issued by a trustworthy authority and is protected by cryptographic measures that ensure its integrity and validity.

Verifiable Credential Standards create a uniform language for digital credentials, promoting consistency and interoperability across systems. They define the rules for creating, structuring, and verifying credentials, providing a smooth and secure interchange of information. These standards provide a common ground for various organizations and technologies to comprehend and interact with VCs in a standardized manner.

The W3C has specified one notable set of standards in the form of the Verifiable Credentials Data Model. The structure and behavior of VCs are defined by this comprehensive framework. It explains fundamental concepts such as credential subjects (entities being described), claims (information about the topic), and digital signatures to ensure the credentials' integrity and validity.

The interaction between VCs and DIDs strengthens the user-centric character of digital identity. DIDs serve as a decentralized and user-controlled foundation for the creation and management of VCs. This partnership adheres to the concept of SSI, giving people more control over their digital personas.

Verifiable Credential Standards are gaining traction in a variety of businesses. These standards have the potential to revolutionize how information is exchanged, validated, and trusted in the digital environment, from speeding up identity verification processes in financial services to allowing safe access to healthcare records.

Adoption of Verifiable Credential Standards, like every disruptive technology, is fraught with difficulties. These include assuring public acceptability, resolving privacy issues, and encouraging varied stakeholders to work together. Nonetheless, continual advances and joint initiatives are paving the way to a future in which digital credentials are not just safe and dependable, but also profoundly embedded in our digital interactions.

Verifiable Credential Standards are basic threads in the complicated tapestry of digital trust, creating a story of safe, user-centric, and interoperable identity management. These standards, led by the W3C's Verifiable Credentials Data Model, move us into a future where digital credentials seamlessly integrate into our daily lives, creating a heightened feeling of confidence and security in the ever-changing digital realm.

Key components of verifiable credential standards

Understanding the major components of verifiable credential standards is critical for navigating the complex terrain of DID and assuring the safe and interoperable transmission of digital credentials. These components serve as the building blocks that give individuals control over their digital identities while also providing a solid basis for trust and credibility online. Exploring these factors reveals the complexities that make verified credentials such a powerful force in defining the future of identity management.

Credential subject

In the context of VCs, the term **credential subject** refers to the entity to whom the credential is relevant or belongs. Within the digital environment, this entity can take different forms, including persons, organizations, or any other identifiable entities. Let us go over this in further detail:

- **Individuals**: The credential subject in the case of VCs relating to personal qualities is often a human. Consider a digital driver's license as an example of a VC. The credential subject is the individual (the person who holds the credential). The credential may include information such as the individual's name, date of birth, and driving privileges.
- **Organizations**: VCs can indicate organizational connections or accreditations as well as personal traits. The credential subject in such situations might be a firm, educational institution, or any other organizational body. For example, the credential subject on an accrediting certificate issued to a university.
- **Identifiable entities**: Beyond persons and organizations, the idea of the credential subject encompasses any distinct entity in the digital realm. This might include **Internet of Things** (**IoT**) devices, smart contracts, or any other digital entity that can be identified. A VC attesting to the validity of data acquired by a sensor device, for example, would include the device as the credential subject.

The credential subject is essentially the principal figure to whom the VC offers information or attributes. It serves as the credential's pivot point, delivering information relevant to, and frequently owned or managed by, the stated entity. Because of the freedom in establishing the credential subject, VCs are adaptable to a wide range of use cases across domains, from personal identity verification to organizational certifications and beyond.

Claims

The word **claims** in the context of VCs refers to bits of information about the credential subject, whether a human, organization, or other identifiable entity. To further understand this notion, here is a more extensive explanation:

- **Information representation**: Claims are the building elements used to transmit precise information about the credential subject. Consider the credential subject to be an entity, such as a person, and the claims to be the discrete bits of information or characteristics that collectively define and characterize that thing.

- **Key-value pairs**: Claims are organized as key-value pairs, a common data representation type. A *key* indicates the kind or category of information in this context, and its matching *value* is the actual data or attribute connected with that key. Consider the following claim: the key is `name`, and the value is the person's true name – for example, `John Doe`:

```
{
    "name": "John Doe",
    "age": 30,
    "organization": "XYZ Corporation"
}
```

  Each claim in this JSON-like structure is a key-value pair that provides separate bits of information on the credential subject.

- **Diversity of claims**: Claims might include a wide range of information, from personal facts such as name, age, and residence to more specialized characteristics such as work title, educational degrees, and membership affiliations. VCs can express a full and nuanced profile of the credential subject due to the variety of claims.

- **Personal identifiers**: Some claims may function as personal identifiers, identifying the credential subject with specific characteristics. A claim with the key `SSN` (Social Security number), for example, would have the value indicating the individual's unique identity:

```
{
    "SSN": "123-45-6789",
    "name": "Jane Smith",
    "age": 25
}
```

- **Contextual reference**: Claims are chosen in response to the specific use case or situation for which the VC is provided. A university degree certificate, for example, can comprise assertions such as degree type, graduation date, and institution name.

- **Granular control**: Claims provide you with fine-grained control over the information you disclose. The credential holder can selectively disclose certain claims to verifiers in a privacy-centric architecture, ensuring that only relevant facts are given based on the context or the verifier's requirements:

```
{
    "name": "Alice Johnson",
    "age": 28
}
```

In essence, claims inside VCs are essential pieces of information, expressed as key-value pairs, that provide a rich and adaptable representation of the credential subject when combined. This structured and contextualized method enables the conveyance of multiple qualities, permitting secure and selective information sharing in a variety of circumstances across different domains.

Credential issuer

The **credential issuer** is a key player in the VC ecosystem, representing the company in charge of developing and issuing the credential. The issuer affirms the truth and validity of the information included in the VC in this capacity, which carries important authority. Let us go over this in further detail:

- **Creation and assertion authority**: The entity that commences the process of establishing a VC is known as the credential issuer. This entity is authorized to make certain claims regarding the credential subject. When providing a digital degree credential, for example, a university may function as a credential issuer, asserting facts such as the degree type and graduation date.
- **Diverse issuer entities**: Depending on the nature of the credential, the credential issuer might take numerous forms. It might be a government agency that issues official identity credentials, a university that issues academic degrees, an employer that issues employee ID credentials, or any other reputable organization that is authorized to make representations of the credential subject.
- **Trust and authority**: Trust is an essential component of the VC system. The credential issuer's authority is formed based on the entity's credibility and validity within a certain domain. A VC issued from a well-known and prestigious university, for example, is likely to be generally regarded.
- **Digital signatures**: The credential issuer uses digital signatures to ensure the integrity and validity of the VC. These cryptographic techniques prove that the credential was issued by the stated issuer and that the information contained inside it was not tampered with:

```
{
    "issuer": {
        "id": "did:example:123",
        "name": "ABC University",
        "publicKey": "base64url-encoded public key"
    },
    // Other credential details
}
```

- **SSI context**: Individuals can also serve as issuers of their credentials in the context of SSI. For example, a person may issue a VC including self-asserted claims such as their name and contact information:

```
{
    "issuer": "did:self:123",
    "name": "Alice Johnson",
    // Other self-asserted claims
}
```

- **Revocation authority**: Under specific conditions, the credential issuer may also have the right to withdraw a credential. If a university discovers an error in a degree certificate, or if an employee departs a firm, the issuer may revoke the credentials:

```
{
    "proof": {/* Proof details */},
    "status": "revoked",
    "revocationReason": "Credential Holder no longer meets 
eligibility criteria."
}
```

In summary, the credential issuer, who has the capacity to produce, assert, and digitally sign credentials, plays a critical role in the VC life cycle. This function is critical to creating confidence within the decentralized and user-centric VC paradigm, ensuring that statements about the credential subject are correct and legitimate and come from entities with the necessary authority and trustworthiness.

Credential holder

The **credential holder** is the person or entity who possesses the VC. This position is especially important in the context of SSI because the credential holder has extensive control over the administration and display of their digital credentials. Let's take a closer look at this idea:

- **Holding the VC**: The entity to whom the VC has been issued is known as the credential holder, and they retain ownership and control over it. This entity might be a person, a company, or any recognizable digital entity.
- **Control over presentation**: The credential holder has unprecedented discretion over when, where, and how they offer their VCs under the SSI paradigm. In contrast to traditional identification systems, in which third parties frequently hold and control identity information, SSI allows individuals to manage their own digital identities. Depending on the unique requirements or trust levels of the conversation, the credential holder can pick which credentials to display in various scenarios.
- **Selective disclosure**: Selective disclosure is an important characteristic of SSI. The credential holder can selectively release specific claims or characteristics of their VCs, only releasing the information required for a given transaction or engagement. This improves privacy and reduces the amount of information given, resulting in a more nuanced and user-centric approach to identity verification.

```
{
    "name": "Alice Johnson",
    "age": 28,
    // Other claims
}
```

- **User-centric identity management**: SSI places the credential holder at the center of the identity ecosystem, emphasizing a user-centric approach to identity management. This paradigm change gives people more agency and sovereignty over how their identification information is utilized, minimizing dependency on central authorities.
- **Digital wallets and personal datastores**: Credential holders often store and maintain their VCs via digital wallets or personal data storage. These technologies offer credential holders a safe and user-friendly environment in which to organize, validate, and selectively reveal their credentials as needed.
- **Proof of ownership**: The credential holder can offer cryptographic evidence of ownership and control over the VCs they present. These proofs, which are frequently in the form of digital signatures, attest to the validity of the information being exchanged as well as the credential holder's legitimacy.

```
{
    "proof": {
        "type": "Ed25519Signature2018",
        "created": "2022-01-01T10:00:00Z",
        "proofPurpose": "authentication",
        "verificationMethod": "did:self:123#key-1",
        "jws": "eyJhbGciOiJFZERTQSIsImtpZCI6ImRpZDp..."
    }
}
```

- **Privacy and security**: The SSI paradigm prioritizes privacy and security, giving credential holders greater control over their personal information as they navigate digital interactions. This user-centered strategy adheres to the concepts of empowerment, permission, and minimum disclosure.

In essence, the credential holder is the individual or entity with control and ownership over their digital credentials in the context of VCs, notably in the area of SSI. In the digital era, this trend toward user-centric identity management promotes privacy, security, and a more empowered and autonomous approach to identification.

Credential verifier

The **credential verifier** is an important function in the VC ecosystem since it represents the entity that wants to validate the validity and correctness of the provided VC. This function is critical in a variety of transactions and interactions where certain traits must be proven. Let us go over this in further detail:

- **Purpose of verification**: The credential verifier is a person or organization that demands evidence or confirmation of specified features about the credential subject. This might include validating someone's identity, reviewing educational credentials, confirming job information, or any other situation where trust and verification are required.

- **Role diversity**: Depending on the circumstances, the credential verifier might take several forms. A service provider, employer, government agency, educational institution, or any other body that requires the verification of statements contained in a VC might be this entity.
- **Trust in the credential issuer**: The credential verifier relies on the credential issuer's credibility and authority. The verifier must believe that the information in the VC was properly issued by an entity that has the ability to make statements about the credential subject.

```
{
    "issuer": {
        "id": "did:example:123",
        "name": "ABC University",
        "publicKey": "base64url-encoded public key"
    },
    // Other credential details
}
```

- **Cryptographic verification**: Credential verifiers use cryptographic techniques to validate the VC's validity and integrity. This includes verifying the digital signatures supplied by the credential issuer to ensure that the credential has not been tampered with and came from the stated issuer.

```
{
    "proof": {
        "type": "Ed25519Signature2018",
        "created": "2022-01-01T10:00:00Z",
        "proofPurpose": "assertionMethod",
        "verificationMethod": "did:example:123#key-1",
        "jws": "eyJhbGciOiJFZERTQSIsImtpZCI6ImRpZDp..."
    }
}
```

- **Attribute-specific verification**: The VC's specific properties or assertions are of interest to the credential verifier. For example, while verifying work history, the verifier may concentrate on assertions such as job title, employment dates, and the name of the organization:

```
{
    "employment": {
        "jobTitle": "Software Engineer",
        "organization": "TechCorp",
        "employmentDates": {
            "start": "2020-01-01",
            "end": "2022-12-31"
        }
    }
}
```

- **Selective disclosure**: Credential verifiers may only require certain information for a given transaction. In the context of SSI, the credential holder has the power to selectively divulge relevant claims, ensuring that only essential information is disclosed, boosting privacy, and minimizing data exposure.

```
{
    "name": "Alice Johnson",
    "age": 28
}
```

- **Decentralized and trust minimization**: The job of a credential verifier in DID systems coincides with the notion of trust reduction. Rather than depending on a central authority for verification, the verifier may independently check the authenticity of VCs using cryptographic methods, leading to a more dispersed and trust-reducing ecosystem.

In essence, the credential verifier is the entity looking for proof or verification of specified properties by examining VCs. This position is critical in a variety of circumstances, ranging from identity verification by service providers to employment confirmation by employers, all of which contribute to a trustworthy and secure digital interaction landscape.

W3C Verifiable Credentials Data Model

Through its Verifiable Credentials Data Model, the W3C has played a critical role in standardizing VCs. This model specifies the fundamental principles and structures that ensure the interoperability and security of VCs. Let's look at some of its major points.

Credential schema

The structure and features of a VC are defined by a credential schema. It contains information on the many sorts of claims that may be included in a credential. A credential schema for a university degree, for example, can comprise assertions such as degree type, graduation date, and institution:

```
{
    "@context": "https://www.w3.org/2018/credentials/v1",
    "type": ["VerifiableCredential", "UniversityDegreeCredential"],
    "issuer": "did:example:123",
    "issuanceDate": "2022-01-01T10:00:00Z",
    "credentialSubject": {
        "id": "did:example:456",
        "degree": {
            "type": "BachelorDegree",
            "name": "Bachelor of Science"
        }
    }
}
```

Proofs and signatures

To secure the integrity and validity of VCs, the W3C Verifiable Credentials Data Model emphasizes the significance of cryptographic proofs. Digital signatures are used to generate proofs that prove the authenticity of the information given.

```
{
    "proof": {
        "type": "Ed25519Signature2018",
        "created": "2022-01-01T10:00:00Z",
        "proofPurpose": "assertionMethod",
        "verificationMethod": "did:example:123#key-1",
        "jws": "eyJhbGciOiJFZERTQSIsImtpZCI6ImRpZDpleGFtcGxlOjEyMw..
tUwe8J69G96j..."
    }
}
```

Revocation mechanism

The Verifiable Credentials Data Model covers an important issue: credential revocation. It establishes a way for credential issuers to transmit a credential's revocation status.

```
{
    "proof": {
        // Proof details
    },
    "status": "revoked",
    "revocationReason": "Credential Holder no longer meets eligibility
criteria."
}
```

Presentation exchange

The model describes a standardized way of presenting VCs, allowing credential holders to choose and provide information to verifiers based on their individual needs.

```
{
    "@context": [
        "https://www.w3.org/2018/credentials/v1",
        "https://www.w3.org/2018/credentials/examples/v1"
    ],
    "type": ["VerifiablePresentation"],
    "verifiableCredential": [/* Array of Verifiable Credentials */],
    "proof": {/* Proof details */}
}
```

Finally, the W3C Verifiable Credentials Data Model, with its clear and complete structure, establishes the foundation for a safe and interoperable framework for handling digital credentials. The given JSON examples demonstrate the model's versatility and efficiency, stressing its potential to transform how we manage identity and credentials in the digital era. Adopting these standards offers the possibility of creating a more user-centric, privacy-conscious, and trustworthy digital environment.

Examples of implementing VCs

Starting with realistic implementations, let's look at some real-world situations that demonstrate the smooth integration and deployment of VCs in diverse circumstances:

- **University degree**: Consider a VC as a university diploma. The credential schema describes the structure, including the kind of degree, the date of issue, and the identification of the institution.
- **Employment ID**: A VC might serve as an employee ID in a professional setting. The credential schema defines information such as the employee's name, their job title, and the organization that is granting the credential.
- **COVID-19 vaccination credential**: In the context of global health problems, VCs play an important role in communicating COVID-19 vaccination status. The credential schema contains information such as the immunization type, dosages provided, and healthcare provider.

To conclude, Verifiable Credential Standards usher in a new era of safe, interoperable, and user-centric digital identity management, particularly as expressed by the W3C Verifiable Credentials Data Model. These standards provide a common vocabulary for entities across heterogeneous systems, facilitating information sharing while prioritizing security and privacy. As digital interactions expand, the solid basis established by Verifiable Credential Standards promises a future in which individuals have more control over their digital identities, encouraging trust in an interconnected and dynamic digital world.

Summary

This chapter investigated the notion of privacy by design, highlighting the need for prioritizing privacy at all stages of digital system development. It provided seven principles, including end-to-end security, transparency, and respect for user privacy, as well as implementation methodologies and examples to demonstrate their practical use. It also discussed the importance of protocols and standards in the digital age, emphasizing their role in facilitating interoperability, security, and privacy while presenting instances of their practical use. The chapter also examined the W3C Verifiable Credentials Data Model and its function in standardizing VCs while ensuring compatibility and security. Overall, it emphasized the significance of PbD and the role of standards and protocols in fostering a safe, efficient, and privacy-conscious digital society.

In the next chapter, we shall take a look at DID authentication and the methods to leverage DIDs for passwordless access into systems.

13
DID Authentication

DID authentication, or decentralized identity authentication, is a method of authentication that relies on decentralized digital identities. Decentralized identities are digital identities that are not controlled by any single organization or authority but instead are created and managed by the individuals themselves. They are based on blockchain technology and use public-private key cryptography to secure and verify identity.

In a DID authentication system, individuals create and manage their own digital identities, which are stored on a decentralized blockchain network. When they need to authenticate their identity, they can use their private key to sign a message, which can be verified by a trusted third party using their public key. This allows individuals to prove their identity without having to rely on a centralized authority or give up control of their personal information.

DID authentication has several advantages over traditional authentication methods. First, it provides a high level of security and privacy, as each user's personal information is not stored on a central server that can be hacked or compromised. Second, it gives users more control over their personal data, as they can choose what information to share and with whom. Finally, it provides a more flexible and interoperable authentication method, as users can use their digital identity to access a wide range of services and platforms.

However, there are also some challenges associated with DID authentication. For example, it requires a high level of technical knowledge and understanding of blockchain technology, which may be a barrier for some users. In addition, there is currently a lack of standardization and interoperability between different decentralized identity systems, which can make it difficult for users to use their digital identity across different platforms and services.

Overall, DID authentication is an innovative and promising approach to identity authentication that has the potential to provide a high level of security, privacy, and control for users.

In this chapter, we will cover the following topics:

- Traditional authentication methodologies
- DID authentication protocols and implementation methodologies
- Real-world examples and use cases of DID authentication methods

Traditional authentication

Before getting into the complexities of DID authentication, it's critical to understand the faults in standard authentication techniques. Passwords, once thought to be the basis of digital security, now have a slew of flaws:

- **Password vulnerabilities**: Passwords are vulnerable to a variety of attacks, ranging from readily guessable weak passwords to the possibility of password reuse across numerous platforms, exposing users to possible security issues. The human instinct to select easily remembered passwords frequently leads to compromised security.
- **Phishing attacks**: Phishing attacks, in which hostile actors deceive users into disclosing their credentials, take advantage of the confidence that users place in traditional login protocols. Despite attempts to educate and raise awareness, consumers might still fall victim to well-crafted phishing attacks.
- **Centralized data breaches**: The centralization of user passwords in databases is a big danger. Millions of usernames and passwords have been revealed as a result of large-scale data breaches, allowing unauthorized access and identity theft.

Securing users' access to various programs is a significant concern for the IT department. Users frequently require access to several systems, which is why authentication procedures, which are usually open standards, are so important. People frequently ask, "*Can you help me find the right authentication protocol for my specific use case?*" This emphasizes the necessity of selecting a suitable authentication mechanism depending on the unique requirements. The DID authentication protocols are probably the recommended ones. But prior to diving into DID authentication, let's take a peek at traditional authentication methods and mechanisms in play today and the challenges or pitfalls with these existing methodologies or approaches.

Lightweight Directory Access Protocol

Lightweight Directory Access Protocol (**LDAP**) acts as a software protocol that assists users in locating entities such as organizations, people, and other resources, such as devices and data. Whether on the vast terrain of the public internet or inside the boundaries of a company intranet, LDAP is critical to enabling effective network travel. In a nutshell, it serves as a directory service, storing and granting access to user account information such as usernames and passwords. LDAP authentication, with its ubiquity, is not without its issues. The following diagram illustrates the flow from applications to a directory service:

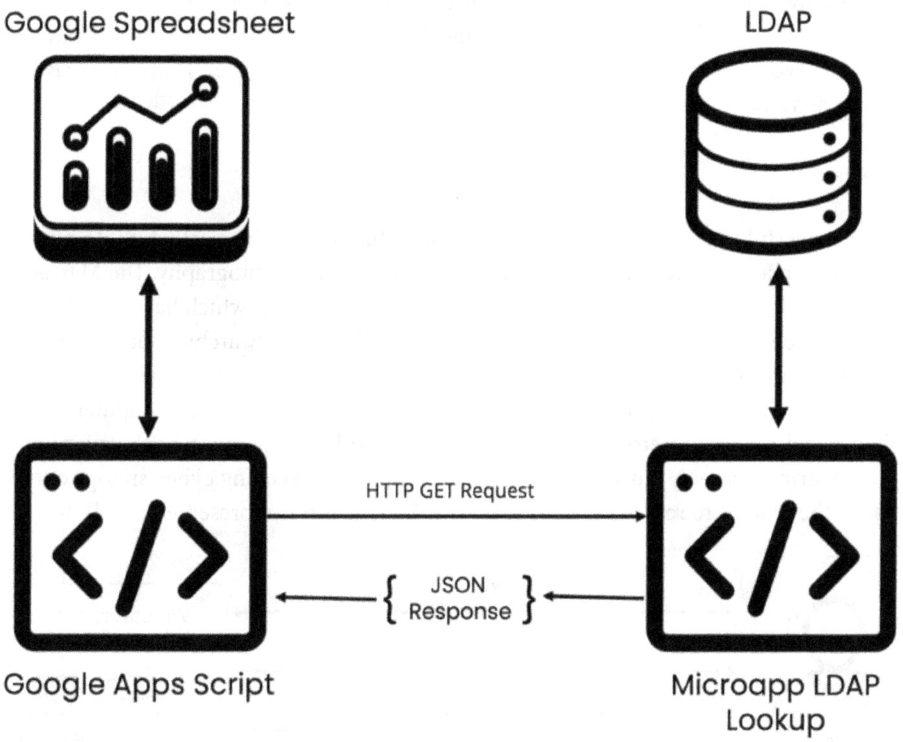

Figure 13.1 – LDAP query structure

One key challenge is the risk of sending sensitive information, such as passwords, over the network. LDAP often uses unencrypted connections, which exposes this data to hostile actors who may intercept and manipulate it. To overcome this, organizations frequently incorporate extra security measures such as SSL/TLS encryption; however, configuring and managing these security features can be difficult.

Another issue is the centralized nature of LDAP. Because it functions as a central directory, any disruption or failure in the LDAP server might hamper authentication for all associated systems. This single point of failure can jeopardize network service availability significantly. Furthermore, expanding LDAP to handle an increasing number of users and devices can be difficult and may necessitate considerable infrastructure changes.

Furthermore, managing and updating user information within LDAP directories may be time-consuming, particularly in big organizations with dynamic user hierarchies. Ensuring the quality and timeliness of user data is a continual problem that, if not properly managed, can lead to authentication difficulties or security vulnerabilities.

In summary, while LDAP authentication is a widely used and effective method for managing user identities, it does have some drawbacks, such as potential security risks during data transmission, the risk of centralization leading to a single point of failure, and the complexities associated with scaling and maintaining accurate user data. To provide a safe and dependable authentication process, addressing these difficulties necessitates the careful consideration of security measures, resilient infrastructure, and efficient management practices.

Kerberos

Kerberos is a powerful network authentication protocol that was methodically built to offer strong authentication for client/server applications by utilizing secret key cryptography. The Massachusetts Institute of Technology provides a free implementation of this protocol, which has received extensive acceptance and appears in a variety of commercial devices. This security architecture provides a high degree of protection for digital interactions, hence increasing the overall resilience of networked systems. Kerberos is an authentication system that, like a secret code for your computer, allows you to safely log in and access numerous network services. Consider Kerberos to be the guardian at the digital gate, ensuring that only authorized users gain access while keeping cyber-snoops out. A flow diagram from the request to response for Kerberos authentication is represented as follows:

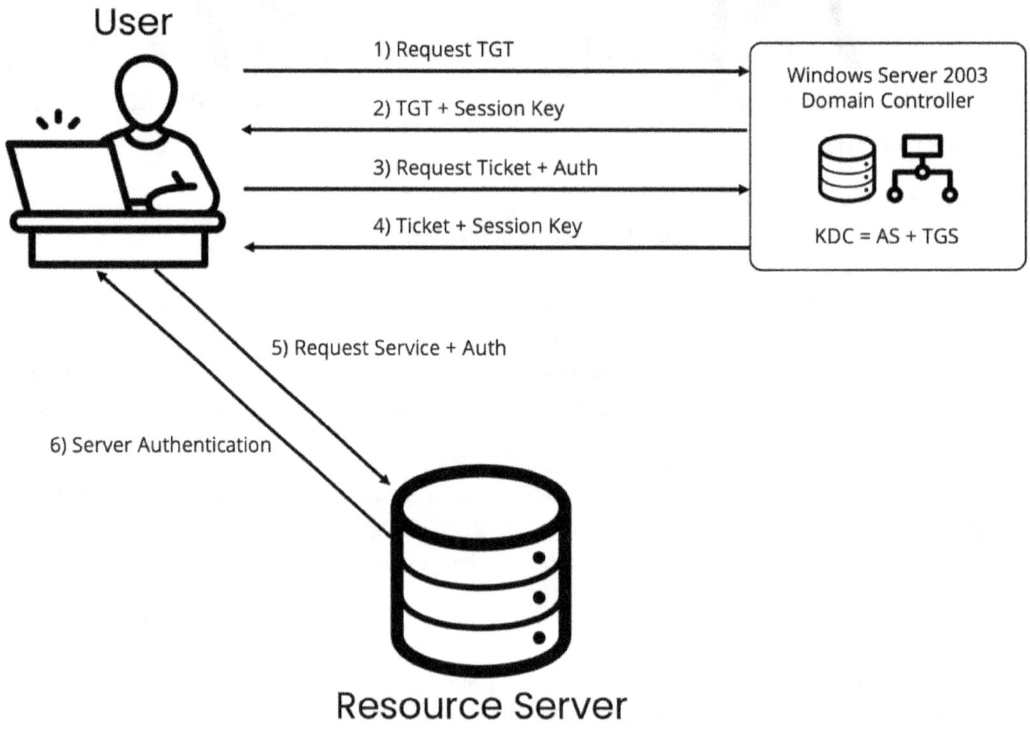

Figure 13.2 – Kerberos ticketing service

How Kerberos works

Consider this—you wish to use a business server to access your email. Instead of just giving your password to the server (which is dangerous), Kerberos steps in to make things safer. It functions as a reliable buddy who stands up for you. This is how it works:

- **Requesting a ticket**: You inform Kerberos that you wish to access your email. Kerberos then offers you a unique ticket, similar to a VIP pass, stating that you are permitted.
- **Authentication server** (**AS**): Kerberos has an authentication server, which acts as a security guard. It checks to see if you're on the guest list and, if so, offers you the ticket.
- **Ticket-granting server** (**TGS**): You now proceed to the ticket-granting server armed with the ticket. It's similar to another checkpoint. You show them your ticket, and they issue you a special ticket (TGT) that allows you to access certain services.
- **Service server**: You eventually go to the service server, where your email is stored, with the TGS-issued ticket or token. They let you in when you show them the TGS-issued token. It's like displaying your VIP card to get to the front row at a concert.

In a nutshell, Kerberos acts as a security guardian, allowing users to safely access services. When you wish to use a server, Kerberos sends you a unique ticket that indicates authorization. The procedure includes authentication servers and ticket-granting servers, which ensure that you have the necessary credentials to access certain services, similar to acquiring VIP tickets for various checkpoints. It improves security by reducing direct password disclosure and serving as a safeguard for digital interactions.

Challenges and pitfalls

Kerberos is a superhero for safe logins, but even superheroes face difficulties. Let's take a look at various Kerberos pitfalls:

1. **Single point of trust**: The **Key Distribution Center** (**KDC**) is the primary figure in Kerberos. If someone gains control of this center, they have the capacity to disrupt the entire authentication process. It's like having a single superhero guarding the city; if they're defeated, the city becomes vulnerable.
2. **Initial authentication vulnerability**: Kerberos issues you a temporary TGT when you initially log in. If someone intercepts this ticket on its way from the authentication server to you, they may be able to pose as you for a brief period. It's the equivalent of a thief stealing your concert VIP pass before you get to the event.

3. **Complicated configuration**: Kerberos isn't the simplest superhero to create. Configuring it across several platforms can be difficult. If not done correctly, it may result in authentication issues or even security flaws. It's like having a very great gadget that's a little difficult to put together.
4. **Not ideal for every scenario**: Kerberos thrives in certain settings but may not be the greatest match in others. Its overhead may feel like utilizing a rocket to travel down the block in smaller settings or less sophisticated networks. Simpler solutions are sometimes more practical.

Kerberos offers an extra layer of security to your digital activities with its VIP passes and superhero-like protection. However, it is not invincible, just like any other superhero. Understanding its difficulties allows us to utilize it intelligently, ensuring that our digital borders stay secure and only accessible to the rightful heroes.

OAuth 2 and OIDC

OAuth 2.0 and **OpenID Connect** (**OIDC**) are linked; however, they fulfill different functions in terms of digital authentication and authorization.

OAuth 2.0

OAuth 2.0, a widely used authorization framework, is critical in providing safe and regulated access to resources across several applications. This introduction delves into the core ideas and procedures of OAuth 2.0, offering insight into its role in current authentication and authorization processes:

- **Purpose**: OAuth 2.0 serves largely as an authorization framework. It allows third-party programs to access a user's resources on a resource server (for example, user data on a server) without revealing the user's credentials.
- **Usage**: It is widely used to let a mobile app access a user's data on a social networking site without requiring the user to provide their credentials with the app.
- **Endpoints**: It defines a number of endpoints, including the authorization endpoint, token endpoint, and others, to help the flow of authorization between the user, client (application), and resource server.

OIDC

OpenID Connect (**OIDC**) is a popular authentication layer built on top of OAuth 2.0 that aims to improve user identity verification in online applications. This introduction digs into the fundamentals of OIDC, explaining its function in providing a consistent and secure framework for user identification across several online platforms:

- **Purpose**: OIDC is an identification layer that sits on top of OAuth 2.0. It adds an authentication layer, giving clients a consistent means of getting user identification information.

- **Usage**: OAuth 2.0 handles permissions (access to resources), whereas OIDC handles authentication (verification of the user's identity). It enables programs to determine who the end-user is and collect basic personal information.
- **Endpoints**: It adds new endpoints such as the `UserInfo` endpoint and the ID token to offer identification information with the access token.

Differences between OAuth 2.0 and OIDC

- **Authentications vs authorization**: OAuth 2.0 focuses on permission, which grants access to resources. OIDC expands OAuth 2.0 by including authentication and end-user information.
- **Tokens**: OAuth 2.0 provides an access token for resource access. OIDC introduces an ID token that stores user identity information.
- **Endpoints**: While both employ common APIs, such as the authorization endpoint, OIDC adds authentication and identity-specific endpoints.

Similarities between OAuth 2.0 and OIDC

- **Foundation**: OIDC is developed on top of OAuth 2.0, inheriting its fundamental ideas and concepts
- **Web-based**: Both frameworks are intended for web-based authentication and authorization, making them ideal for current online and mobile applications

In summary, OAuth 2.0 focuses on resource access, whereas OIDC adds an authentication layer, making it more suitable for identity-related applications. They frequently work together to give a comprehensive solution for secure and user-friendly authentication and authorization. OIDC provides an identity layer to OAuth, allowing apps to authenticate user identities and retrieve basic profile information. OIDC improves security by employing OAuth's token-based authorization procedure, which ensures that only authorized users have access to protected resources. Understanding OAuth is critical for understanding the basic concepts of authorization, whereas OIDC extends OAuth to provide a full authentication method, making both protocols necessary for designing safe and trustworthy identification solutions in modern applications. The following section focuses on understanding the OAuth framework, as it's foundational to OIDC.

Understanding the OAuth 2.0 authorization model

An abstract OAuth2.0 flow can be represented by the following diagram:

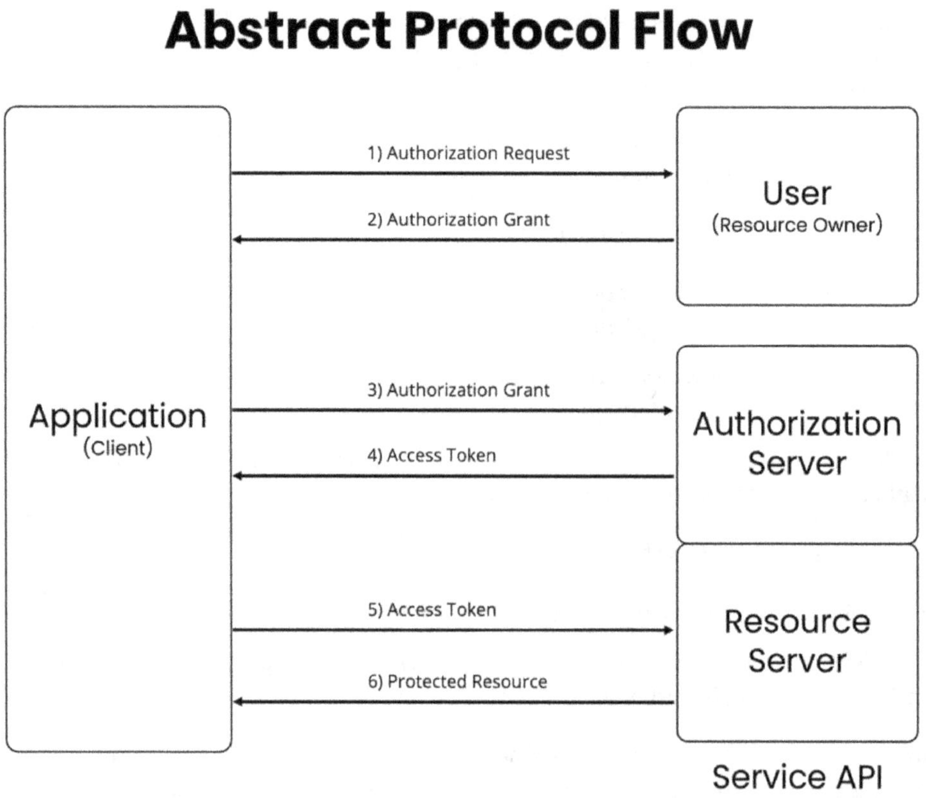

Figure 13.3 – Authorization protocol flow

Assume you have a magical key that can open several doors without disclosing your secret password each time. In the digital world, OAuth 2.0 allows you to access numerous services without providing your password with each app. This magical key, however, comes with its own set of difficulties.

Consider OAuth 2.0 to be a trusted intermediary between you (the user), the app you want to use, and the service you want to utilize. This is how the magic works:

1. **User accesses a web application**: You opt to use an app, such as a music streaming service. Instead of entering your password into the app (which might be dangerous), the app requests access to your account using OAuth 2.0.
2. **Redirect to an OAuth2 provider**: The magic messenger, OAuth 2.0, transfers you to your account provider (such as Google or Facebook). It's similar to the messenger escorting you to the castle where your magical key is stored.

3. **Obtain user consent**: Your provider asks if you're comfortable with the app accessing your account at the castle. If you answer yes, the supplier gives the app a particular key known as a token. This token functions similarly to a key to a certain door; it only unlocks the door you permit.
4. **Grant access**: With the token in hand, the app returns to OAuth 2.0. Now, any time the app needs to access your account (for example, to fetch playlists), it displays the token as proof. It's similar to displaying the magic key to unlock the door to your music.

OAuth 2.0 confronts hurdles even with its magic. Let's look at a few of them:

- **Limited control**: When you provide permission, it is frequently all or nothing. OAuth 2.0 may not provide fine-grained control if you want an app to access one portion of your account, such as viewing your playlists but not publishing. It's like giving someone full access to your castle when all you wanted them to see was the library.
- **Token lifespan**: Tokens have a life cycle. When they expire, the app must obtain a new one. This ongoing renewal procedure, like replacing your magic key on a regular basis, adds a degree of intricacy.
- **Security risks**: Token mishandling may be dangerous. Someone who obtains your token has access to your account. As a result, it is critical that programs handle these tokens with caution. Consider the possibility of someone snatching your magical key and utilizing it to open your doors.
- **Provider dependency**: OAuth 2.0 is based on account providers (such as Google or Facebook). If these providers have problems, you may be unable to utilize apps that rely on OAuth 2.0. It's almost as if your magical key only works when the castle's guards are on duty.
- **Deployment and development complexity**: OAuth 2.0 implementation in apps might be difficult for developers. The procedure has several phases and needs careful management of tokens. It's like putting together a complicated puzzle: if one piece is out of position, the whole thing may not function.

The magic key in OAuth 2.0 streamlines how we access services, but recognizing its difficulties allows us to utilize it carefully. Being mindful of possible hazards as we explore the digital realm ensures that our magic keys stay secure, providing access only where and when needed.

Security Assertion Markup Language

Security Assertion Markup Language (**SAML**) is a method of transferring authentication and authorization data between parties that use an XML-based, open-standard approach, most notably the identity supplier and the service provider. SAML is an OASIS Security Services Technical Committee patent.

SAML appears as a heroic guardian in the enormous terrain of digital interactions, enabling safe passage for users across numerous online domains. Consider SAML to be a trustworthy guide that allows you to enter secure domains without providing important information at each gateway. Even in its laudable endeavor, SAML faces problems and dangers that necessitate our comprehension in order to traverse the complexity of this digital voyage. The following UML diagram represents the interactions between the service and the identity provider in a SAML access request flow:

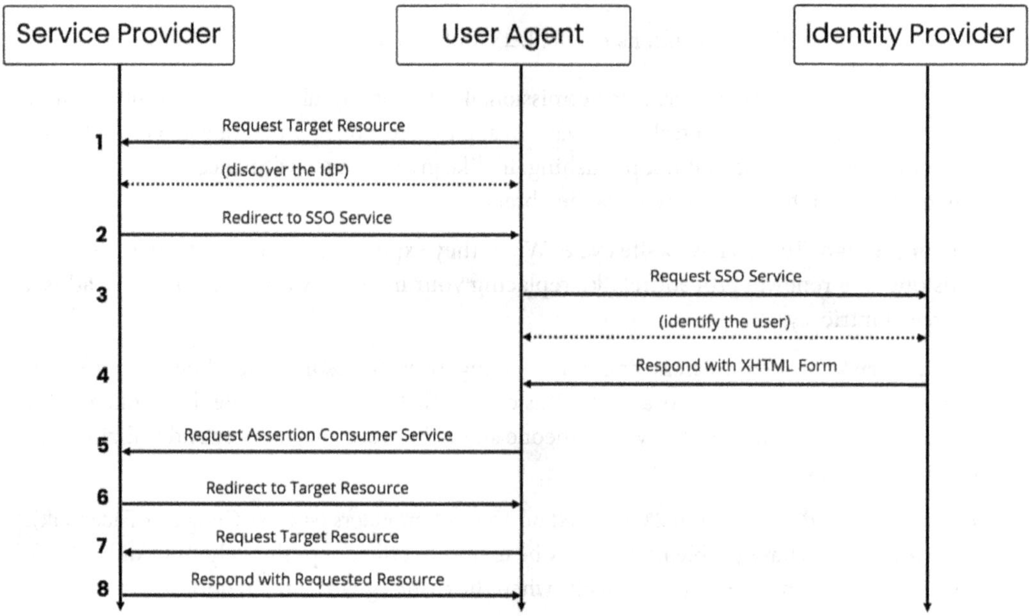

Figure 13.4 – SAML UML flow diagram

Protocols and standards

Users seeking access to protected services go on a trip comparable to a medieval quest in the enchanting region of SAML. Here's how this authentication mechanism works:

1. **Requesting entity**: Consider yourself as a user who wishes to enter a gorgeous castle, which is the service you wish to utilize. You make a modest request to the castle's protector, commonly known as the **identity provider** (**IdP**), rather than storming the gates with a password.
2. **Credential issuance**: The IdP, like a clever wizard, confirms your identity and provides you with a token containing vital information. This credential, similar to a royal seal, confirms your validity without divulging critical information.
3. **Token presentation**: You approach the castle gates now, armed with the token. Instead of disclosing your identity to the service (**service provider**, or **SP**), you submit the token—a mystical emblem attesting to your credibility.

4. **Token verification**: The SP checks the token while working as the castle gatekeeper. The gates swing open, enabling you access to the safeguarded riches within if it bears the unmistakable mark of the IdP, indicating your identity.
5. **Information exchange**: SAML facilitates the safe sharing of information during this route. Your sensitive information is kept safe within the token, away from prying eyes on the treacherous journey across the digital terrain.

Challenges in implementing SAML

While on this laudable effort to safeguard digital identities, SAML faces hurdles that add twists to its narrative. Let us look at the challenges that put our brave guardian to the test:

- **Complex configuration**: SAML, like a mythological spell, necessitates complex setups. Implementing and monitoring these setups can be equivalent to unraveling a mystical spell, especially in complicated systems. Mistakes can cause authentication issues and disrupt the smooth flow of digital experiences.
- **Single point of trust**: SAML is frequently based on a single point of trust—the IdP. If this trustworthy entity fails or is influenced by bad powers, the entire authentication ecosystem is vulnerable. It's like putting the fate of the entire kingdom in the hands of a single guardian, who is vulnerable to external forces.
- **Session management challenges**: SAML controls sessions in order to keep users authenticated throughout their digital journeys. However, efficiently managing sessions across several apps and platforms might be compared to choreographing a complicated dance. Problems may emerge, resulting in unexpected expirations or chronic access issues.
- **Limited standardization**: While SAML complies with certain standards, there are differences in implementations and interpretations. This lack of total standardization is analogous to dialects spoken in various parts of a kingdom, posing possible interoperability issues.
- **Limited user experience**: The major focus of SAML on security might occasionally result in a trade-off with user experience. The redirections and intricate operations involved in the SAML dance may feel like a maze to consumers, potentially generating dissatisfaction.
- **Token exposure risks**: Tokens, while effective in demonstrating identity, might be insecure if handled incorrectly. If an opponent has access to a user's token, they may use it to gain unauthorized access. Defending these tokens becomes as important as defending the kingdom's most valuable artifacts.
- **Evolutionary pressure**: SAML confronts the problem of keeping up with evolving technologies as digital landscapes develop. Keeping up with the changing tides of cybersecurity and user expectations necessitates constant monitoring and improvement, analogous to a never-ending quest.

SAML appears as a noble guardian in the digital sphere, revealing the path to safe authentication. However, its voyage is not without difficulties and perils. Understanding the complexities of SAML's mission allows us to approach these difficulties sensibly, ensuring that our digital kingdom stays safe, accessible, and resilient in the face of increasing threats and ambitions.

The need for protocols and standards

Protocols and standards for the digital society are needed for several reasons:

- **Interoperability**: In a digital society, there are many different systems and devices that need to communicate with each other. Protocols and standards help ensure that these systems can work together seamlessly, regardless of the manufacturer or vendor.
- **Security**: Digital systems are vulnerable to a wide range of security threats, such as hacking, malware, and phishing attacks. Protocols and standards help establish best practices for security and help ensure that systems are designed to protect against these threats.
- **Privacy**: With the proliferation of digital data, there are increasing concerns about the privacy of personal information. Protocols and standards can help ensure that personal data is collected, stored, and used in a way that respects individuals' privacy rights.
- **Efficiency**: Digital systems can help streamline and automate many processes, but this requires standardized protocols to ensure that data is transmitted and processed efficiently.
- **Innovation**: Protocols and standards can help drive innovation by providing a common foundation that developers can build upon. This can help reduce the duplication of effort and enable new ideas and applications to emerge.

Overall, protocols and standards are essential for creating a digital society that is secure and efficient and respects individuals' privacy rights. Without them, digital systems would be more fragmented, less secure, and less interoperable, which would limit their potential to improve our lives and solve important societal challenges.

DID authentication protocols

DID ushers in a paradigm shift in how we create and maintain digital identities, ushering in a more user-centric and secure approach. Each DID user has a unique identification, similar to a digital passport. These DIDs are not attached to any central authority, but rather to decentralized networks, which are frequently aided by blockchain technology.

Consider DIDs to be digital keys that users possess, providing them with power over their online identities. Traditional identification systems are analogous to having a master key held by a central authority, requiring users to rely on this authority for verification. DIDs, on the other hand, provide people with their own keys, allowing them to open multiple digital doors without the need for continual permission from a central body.

Authentication in the domain of DIDs is a complex dance of protocols that ensures only the proper key holder has access to certain digital worlds. The **decentralized key management system (DKMS)**, which orchestrates the development, storage, and use of cryptographic keys associated with DIDs, is one significant protocol in this dance.

The DKMS may be thought of as a digital locksmith. When a user establishes a DID, the DKMS generates a pair of cryptographic keys: a public key, which acts as the visible lock, and a private key, which acts as the secret key that unlocks the door. The public key is shared with others so that they may verify the user's identity, while the private key is kept secure by the user.

The authentication protocol, a collection of rules defining how users confirm ownership of the private key, now enters the picture. The **Elliptic Curve Digital Signature Algorithm** (**ECDSA**), which is similar to the unique signature a person uses to authenticate papers, is one frequently used technique. When a user needs to access a service, they use their private key to sign a digital message, establishing a cryptographic proof. The service may validate this proof using the user's public key, guaranteeing that only the legitimate owner of the private key obtains access.

There's also the verifiable credentials protocol in the DID authentication dance, which allows users to give their digital credentials without disclosing unneeded information. It's the equivalent of handing a bouncer at a bar a digital ID card proving your age without exposing your precise age or home location. This protocol protects users' privacy by allowing them to choose to divulge particular claims, resulting in a more complex and secure authentication procedure.

While the dance of DID authentication protocols has a bright future, it is not without its difficulties. Scalability, for example, is an important factor. As the number of DIDs increases, creative solutions are required to provide a smooth and effective authentication procedure. The establishment of standards and best practices becomes critical, analogous to setting dance rules.

Another difficulty in the DID authentication landscape is interoperability. To enable a smooth flow of authentication information, different systems and applications must communicate in a standard language. Consider a dance in which each participant follows their own set of steps; anarchy would result. By establishing universal standards, DIDs may be authenticated smoothly across several systems.

Furthermore, usability is an important consideration. The dance must be user-friendly in order for DID authentication to be widely used. Individuals should be able to maintain their DIDs and negotiate the authentication processes with ease without having to execute difficult choreography. This necessitates user-centered design and simple interfaces.

Looking ahead, continued DID authentication research and development promises ever more robust and user-friendly procedures. Zero-knowledge proofs, which allow users to verify ownership of certain information without exposing the information itself, add an added element of intricacy to the dance. Like dancers using masks, advances in privacy-preserving technology attempt to improve the secrecy of the authentication process.

Finally, the realm of DID authentication protocols is an enthralling dance in which consumers control their digital identities. It entails the coordination of protocols such as DKMS, ECDSA, and verifiable credentials in order to provide a safe and user-centric authentication experience. Addressing difficulties and embracing advances will define the future landscape of DID authentication, paving the way for a more secure and empowered digital identity frontier as this dance progresses.

Implementing DID authentication

DID authentication is a paradigm change, reinventing how people assert and control their digital identities. DID authentication, at its heart, employs decentralized IDs, blockchain technology, and cryptographic protocols to create a safe, privacy-preserving, and user-centric authentication experience. The following three essential components are transforming the landscape of contemporary authentication:

- **Decentralized identifiers**: DIDs provide the foundation of this new authentication paradigm. They are distinct IDs embedded in decentralized networks, which are frequently assisted by blockchain technology. DIDs, as opposed to standard usernames that are related to centralized entities, are self-owned and managed by the individual.
- **Cryptographic protocols**: DID authentication employs public-key cryptography in its cryptographic dance. Users have two cryptographic keys: a public key that is publicly shared and a private key that is securely maintained by the user. Cryptographic systems such as ECDSA ensure safe authentication by validating the ownership of these keys.
- **Verifiable Credentials**: Verifiable credentials supplement the DID authentication process by allowing users to submit digital credentials while avoiding unwanted information sharing. Users can communicate certain claims selectively, increasing privacy while developing trust in the legitimacy of their identity.

The adoption of DID implies a radical change toward user-centric and safe authentication in the changing world of digital identity. Implementing DID authentication entails traversing a complicated ecosystem of approaches and tactics, which frequently use cryptographic protocols, decentralized networks, and novel verification processes. This section digs into the fundamental concepts, methodology, and case studies that explain the way to a decentralized and user-controlled digital identity frontier.

Core methodologies in DID authentication

The idea of DIDs is central to DID authentication. Individuals in the digital sphere use DIDs as unique and self-owned identifiers. Because these identities are not related to any central authority, people have complete control over their digital appearance. DID implementation entails the creation and administration of these unique identifiers, which are frequently assisted by blockchain technology.

In practice, creating a DID requires the use of DKMS. DKMS is in charge of producing a pair of cryptographic keys, one public and one secret. The public key is freely distributed and serves as the user's digital address, while the private key is kept safely in the user's possession. This process assures that the person retains ownership and control of the DID.

During the authentication process, cryptographic techniques play a critical role in authenticating the validity of DIDs. ECDSA is a crucial protocol in this domain. It allows a user to use their private key to sign digital communications, resulting in cryptographic evidence of ownership. This proof may be validated by reliant parties that have access to the user's public key, guaranteeing that only the rightful owner has access.

The verifiable credentials protocol extends the cryptographic dance by offering a means for presenting digital credentials while avoiding excessive information exposure. This selective sharing protects privacy while adhering to self-sovereign identification standards. Users can show valid credentials, similar to digital ID cards, while maintaining control over the specifics of the information given.

It has become difficult to provide verified information online, such as proof of age, financial information, healthcare data, and educational qualifications. The architecture provided here seeks to overcome this issue by defining long-term goals for DID verifiable assertions. These objectives include giving claim holders entire control over their identities and related claims, improving website usability, decreasing fraud by standardized qualifying transfer, and assuring maximum privacy in claim sharing. The architectural goals include separating control over identifiers and claims, distinguishing control over claims sharing from claim production, setting standards for cross-role interactions, and using existing protocols where appropriate. Key language includes *claim* as a declaration for the topic, *subject* as the entity showing the claim, and *verifiable claim*, indicating that the authenticity, integrity, and non-repudiation of the authorship can be verified.

A set of verifiable claims is made up of four parts, as shown in the following figure:

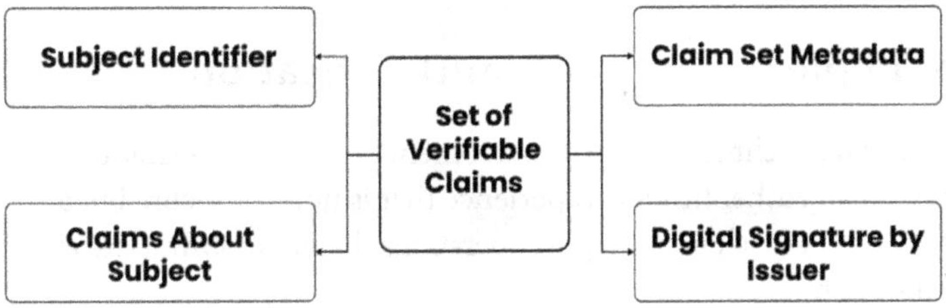

Figure 13.5 – Verifiable claims components

A key framework for verifiable claims must differentiate the critical roles of core players and their connections—how they interact. Roles can be construed in a variety of ways, and identifying them reveals potential limits and standardized standards. The DID Verifiable Claims Architecture consists of the following roles:

- **Holder**: Acquire credible claims from the issuer and selectively distribute them to inspectors. The holder is frequently the subject of the claims.
- **Issuer**: Issue credible claims to the claim holders.
- **Inspector**: Request credible claims from holders to authenticate them.
- **Identifier registry**: Mediate the creation and verification of globally unique IDs.

The following diagram depicts a verifiable claims model:

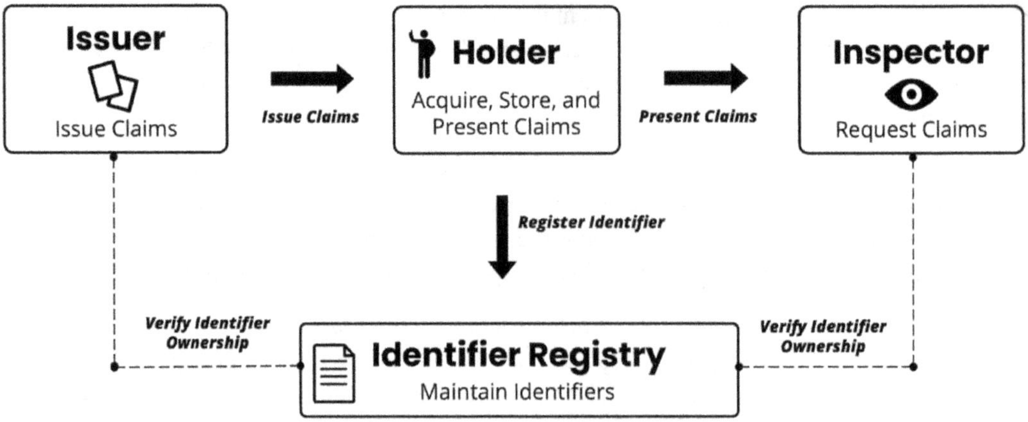

Figure 13.6 – Verifiable claim model

In summary, the basis of DID authentication is DIDs, which provide people with unique, self-owned digital identities that are independent of central authority. DIDs, when combined with cryptographic keys and protocols such as ECDSA, provide ownership control, while the verifiable credentials protocol improves privacy by allowing for selective distribution of digital credentials. The architecture tackles the issues of delivering verified online information by establishing long-term objectives for DID-verifiable assertions that empower users, improve usability, decrease fraud, and optimize privacy.

Strategies for implementing DID authentication

User-centric design is a vital technique in the development of DID authentication. The objective is to provide consumers with an authentication experience that is not just secure but also accessible and straightforward. The complexity of cryptographic operations should be abstracted, allowing consumers to participate in a smooth manner.

Real-world implementations, such as those shown in Sovrin, place an emphasis on user-centricity. DIDs and verifiable credentials are used by Sovrin, a worldwide public utility for self-sovereign identification, to enable users to maintain and share their digital identities. The approach here is to create a user-friendly interface that allows people to interact with their DIDs without being bogged down in the complexities of cryptographic protocols.

Interoperability is a critical technique for DID authentication's widespread adoption. DIDs should be able to interact fluidly across several systems and apps, offering a universal and consistent authentication experience. Standards and best practices become the pivotal point in promoting interoperability.

The **Decentralized Identity Foundation** (**DIF**) is at the forefront of standards development in the DID domain. DIF focuses on developing standards that allow diverse systems to communicate in a common language. The DID communication protocol, for example, provides a standard for safe communication between DID entities. This standardization makes it easier to integrate DIDs into a wide range of applications and services.

Privacy concerns are critical in the deployment of DID authentication. Users should be able to manage the information they reveal during authentication processes. In this setting, zero-knowledge proofs emerge as a strategic tool, allowing users to demonstrate ownership of certain information without exposing the information itself.

The Aztec protocol is a real-world example of the integration of privacy-preserving technology. While largely used in financial transactions, Aztec's use of zero-knowledge proofs demonstrates the possibility of improving privacy in a variety of digital interactions. Using similar tactics in DID authentication allows users to assert their identities while protecting critical information.

Real-world examples and case studies

Microsoft has been a trailblazer in the exploration and implementation of decentralized identification systems. Their work with the DIF and contributions to the open source project *Ion* demonstrate their dedication to the topic.

DIDs and Sidetree, a scaling solution for decentralized identifiers, are used by Ion, a decentralized identity network built on the Bitcoin blockchain. Creating a decentralized and interoperable identification network that transcends traditional silos is the approach here. Microsoft's venture into decentralized identification demonstrates the practical implementation of DIDs in a real-world setting, with an emphasis on scalability and broad acceptance.

Sovrin, a key decentralized identity player, acts as a worldwide public utility for self-sovereign identification. The Sovrin Network makes it easier to issue and verify DIDs and verifiable credentials while emphasizing user autonomy and privacy.

The idea of Sovrin's implementation is to build a strong and scalable network for decentralized identification. Sovrin empowers users to manage their digital identities freely by utilizing DIDs and verifiable credentials. The network's design is consistent with self-sovereign identity principles, offering a look into the possibility of decentralized identity ecosystems.

uPort, a blockchain-based identity platform, is an example of DID authentication in action, providing users with portable identities. To promote safe and user-controlled identification interactions, uPort employs Ethereum-based DIDs and smart contracts.

uPort's strategy focus is on providing users with portable and interoperable identities. uPort guarantees a decentralized and tamper-resistant basis for DIDs by leveraging Ethereum's blockchain. This solution demonstrates blockchain technology's adaptability in supporting DID authentication approaches.

1Kosmos BlockID, a trailblazing solution that redefines identity verification via the lens of decentralized principles, has been added to the roster of real-world instances in the landscape of DID authentication. BlockID incorporates DIDs into its structure, prioritizing user autonomy and privacy. 1Kosmos BlockID, by using DIDs, provides individuals with a self-owned digital identity that crosses traditional borders, providing a safe and efficient way of authentication. The 1Kosmos BlockID implementation approach is heavily anchored in preserving the privacy and security of user identities. The platform makes use of cutting-edge cryptographic protocols that adhere to the concepts of self-sovereign identification. Users retain control over their DIDs, instilling more confidence and security in the authentication process.

1Kosmos BlockID prioritizes user-centric design by abstracting the intricacies of cryptographic procedures for a seamless and intuitive user experience. The approach relies on making DID authentication accessible to a large user base, as well as demystifying the complexities of decentralized identity for users across several domains. As an example, 1Kosmos BlockID is making significant progress in industries where identity verification is crucial, such as labor management and access control. The platform's application demonstrates how DID authentication may improve security, privacy, and efficiency in real-world applications.

1Kosmos BlockID is a testament to the shifting environment of identity verification, where user-centric design, privacy, and security are vital, by carefully incorporating decentralized identity concepts. This example adds to the larger story of how DID authentication is actively affecting the future of digital identities across businesses.

DID authentication implementation examples

Of the several use cases and examples mentioned previously here are a few implementation examples of DID authentication. Implementing login with Ethereum, BlockID, and WalletConnect necessitates the use of many protocols and technologies.

Logging in with Ethereum

When logging in using Ethereum, DIDs and authentication methods based on Ethereum are commonly used. User authentication is often performed using Ethereum wallets such as MetaMask.

Implementation method:

1. **Integrate web3.js**: Make sure your app has `web3.js`, a JavaScript library for interfacing with the Ethereum blockchain:

```
<script src="https://cdn.jsdelivr.net/npm/web3@1.5.2/dist/web3.
min.js"></script>
```

2. **Request an Ethereum wallet address**: Encourage people to link their Ethereum wallets using MetaMask or a comparable service, as follows:

```
const ethereumButton = document.querySelector('.
enableEthereumButton');
```

```
ethereumButton.addEventListener('click', () => {
  getEthereum();
});

async function getEthereum() {
  if (window.ethereum) {
    try {
      // Request account access
      await window.ethereum.request({ method: 'eth_
requestAccounts' });
    } catch (error) {
      console.error('Error requesting Ethereum access:', error);
    }
  } else {
    console.error('No Ethereum provider detected. Please install
MetaMask.');
  }
}
```

3. **Retrieve Ethereum address**: After successful authentication, obtain the user's Ethereum address, like so:

```
async function getUserAddress() {
  const accounts = await window.ethereum.request({ method: 'eth_
accounts' });
  const userAddress = accounts[0];
  console.log('User Ethereum Address:', userAddress);
}
```

In conclusion, integrating login with Ethereum not only improves security through blockchain-based authentication but also corresponds with the decentralized ethos, delivering a strong and trustworthy user experience in the digital domain.

Logging in with BlockID

BlockID is a solution for authentication that leverages DIDs. The implementation may differ depending on the BlockID platform's individual APIs and SDKs.

Implementation method:

1. **Integrate BlockID SDK**: Include the BlockID SDK as part of your project:

```
<!-- Example script tag; actual integration may vary based on
BlockID documentation -->
<script src="https://blockid-sdk.example.com/sdk.js"></script>
```

2. **Initialize BlockID**: Set up the BlockID SDK using your application credentials:

```
const blockID = new BlockID({
  apiKey: 'your-api-key',
  // other configurations...
});
```

3. **Authenticate with BlockID**: Create a function to start the BlockID authentication procedure:

```
async function loginWithBlockID() {
  try {
    const userData = await blockID.authenticate();
    console.log('User Data from BlockID:', userData);
  } catch (error) {
    console.error('BlockID Authentication Error:', error);
  }
}
```

To summarize, combining login with BlockID not only strengthens security measures but also reflects the ideals of decentralized identity management, giving users a safe and simplified authentication experience across digital platforms.

Logging in with WalletConnect

WalletConnect is a protocol that allows decentralized apps to communicate with mobile wallets. Users may engage with DApps on their mobile devices using their favorite wallets.

Implementation method:

1. **Integrate WalletConnect SDK**: WalletConnect SDK should be included in your project:

```
<!-- Example script tag; actual integration may vary based on
WalletConnect documentation -->
<script src="https://cdn.walletconnect.org/1.3.3/walletconnect.
min.js"></script>
```

2. **Initialize WalletConnect**: Set up WalletConnect using your application information:

```
const connector = new WalletConnect({
  bridge: 'https://bridge.walletconnect.org', // replace with
your bridge URL
});
```

3. **Initiate wallet connection**: Create a method that starts the connection between your app and the user's wallet:

```
async function connectWithWallet() {
  if (!connector.connected) {
```

```
        await connector.createSession();
    }
}
```

4. **Handle WalletConnect events**: To manage the connection and user interactions, listen for events:

```
connector.on('connect', (error, payload) => {
  if (error) {
    throw error;
  }
  // Handle connection success
});

connector.on('session_update', (error, payload) => {
  if (error) {
    throw error;
  }
  // Handle session update
});
```

Finally, the WalletConnect integration procedures open the way for a safe and seamless link between decentralized applications and users' mobile wallets, improving the user experience and supporting the blockchain ecosystem's user-controlled identification principles.

Paving the way for a decentralized identity frontier

The old methodology of username and password authentication is rapidly proving unsuitable in the ever-changing context of digital security. Passwords are vulnerable to a variety of security concerns, including breaches, phishing attempts, and ineffective credential management practices. The search for a more secure and user-friendly authentication paradigm has resulted in the rise of DID authentication, a game-changing technique that holds the promise of ushering in a password-free society.

DID authentication prioritizes user-centric design, with the goal of simplifying the authentication process. Users begin by generating their DIDs and controlling their cryptographic keys, giving them authority over their digital identities. The following code snippet demonstrates the use of a `DIDAuthentication` class to manage DIDs in a system:

```
const didAuth = new DIDAuthentication();
const userDid = didAuth.createUserDID();
const userPublicKey = didAuth.getUserPublicKey(userDid);
```

DID authentication is built with privacy in mind. Users can selectively reveal information using verified credentials, limiting the danger of oversharing sensitive information. The emphasis on user permission and minimum disclosure is consistent with privacy-by-design concepts. The following code snippet demonstrates the creation and signing of a verifiable credential using a DID and a `DIDAuthentication` class:

```
const verifiableCredential = {
  '@context': 'https://www.w3.org/2018/credentials/v1',
  type: ['VerifiableCredential', 'EmailCredential'],
  issuer: userDid,
  issuanceDate: new Date().toISOString(),
  claim: {
    email: 'user@example.com',
  },
};
const signedCredential = didAuth.signCredential(verifiableCredential,
userDid);
```

DID authentication reduces phishing threats by transferring authentication from centralized entities to user-controlled DIDs. As users authenticate directly with their decentralized IDs, phishing attempts targeting centralized login credentials become outdated. The following code snippet depicts an authentication method with a DID by leveraging a QR code:

```
const qrCodeData = didAuth.generateAuthenticationQRCode(userDid);
const userScansQRCode = didAuth.scanAuthenticationQRCode(qrCodeData);
```

DIDs and their associated credentials benefit from the immutability and tamper-resistant features of the decentralized ledger when using blockchain technology. The blockchain's distributed and trustless nature improves the overall security of the authentication process. The following code snippet depicts a method to store a user's DID on the blockchain:

```
const blockchainTransaction = didAuth.storeDIDOnBlockchain(userDid);
```

While the concept of a password-free society driven by DID authentication is appealing, limitations and acceptance barriers must be recognized and solved. The lack of standardized protocols, as well as compatibility issues, might stymie wider adoption. Collaboration between industry standards groups, as well as the creation of interoperable specifications, is critical. To move users from familiar but unsafe password-based systems to DID authentication, strong education and user-friendly onboarding experiences are required. To manage a worldwide user population effectively, the scalability and performance of DIDs and associated blockchain networks must be continuously improved.

Continuous advances and initiatives define the future of digital identification as the DID authentication environment changes. Zero-knowledge proofs, advances in cryptographic algorithms, and improved user experiences all contribute to the ongoing development of DID authentication:

- **Zero-knowledge proofs**: Users can use zero-knowledge proofs to demonstrate ownership of specified information without exposing the information itself. The use of zero-knowledge proofs improves DID authentication privacy and secrecy.
- **Enhanced user experience**: Usability enhancements, intuitive interfaces, and user-friendly interactions are critical for DID authentication's widespread acceptance. Ongoing developments in decentralized identity systems are aimed at improving the overall user experience.
- **Industry collaboration**: Collaboration among industry players and standardization organizations is critical in influencing the future of DID authentication. The establishment of universal standards and protocols guarantees that the ecosystem is coherent and compatible.

DID authentication appears as a lighthouse, blazing the route to a password-free society. DID authentication reshapes the mechanisms of digital identity by emphasizing user autonomy, privacy, and security. Overcoming difficulties, developing industry collaboration, and accepting continuous developments are all part of the route to universal acceptance. As the world moves away from traditional authentication techniques, the goal of a password-free society powered by DID authentication is becoming a reality, providing a more secure, user-centric, and privacy-preserving digital future.

Security and privacy considerations

DID authentication has emerged as a revolutionary paradigm in the quickly expanding environment of digital identification, offering greater security, privacy, and user control. This radical transition, however, comes with its own set of obstacles and considerations. Next, we delve into the complicated tapestry of security and privacy issues in DID authentication in this detailed examination, unraveling the methods and techniques that enable the protection of the digital identity frontier.

The pillars of DID security

The resilience of cryptographic concepts is at the basis of DID security. Not only does cryptography enable secure communication but it also serves as the foundation for the creation, maintenance, and validation of DIDs and related credentials.

The usage of public and private key pairs provides users with a safe way to prove ownership of their DIDs. The identification is provided by public keys, while the user's identity is validated using private keys. The cryptographic strength of these keys, which are frequently implemented using techniques such as ECDSA, strengthens the basis of DID security.

The following is an example implementation:

```
const didAuth = new DIDAuthentication();
const userDid = didAuth.createUserDID();
const userPublicKey = didAuth.getUserPublicKey(userDid);
```

Decentralization is an essential concept of DID authentication since it eliminates the possibility of a single point of failure. Traditional authentication solutions sometimes depended on centralized databases, making them appealing to bad actors. The risk of large-scale breaches is reduced when DIDs are kept on decentralized networks.

Blockchain is the foundation of decentralized identity ecosystems. The blockchain's immutability and distributed nature increase the security of DIDs and related credentials. Each transaction on the blockchain, such as the creation or change of a DID, is cryptographically protected, which contributes to the system's tamper-proof character.

The following is an example implementation:

```
// Storing user DIDs on a blockchain
const blockchainTransaction = didAuth.storeDIDOnBlockchain(userDid);
```

Finally, the pillars of DID security provide a solid foundation for a trustworthy and user-centric decentralized identification system. DID security ensures control, privacy, cryptographic integrity, and verifiable credentials, paving the path for a secure digital ecosystem in which individuals may confidently manage and validate their identities.

Privacy-first design

Privacy concerns are woven into the very structure of DID authentication. Verifiable credentials play a critical role in allowing users to selectively disclose information, allowing them to offer exactly the facts they need without jeopardizing their entire identity.

Users can issue verified credentials containing certain assertions, such as age or affiliation, without disclosing any other information. This limited disclosure principle guarantees that consumers keep control over their personal data.

The following is an example implementation:

```
const verifiableCredential = {
  '@context': 'https://www.w3.org/2018/credentials/v1',
  type: ['VerifiableCredential', 'EmailCredential'],
  issuer: userDid,
  issuanceDate: new Date().toISOString(),
  claim: {
    email: 'user@example.com',
  },
```

```
};
const signedCredential = didAuth.signCredential(verifiableCredential,
userDid);
```

A fundamental element of DID authentication is user empowerment in controlling their DIDs. Users may create, amend, and cancel their DIDs, ensuring that their digital identity stays entirely under their control.

The user-centric design prioritizes the provision of interfaces and experiences that enable users to manage their DIDs effortlessly. The adoption of user-managed DIDs is supported by offering intuitive tools and interfaces.

The following is an example implementation:

```
const didAuth = new DIDAuthentication();
const userDid = didAuth.createUserDID();
```

In conclusion, the strength of cryptographic principles supports the security of DIDs, providing not only safe communication but also the integrity of identity management. Users may firmly declare their ownership of DIDs by using public-private key pairs, while blockchain technology facilitates decentralization, which reduces the dangers associated with centralized systems, creating a trustworthy and user-centric decentralized identity ecosystem.

Security challenges

While the decentralized nature of DIDs is a virtue, it also creates standardization and interoperability issues. The lack of common protocols might make it difficult for different decentralized identification systems to communicate with one another.

To build universal protocols, active collaboration between industry standards and organizations is required. Initiatives such as DIF help to produce standards that facilitate interoperability.

Credential revocation and expiration are important aspects of DID security. To avoid the abuse of obsolete or hacked credentials, ensure that credentials are revoked when appropriate and have established expiry periods.

Implementing revocation lists or registries enables users and issuers to communicate the status of credential revocation. Periodic checks against these lists aid in the rapid identification and invalidation of compromised credentials.

This is an example implementation:

```
const didAuth = new DIDAuthentication();
const credentialRevocation = didAuth.revokeCredential(userDid,
credentialId);
```

Security is like constructing an impenetrable castle for your secrets in the enchanting land of digital identity. However, every fortress has its own set of difficulties. Let us take a voyage inside this beautiful domain to better grasp the difficulties and solutions.

Consider your castle to have two magical keys, one that everyone knows and one that only you have. These keys function similarly to the keys of your digital castle. The problem is ensuring that these keys are truly magical and cannot be reproduced. We want to make sure that only the proper owner (you) has access to the castle gates.

The castle will not open if a fraudulent key is used!

Instead of one large castle where everyone's secrets are held, our kingdom now has several smaller castles (similar to little forts). This is an example of decentralization. However, having several castles creates a challenge: they must all communicate smoothly with one another. Consider each castle speaking a similar language so that they may collaborate. This language is responsible for ensuring the security of our digital secrets, regardless of which castle they are held in.

Even if one castle collapses, the others hold firm!

Our digital domain is teeming with cunning animals—hackers out to steal our precious keys. The difficulty is keeping them out. We want magical barriers (firewalls and defenses) capable of detecting and stopping these entities. Consider it similar to having guards that patrol your fortress at all hours of the day and night to ensure no one slips in.

The guards raise the alarm whenever a devious monster tries to scale the walls!

Our kingdom is large, with several lands (platforms and services). The problem is ensuring that all of these countries comprehend each other's rules. We want to make sure that when we go from one place to another, our magical keys and secrets continue to function properly.

When you go to a friend's castle, your magical key should also open the gates!

To keep our secrets hidden in our magical realm, we utilize spells (cryptography). The challenge is to make these spells indestructible. We don't want any witches or wizards attempting to decode our mystical signals.

Even if someone hears, they will be unable to comprehend the secret magic we are employing!

Overcoming these hurdles in the enchanted journey of digital identification involves creating a kingdom where everyone's digital secrets are safe and only the genuine owners have the keys to their magnificent castles. So, let the adventure continue as we work to make our digital domain the safest and most secure place on the internet for everybody!

Privacy challenges

The deployment of DID authentication faces major challenges in terms of user education and getting informed permission. Users must understand the concepts of DID and actively participate in the administration of a user's DIDs.

Creating user-friendly onboarding experiences, in conjunction with training resources, helps to bridge the knowledge gap. This involves assisting users through the creation of DIDs, key management, and comprehending the consequences of verified credentials.

Scalability and performance become increasingly important as the user base for decentralized identification systems expands. It is a constant problem to ensure that DIDs and associated networks can manage a worldwide user population without sacrificing speed.

Continuously optimizing blockchain networks, investigating scalability options such as sharding, and exploiting layer-2 solutions all help to overcome scalability issues.

The following is an example implementation:

```
// Optimizing blockchain network for scalability
const didAuth = new DIDAuthentication();
const optimizedNetwork = didAuth.optimizeBlockchainNetwork();
```

Ongoing breakthroughs in zero-knowledge proofs offer the prospect of improving DID authentication privacy even more. Users can use zero-knowledge proofs to confirm the validity of information without exposing the information itself.

Using zero-knowledge proofs, users may choose to prove statements without revealing the real facts. This guarantees that anonymity is preserved even in instances when certain traits must be proven.

The following is an example implementation:

```
const didAuth = new DIDAuthentication();
const zeroKnowledgeProof = didAuth.generateZeroKnowledgeProof(userDid,
claim);
```

The continuous improvement of user experiences in decentralized identification systems drives mainstream adoption. Efforts to improve user experiences include efforts to speed up user interactions, simplify key management, and seamlessly integrate into current systems.

User adoption is increased by creating intuitive interfaces that abstract complicated cryptographic operations and give explicit instructions on DID management. User testing and feedback are critical.

The emphasis on user agreement and the notion of data minimization are critical features of decentralized identification systems. While these notions help to protect privacy, they also bring complex issues with regard to achieving the correct balance between user control and sensitive data protection.

In order to solve privacy concerns, granular consent procedures must be implemented. Users must be able to offer explicit authorization for certain data exposures in order for their preferences to be honored in a variety of authentication circumstances.

The following is an example implementation:

```
const didAuth = new DIDAuthentication();
const userConsent = didAuth.requestUserConsent(userDid, dataTypes);
```

As decentralized identification systems arise, they must live with legacy systems that may not follow the same privacy rules. Bridging the gap between decentralized identification and traditional systems creates privacy problems during cross-system interactions.

It is a huge difficulty to provide privacy-bridging solutions that allow smooth yet privacy-respecting communication between decentralized identification systems and older platforms. Encapsulating and translating privacy-preserving techniques for interoperability is required.

The following is an example implementation:

```
const didAuth = new DIDAuthentication();
const privacyBridge = didAuth.createPrivacyBridge();
```

Maintaining user identification in certain settings while maintaining overall anonymity is a complex privacy topic. Finding the correct balance guarantees that users may engage in attribution-required services without jeopardizing their identity throughout the whole digital landscape.

Implementing pseudonymous identifiers allows users to interact with ascribed identities while being anonymous overall. This necessitates careful study in order to avoid unintentional linkage between pseudonymous and totally anonymous interactions.

The following is an example implementation:

```
const didAuth = new DIDAuthentication();
const pseudonymousIdentifier = didAuth.
generatePseudonymousIdentifier(userDid, context);
```

Inadvertent data leakage can occur in the decentralized identity ecosystem owing to a variety of circumstances, including poor credential usage or inadvertent exposure. Mitigating the risk of unintentional data leakage necessitates effective procedures for detecting, preventing, and correcting unintended disclosure.

Organizations and users can track and audit the use of credentials by using runtime monitoring and auditing technologies. Automated warnings and real-time monitoring help to spot irregularities quickly, allowing for quick action to minimize inadvertent exposure.

The following is an example implementation:

```
const didAuth = new DIDAuthentication();
const runtimeMonitoring = didAuth.setupRuntimeMonitoring(userDid);
```

These solutions are critical in supporting a privacy-first strategy in negotiating the complicated landscape of privacy problems in DID authentication. Continuous refinement and flexibility to emerging issues will be critical in realizing the goal of a safe, private, and user-centric digital identity environment as decentralized identity systems emerge.

Ongoing developments and initiatives

The world of digital identification and passwordless authentication is constantly changing, with a slew of ongoing breakthroughs and ground-breaking efforts promising to reshape how we safeguard and maintain our online personas. As we go through the maze of innovation, many major themes emerge, each adding to the story of a more secure, user-centric, and privacy-respecting digital world.

Pioneering zero-knowledge proofs

The incorporation of zero-knowledge proofs, cryptographic miracles that allow users to confirm the validity of particular information without disclosing the information itself, is at the forefront of continuing advances. This approach fits in perfectly with the decentralized identity ethos, where anonymity is vital. Individuals may show possession of specified qualities without releasing the actual data by using zero-knowledge proofs, providing an unsurpassed degree of privacy guarantee. Initiatives are underway to further improve and standardize the use of zero-knowledge proofs across decentralized identity systems, guaranteeing a consistent approach to privacy-preserving interactions.

Advancements in cryptographic techniques

As digital security guards continue to harden the walls against possible intrusions, continuing efforts focus on developing cryptographic algorithms. Post-quantum cryptography, homomorphic encryption, and secure multi-party computation are establishing themselves as sturdy guardians of digital secrets. These cryptographic advances not only increase the security of decentralized identification systems but also open the door for more robust and resilient passwordless authentication procedures. The continuing research of cryptographic frontiers seeks to keep us one step ahead of possible dangers, offering a defense against an ever-changing world of cyber enemies.

Usability and user experience refinement

While cryptographic concepts serve as the backbone of decentralized identification, the continuous narrative lays a strong emphasis on improving usability and user experiences. Intuitive interfaces, simple onboarding procedures, and seamless integration with current apps are all becoming increasingly important. Initiatives are being launched to simplify the often-complex realm of cryptographic key management, guaranteeing that users, regardless of technical skill, can browse and handle their decentralized identities with ease. The objective is to close the gap between cryptographic robustness and user-friendly interactions, hence encouraging wider use and acceptance.

Industry collaboration and standardization

A common language is required for the harmonic symphony of decentralized identity, and continuing initiatives in industry collaboration and standardization attempt to provide precisely that. Initiatives to build common standards and protocols are led by organizations such as DIF and the **World Wide Web Consortium** (**W3C**). The goal is to build an interoperable environment in which diverse decentralized identity systems may connect with one another, enabling a unified and widely acknowledged approach to digital identification. Standardization initiatives encompass both technical specifications and usability principles, enabling an environment in which individuals and organizations can confidently and clearly traverse the digital identification domain.

Convergence of decentralized identity and blockchain

Blockchain technology continues to be a reliable ally in the ongoing drama of decentralized identification. The intersection of these two spheres creates a symbiotic connection in which blockchain's immutability and decentralized nature improve the security and integrity of decentralized identification systems. Ongoing projects investigate innovative consensus processes, scalability solutions, and interoperability protocols inside blockchain networks to meet the rising worldwide need for decentralized identification. The objective is to build a strong, tamper-proof foundation that supports the ideals of self-sovereignty and user control.

Integration with emerging technologies

The continual growth of decentralized identity is intertwined with the larger tapestry of developing technologies. Initiatives look at how AI, edge computing, and IoT might be used to improve the capabilities of decentralized identity systems. These integrations foresee a future in which digital identity easily aligns with the complexities of our networked and technologically evolved lives, from smart gadgets handling cryptographic keys to the AI-driven customization of authentication procedures.

Fostering the passwordless paradigm

Continuing projects are pioneering novel authentication techniques that reduce the need for standard passwords in the pursuit of a passwordless future. In this story, biometric authentication, secure multi-factor authentication, and device-based authentication are emerging as leaders. The emphasis is not just on improving security, but also on streamlining the user experience by minimizing the burden of password management. Users may look forward to a future where their digital interactions are not hampered by the challenges of remembering and preserving passwords as these passwordless paradigms gain traction.

As we traverse the maze of continuous innovations and activities in the DID authentication and passwordless sector, our collective objective remains unwavering: to create a digital identity ecosystem that prioritizes privacy, security, and user control. The path has been distinguished by cooperation, creativity, and an unshakable commitment to crafting a future in which our digital selves thrive in an environment of trust and resilience.

Summary

This chapter explored the problems of delivering verified information online and offered a decentralized architecture to solve them. It described DID verifiable assertions' long-term goals, which include increased user control over identities and claims, greater website usability, standardized transfer criteria, and claim-sharing privacy. The design intends to divide control over identifiers and claims, define standards for cross-role interactions, and use existing protocols when possible. The chapter also emphasized the necessity of protocols and standards in the digital society, such as SAML and Kerberos, for safe authentication and secure digital interactions. It also discussed the issues and drawbacks associated with these protocols, such as single points of trust, session management, and insufficient standardization. Overall, the chapter underlined the need for safe and interoperable digital identification systems, as well as the problems that come with accomplishing that aim. The next chapter digs into the broad field of identity verification, looking at the various methodologies, tools, and frameworks used to verify the identities of people and entities in digital ecosystems.

14
Identity Verification

Identity verification is the process of confirming that a person's claimed identity matches their actual identity. It involves gathering and verifying information about an individual, such as their name, date of birth, social security number, and other personal identifying information. The goal of identity verification is to prevent identity theft, fraud, and other types of malicious activity by ensuring that the person accessing a system or service is who they claim to be.

Identity verification can be accomplished using a variety of methods:

- **Knowledge-based verification**: This method involves asking the user a series of questions to verify their identity, such as their mother's maiden name or the name of their high school
- **Document verification**: This approach entails determining the legitimacy of a government-issued ID, such as a passport or driver's license
- **Biometric verification**: This method involves verifying the user's identity using their unique physical or behavioral characteristics, such as fingerprints or facial features
- **Database verification**: This method involves checking the user's information against a database of known identities, such as credit bureaus or government databases

Identity verification is used in a variety of contexts, such as online account creation, financial transactions, and government services. By verifying a person's identity, organizations can help prevent identity theft and fraud, protect sensitive information, and ensure that their systems and services are used only by authorized individuals.

In this chapter, we are going to cover the following topics:

- History of identity verification
- Challenges in traditional identity verification methods or systems
- Technological innovations in the identity management space
- Biometric identity verification

Historical evolution of identity verification

As we traverse the maze of continuous innovations and activities in the DID authentication and password-less sector, our collective objective remains unwavering: to create a digital identity ecosystem that prioritizes privacy, security, and user control. The path has been distinguished by cooperation, creativity, and an unshakable commitment to crafting a future in which our digital selves thrive in an environment of trust and resilience.

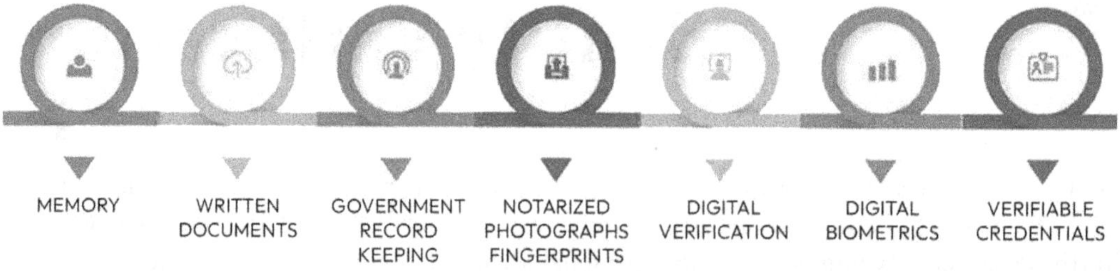

Figure 14.1 – Evolution of identity verification

The birth of trust and recognition

Personal recognition and trust provided the primitive basis for identification verification in the infancy of early civilizations. Individuals in close-knit groups were known by their faces, names, and reputations. The village elder's recognition of each member became an informal way of identification verification. Personal seals and symbols were the earliest attempts to create more tangible forms of identification as civilizations grew. Clay tokens with unique symbols, for example, were used as antecedents to the distinguish markings and seals we identify with identity in ancient Mesopotamia.

Seals, signatures, and scrolls

As civilizations progressed, so did the necessity for more formalized identification verification. To authenticate papers and transactions, the ancient Egyptians, known for their bureaucracy, used hieroglyphic seals on papyrus scrolls. Seals, which were frequently personalized and one-of-a-kind, were a sign of power and identification. The wax seal acquired significance in the Graeco-Roman civilization, denoting the validity of written agreements and declarations. Personal signatures arose in a separate manner, particularly during the Middle Ages, as people sought to append their own mark as a kind of identification proof.

Medieval guilds and the advent of credentials

The growth of trade guilds and the intricacies of business needed more sophisticated means of identification in medieval times. Guilds, which are groups of skilled artisans and merchants, provided membership and expertise certifications such as diplomas and badges. Although not as standardized as they are now, these early kinds of credentials indicated a shift from personal recognition to formalized documentation. The notarial system, in which trustworthy persons functioned as official witnesses, rose to popularity as well, offering some form of validation for legal transactions.

Renaissance and early modern period

The Renaissance saw a revival of interest in individual identity and portraiture. Portraits evolved to become a means of personal identity, reflecting not just physical characteristics but also social position. The Medicis in Florence, for example, established a primitive kind of identification authentication by creating an immense archive of portraits. Passports developed from informal papers provided by monarchs to verify the identification and loyalty of persons traveling across borders to formalized documents issued by monarchies to authenticate the identity and allegiance of individuals traveling across borders.

The birth of photography

The Industrial Revolution heralded a period of greater urbanization and movement, needing more sophisticated identification verification procedures. In the mid-nineteenth century, photography developed as a revolutionary technique, providing a visual record of persons. In the United States, organizations such as the Pinkerton National Detective Agency used early photographic identity cards. These photos, which were frequently accompanied by physical descriptions, paved the way for the incorporation of photographs into official papers such as passports and driver's licenses.

The rise of identification documents

With the widespread deployment of official identification cards in the twentieth century, identity verification underwent a paradigm change. Governments all across the globe have implemented standardized forms of identification, such as national ID cards, driver's licenses, and social security numbers. The emergence of these papers signaled a deliberate attempt to create a technique of identity verification that was systematic and widely recognized. Furthermore, advances in fingerprinting technology during this time period allowed for a more scientifically sound approach to authentication.

The digital age

The digital era resulted in a significant shift in identity verification. In the 1970s, the advent of magnetic stripe cards transformed how personal information was stored and accessible. This age also saw the introduction of computer databases, which allowed for more efficient retrieval of identification records. As the internet grew more prevalent in daily life, online identity verification mechanisms such as usernames, passwords, and security questions evolved.

The use of biometric technology in identity verification systems marked a substantial leap in the late twentieth and early twenty-first centuries. Fingerprint scanners, iris recognition, and face recognition technology made authentication safer and more convenient. These technologies, which were primarily identified with high-security contexts, progressively made their way into common applications, ranging from smartphone unlocking to airport security.

Self-sovereign identity in the digital landscape

The notion of **self-sovereign identity** (**SSI**) is gaining traction in the modern landscape. Individuals are placed at the center of their digital identities by SSI, which emphasizes user control and permission. Blockchain technology, which is recognized for being decentralized and tamper-resistant, is being investigated as a basis for SSI. DID systems let users control and display their credentials without relying on a central authority. This represents a trend away from old, centralized strategies and toward more user-centric, privacy-protecting alternatives.

Challenges and opportunities

The adventure of identity verification continues as we stand at the crossroads of history and innovation. In a world of ever-changing technology, the issues of assuring security, privacy, and user empowerment continue. The continuous research into biometrics, artificial intelligence, and decentralized systems has the potential to reshape how we verify identification in the coming years. Through the epochs, the identity verification story reflects not just technological improvements but also cultural transformations in our perception of trust, privacy, and the individual's responsibility in maintaining their digital identity. The chapters of history unfold as we travel this dynamic terrain, showing the continuing tale of who we are and how we prove it.

Challenges in traditional identity verification

Traditional techniques of identification verification have long been the sturdy protectors of security and confidence. These age-old practices have formed the underpinning of establishing and validating individual identities, from handwritten signatures to tangible identity cards. However, a tour through the historical tapestry of traditional identity verification reveals a maze of obstacles, each of which echoes the need for innovation and change in our approach to safeguarding and certifying human identification.

The vulnerability of physical papers to fabrication and counterfeiting is one of the persistent issues in traditional identity verification. Paper-based identifying documents, such as passports and driver's licenses, are vulnerable to adept counterfeiters who can copy complex characteristics with frightening accuracy. This problem extends to signatures, as expert forgers may imitate the distinctive strokes and loops, compromising the trustworthiness of handwritten verification.

For identity verification, traditional approaches frequently rely on static data such as names, dates of birth, and addresses. While this information may have been useful in simpler times, it no longer serves its function in the complex, linked digital world. Static data offers a substantial danger when it is hacked or easily accessible. Hackers with stolen databases can take advantage of this weakness, making traditional identity verification vulnerable to breaches.

Manual verification processes, which are inherent in many older approaches, cause inefficiencies and delays. Manually scrutinizing papers, checking records, and validating identities can result in lengthy wait times and higher operating expenses. This not only causes hardship for individuals seeking fast services, but it also puts strain on organizations dealing with the requirement for quick and precise identification verification.

While paper trails were previously the hallmark of bureaucratic processes, they are today a cause of irritation for individuals undertaking identity verification. Long forms, photocopying, and in-person visits all add to an inefficient user experience. The desire for digital services and seamless interactions collides with the antiquated nature of paper-based verification, resulting in an uncomfortable and fragmented trip for individuals.

Many traditional identity verification systems rely on centralized information stores, resulting in single points of failure. If centralized databases are breached, they can disclose a plethora of sensitive information, jeopardizing people's privacy and security. Furthermore, the concentration of data in a single authority raises worries about unauthorized access, which might result in a loss of control over personal information.

Traditional identity verification has worldwide compatibility issues in a linked world where people routinely traverse borders for business, tourism, or housing. Varying document standards, disparate authentication practices, and a lack of a standardized strategy impede seamless identity verification across geographies. This presents difficulties not just for people but also for organizations attempting to establish a consistent and internationally relevant verification method.

Identity theft and impersonation continue to cast long-lasting shadows over the traditional identity verification environment. Stolen documents, whether physically stolen or obtained via cyber intrusions, allow malevolent actors to assume fraudulent identities. The reliance on static information and the lack of adequate verification procedures makes people vulnerable to identity theft, with potentially severe implications for victims.

Traditional identity verification methods face the weight of compliance duties in an era of increasing data protection and privacy rules. From the **General Data Protection Regulation** (**GDPR**) to regional data protection regulations, organizations must traverse a complicated web of legal responsibilities. Ensuring compliance with these requirements adds another degree of complication to an already difficult identity verification procedure.

Traditional approaches may need substantial resources, both in terms of staff and infrastructure. The requirement for qualified workers to physically check papers, as well as the upkeep of physical archives, adds to operating costs. For organizations that rely on conventional identity verification, striking a balance between guaranteeing effective security measures and reducing operating expenditures becomes a difficult act.

Traditional identity verification techniques have the difficulty of adapting to the digital transformation age as the digital world advances. Legacy systems may fail to integrate seamlessly with current, technology-driven processes since they are strongly ingrained in established practices. The unwillingness or inability to adopt digital innovations can lead to a growing disparity between conventional techniques and the aspirations of a technologically empowered populace.

Traditional approaches, which frequently include the collecting and storage of considerable personal information, generate legitimate privacy concerns. The centralization of data and the possibility of unauthorized access create a conflict between the need for strong security and the preservation of individual privacy. These issues must be balanced in order to develop a trustworthy and user-centric identity verification platform.

Individuals have little control over how their information is used and shared in many traditional identity verification systems. The idea of self-sovereign identification is undermined by a lack of transparency and user-centric control. Individuals are frequently compelled to give up control of their personal data without clear knowledge of how it will be used, adding to a sense of disempowerment in the identity verification process.

The urge for innovation and revolutionary ideas reverberates as we traverse the obstacles contained in the traditional identity verification environment. The need to overcome these difficulties is more than just a desire for efficiency; it is a fundamental change toward identity verification systems that are in sync with the dynamic, networked, and privacy-conscious needs of today's society. The maze of obstacles provides an engrossing backdrop for the investigation of alternative ways and the continuous search to rethink how we verify and safeguard our identities in the digital era.

Technological innovations in identity verification

Technological breakthroughs have emerged as guiding lights in the ever-changing environment of identity verification, transforming the terrain and reinventing the way we verify individuals. From the origin of biometrics to the dawn of blockchain-based self-sovereign identification, the path through these technologies reveals a story of development, hurdles, and the critical role regulatory institutions play in supporting advancements. Let's go on a journey across the realms of technological advancements and legislative constellations that have moved identity verification forward.

Mapping identity in unique traits

The biometric revolution stands as a cornerstone in the history of identity verification, transforming how people are validated. To establish and validate identification, biometrics use distinctive physical or behavioral attributes such as fingerprints, facial features, iris patterns, or voiceprints. This technological advancement not only improves security by relying on unique biological identifiers but also simplifies user experiences by removing the need for passwords or PINs.

The **National Institute of Standards and Technology** (**NIST**) is a critical regulatory organization that has greatly contributed to the standardization of biometric technology. NIST, which is part of the U.S. Department of Commerce, is critical in defining and maintaining standards that support the accuracy and reliability of biometric systems. The NIST's **Biometric Evaluation and Testing** (**BEAT**) program, for example, tests and validates biometric technologies, ensuring conformity with established criteria and building faith in their usefulness.

A pocketful of identity

Smartphone ubiquity has ushered in a fundamental change in identity verification, bringing about the era of mobile identification. This breakthrough adds multi-factor authentication solutions, such as biometrics and device-based authentication, by using the capabilities of mobile devices. The seamless integration of identity verification into ordinary mobile interactions has not only improved security but has also become a driving factor in the broad use of digital services.

Regulatory entities such as the **Canadian Payments Association** (**CPA**) have played an important role in creating the mobile authentication environment. The CPA's commitment to promoting payment innovation includes the creation of standards and guidelines for safe mobile transactions. Entities such as the CPA help in building confidence in this expanding arena by providing a regulatory framework that protects the security and reliability of mobile authentication techniques.

Empowering individuals in the digital realm

Blockchain technology, which was initially designed to safeguard Bitcoin transactions, has found new applications in identity verification. The blockchain-powered notion of SSI puts individuals in control of their digital identities. Without relying on centralized authority, SSI allows people to own and maintain their identification information, selectively sharing it as needed.

The **Decentralized Identification Foundation** (**DIF**), a collaborative initiative led by industry professionals, is critical to the advancement of SSI concepts. DIF creates standards and protocols that allow diverse decentralized identity systems to communicate with one another, promoting a unified approach. As the digital world moves toward decentralized models, the DIF's contributions become increasingly important in developing a standardized, user-centric approach to identity verification.

eID and government-backed initiatives

Electronic identity (**eID**) projects supported by the government have emerged as a key factor in modernizing identity verification. eID systems, which are frequently connected to official government papers, provide safe and standardized means of online authentication. Digital passports, digital driver's licenses, and other government-issued eID credentials lay the groundwork for safe online interactions.

The **Digital Identity and Authentication Council of Canada** (DIACC) is an example of a collaborative institution that promotes eID improvements. DIACC brings together partners from the public and business sectors to create a trust framework for digital identification in Canada. DIACC helps in the creation of eID systems that are aligned with the demands of individuals and companies by defining standards and principles for safe, privacy-respecting digital identification.

Regulation frameworks and standardization

Regulatory frameworks and standardization organizations act as guiding lights in the enormous universe of identity verification, providing the required structure and coherence. Industry standards guarantee that identity verification technology fulfills established security, interoperability, and privacy norms.

The **European Union Agency for Cybersecurity** (**ENISA**) is a regulatory body devoted to creating a safe and trustworthy digital environment. ENISA's work in defining standards and recommendations for electronic identity and trust services helps to build a strong regulatory environment. Conformance to such standards guarantees that identity verification systems follow a shared set of principles, enabling cross-border trust and interoperability.

Catalyzing cross-border collaboration

Cross-border collaboration is critical for the efficiency of identity verification in the linked world of digital interactions. Global efforts strive to harmonize standards, promote best practices, and stimulate worldwide collaboration, recognizing that identity verification is not geographically limited.

The **International Civil Aviation Organization** (**ICAO**), a United Nations specialized body, is critical in the worldwide panorama of identification verification. The ICAO establishes standards for machine-readable travel documents, such as biometric passports. By defining these worldwide standards, ICAO helps to improve the interoperability and reliability of identity verification procedures across borders, hence improving international travel security.

Open standards and interoperability

In the context of identity verification, the necessity of open standards and interoperability cannot be overlooked. Open standards enable various systems to interact in real time, facilitating the development of interoperable solutions that span silos. These standards serve as the gravitational forces that bind a diverse technological galaxy together.

An industry organization, the OpenID Foundation, shows the value of open standards in the domain of identity verification. The foundation's OpenID Connect is an identity layer that sits on top of the OAuth 2.0 protocol, offering a standardized and secure framework for identity verification. By facilitating interoperability across identity providers and dependent parties, the OpenID Foundation helps in building a unified and user-centric identity ecosystem.

Continuous authentication

The changing nature of online dangers needs a transition in identity verification methodologies from static to dynamic. A unique paradigm—continuous authentication—examines user behavior and applies multi-factor authentication mechanisms during a session. This adaptable strategy improves security by responding to changing risk factors and possible attacks.

The **National Cybersecurity Centre of Excellence** (**NCCoE**), which is part of NIST, is critical to the advancement of continuous authentication systems. NCCoE works with industry partners to create realistic, standards-based solutions to cybersecurity concerns. The NCCoE contributes to the adaptable and resilient character of modern identity verification by offering guidance and resources for the deployment of continuous authentication.

The cognitive revolution

Machine learning (**ML**) and **artificial intelligence** (**AI**) introduce capabilities that imitate human cognitive functions, ushering in a cognitive revolution in identity verification. ML algorithms analyze massive datasets, discover patterns, and adapt to new data, improving the accuracy and efficiency of identification verification operations. From facial recognition to anomaly detection, AI-powered solutions are reshaping the authentication market.

Recognizing the transformational potential of AI in identity verification, regulatory institutions contribute to the development of ethical and responsible practices. The **Office of the Privacy Commissioner** (**OPC**) of Canada offers guidelines and recommendations on the ethical use of AI, ensuring that technological breakthroughs comply with privacy values. As AI continues to influence the identity verification environment, regulatory guidance is becoming increasingly important in navigating the ethical concerns connected with new technologies.

Digital signatures and cryptography

Digital signatures and encryption emerge as virtual seals of legitimacy in the digital arena, where physical signatures have no sway. Cryptographic techniques are used in digital signatures to secure the integrity and validity of electronic documents. As transactions and interactions grow more digital, reliance on cryptographic technologies for safe identity verification becomes critical.

In the United States, the **Electronic Transactions Association** (**ETA**) represents an industry ecosystem that promotes the use of secure electronic transactions. ETA advocates for the use of digital signatures and cryptographic technologies to improve the security and dependability of online transactions. By advocating for these technologies, ETA helps to create a safe and trustworthy digital identification ecosystem.

Quantum computing and beyond

Emerging technologies, such as quantum computing, bring both possibilities and problems for identity verification as we look to the future. With their unmatched computational capacity, quantum computers have the potential to disrupt established cryptography approaches. To ensure the integrity of identity verification in the quantum age, the race to create quantum-resistant cryptographic algorithms becomes critical.

Regulatory bodies, such as the **Cloud Security Alliance**'s (**CSA**) Quantum-Safe Security Working Group, are addressing the difficulties posed by quantum computing. This working group is primarily concerned with producing standards and best practices to guarantee that identity verification systems remain safe in the face of quantum threats. As quantum technologies evolve, regulatory agencies' coordinated efforts become critical in navigating this new frontier.

Navigating the cosmic seas of identity verification

Technological advances and regulatory agencies act as celestial navigators in the vast cosmic seas of identity verification, steering the route of progress and determining the future of identification. From the biometric revolution to blockchain-based self-sovereign identification, each breakthrough marks a step forward toward more secure, user-centric, and internationally interoperable identity verification.

Regulatory institutions play an important role in promoting breakthroughs, maintaining ethical practices, and harmonizing norms, acting as the gravitational forces that hold technical galaxies together. Whether it's the NIST overseeing biometric technology development, the DIACC molding the landscape of digital identification in Canada, or the ICAO establishing worldwide standards for travel papers, these organizations act as beacons of stability in a fast-developing digital cosmos.

As we navigate the vastness of identity verification, the convergence of technological innovations and regulatory frameworks becomes the key to unlocking a future in which individuals have greater control over their identities, transactions are secure and seamless, and trust transcends borders. The constellations of development continue to shift, and the road toward a more sophisticated, inclusive, and resilient identity verification ecosystem is a time-and-space odyssey.

Digital ID verification is a war on identity theft

According to Symantec's 2019 *Internet Security Threat Report*, identities are now the key target for hackers and criminals. To decrease fraud and combat identity theft, companies and websites are gradually using real-time identity verification to verify that their clients truly are who they claim to be. In short, identity verification is increasingly obligatory at a time when identity theft is a top mark for criminals.

Where mice play, the cats are always there to swat them down, and digital ID verification systems are also on the front line of this war. In a world where government identity documents are traded on the dark web, the novel identity-verification-technology-based approach of bringing a physical person into the identification process saves us from criminals using stolen identity documents to pledge fraud in our names.

The war on identity theft has become a defining battleground in the ever-expanding digital domain, where information transmission is quick and frictionless. In this fight, digital identity verification is a strategic assault, utilizing cutting-edge technology to build defenses against the never-ending onslaught of cyberattacks. This paradigm change from old approaches to digital identity verification not only improves security but also speeds up the verification process, ushering in a new age in the fight against identity theft.

Identity theft requires dynamic and robust defense against a shapeshifting opponent that is competent at exploiting holes in existing verification techniques. Individuals depended on tangible papers, signatures, and personal recognition to verify their identity in the analog age. However, the digital era introduced a new species of risk, with hackers employing sophisticated tactics to steal personal information and assume fraudulent identities. As a result, the battlefield shifted to the digital frontier, necessitating a proactive and technologically advanced approach to identity verification.

Identity verification technology has matured into a strong weapon in the fight against identity theft. Biometric authentication, which uses unique physiological and behavioral attributes such as fingerprints, facial features, and speech patterns, is at the vanguard of this arsenal. Not only does biometrics provide a highly secure means of identity verification but they also eliminate the weaknesses associated with traditional passwords or PINs.

In modern digital warfare, AI emerges as a force multiplier, enabling computers to analyze massive databases, detect abnormalities, and respond to new threats in real time. A subset of AI, machine learning algorithms, constantly enhance the understanding of user behavior, discriminating between lawful access and probable security breaches. This cognitive revolution offers a dynamic defense against identity thieves' adaptive techniques.

Blockchain technology, known for its decentralized and tamper-resistant nature, is emerging as a strategic fortress in the fight against identity theft. The notion of self-sovereign identification, enabled by blockchain, enables individuals to securely control and maintain their digital identities. Blockchain reduces the danger of large-scale data breaches by decentralizing identifying information and eliminating the need for a centralized authority, making it a resilient defense against identity theft.

Biometrics as the sentinel of identity

Biometric authentication, which is based on the uniqueness of human attributes, acts as a sentinel defending the identity gates. The combination of fingerprints, retina scans, and facial recognition converts the human body into a complex cryptographic key, allowing access only to those who have been authorized. This personalized approach not only strengthens security but also removes the weaknesses associated with previous systems that rely on static data.

The strength of visual identity verification is exemplified by facial recognition technology, a significant aspect of biometric identification. It analyzes facial traits to generate a one-of-a-kind biometric profile, offering a nearly infallible means of identification verification. Facial recognition acts as a watchdog in the ongoing struggle against identity theft, from unlocking cell phones to safeguarding access to critical data.

Voice biometrics, another aspect of biometric authentication, are based on the unique qualities of a person's voice. Voice biometrics provide a safe and easy form of identity verification by analyzing spoken patterns and characteristics. This technology's applications range from providing customer service to protecting financial transactions, improving defenses against identity theft.

Leveraging AI as the architect for efficiency

In the fight against identity theft, efficiency in the verification process is critical. AI emerges as the architect of this efficiency, revolutionizing how identity verification is carried out. Machine learning algorithms, which are capable of analyzing massive information at remarkable rates, help to identify legitimate users while quickly warning of possible dangers.

Continuous authentication, a dynamic AI-powered technique, analyzes user behavior during a session and adapts to changing risk variables. Unlike static approaches, continuous authentication maintains vigilance, guaranteeing that access is allowed only when the user's behavior corresponds to predefined patterns. This not only improves security but also simplifies the verification process by removing the need for frequent reauthentication.

The use of AI in document verification speeds up the process, especially in cases where individuals must present proof of identification via official papers. AI algorithms can quickly analyze and validate documents, saving time and resources that would otherwise be required for manual verification. This efficiency is especially important in industries such as banking, healthcare, and travel, where rapid and precise identification verification is required.

Blockchains redefining the battlefield

Decentralization appears as a strategic maneuver in the fight against identity theft, and blockchains are the technology redefining the battlefield. The notion of self-sovereign identification, enabled by blockchains, gives people authority over their digital identities. Users maintain control of their personal information in a decentralized identification system, selectively exposing it as needed without relying on a central authority.

The tamper-resistant feature of blockchains maintains the integrity of identifying information, making them highly resistant to unauthorized changes or data breaches. This decentralized architecture avoids the single point of failure found in centralized systems, where a breach might endanger a massive amount of sensitive data. Blockchains enable a strong defense against large-scale identity theft efforts by spreading identifying information over a network of nodes.

Smart contracts, which are self-executing agreements based on blockchains, help to speed up identity verification processes even more. These programmable contracts may enforce established restrictions, such as identity verification requirements, automatically and without the need for human interaction. This not only speeds up the verification process but also ensures consistency and accuracy in meeting regulatory requirements.

Regulatory constellations

In the fight against identity theft, regulatory organizations serve as celestial navigators, directing the ethical and secure application of identity verification technology. These regulatory constellations guarantee that the deployment of biometrics, AI, and blockchains is consistent with recognized privacy, security, and user rights norms.

The NIST, a US regulatory body, is critical in setting standards for identity verification systems. NIST's biometric standards guideline ensures the accuracy and reliability of biometric systems, encouraging trust in their use. NIST enforces the ethical use of biometric technology in the ongoing war against identity theft by establishing a regulatory framework.

The ENISA acts as a regulatory authority to ensure that identity verification technologies are used securely throughout member states. ENISA's work on defining electronic identity standards and recommendations helps in achieving a unified approach, supporting cross-border confidence and interoperability. Globally, regulatory entities such as ENISA play an important role in uniting efforts against identity theft.

The DIACC is a collaborative organization that is influencing Canada's digital identity ecosystem. It helps in building a robust defense against identity theft by establishing a trust framework and pushing for safe identity practices. DIACC and other regulatory agencies guarantee that technological improvements are consistent with the ethical values underlying the fight against identity theft.

The strategic attack spearheaded by digital identity verification technology reshapes the landscape of security and efficiency as the war against identity theft unfolds in the digital domain. Biometrics, AI, and blockchains serve as foundations of defense, reinforcing the battlefield against the mutating foe of identity theft. In this transformational era, where every digital contact has the potential to be a battlefield, the alliance is essential.

Biometric identity verification

With the introduction of biometric technology, the landscape of identity verification has undergone a seismic change, ushering in a revolutionary period in digital security. Biometrics, which use unique physiological and behavioral features, are at the forefront of the fight against identity theft. This in-depth investigation digs into the concepts, uses, and transformational influence of biometric identity verification, with a particular emphasis on the widespread use of face recognition technology.

The idea of uniqueness is at the heart of biometric identity verification. Biometrics, as opposed to traditional approaches that rely on knowledge-based elements such as passwords or PINs, harness intrinsic traits that are intrinsically unique and impossible to copy. Biometric authentication is based on the uniqueness of physiological attributes such as fingerprints, iris patterns, and facial features, as well as behavioral aspects such as voice patterns and typing dynamics.

Let's take a brief look at the various biometric methodologies available and their usages:

- **Fingerprint recognition**: Fingerprint recognition is one of the oldest and most extensively used techniques in biometric identification. The distinct ridges and valleys on a person's fingers provide a distinguishing pattern that acts as an unmistakable identity. Fingerprint recognition systems record and analyze these patterns, allowing for fast and secure identification verification. Fingerprint recognition has grown pervasive in modern identification, from unlocking cell phones to safeguarding entry to high-security locations.
- **Iris recognition**: The intricate patterns of the iris, the colored region of the eye, are used in iris recognition. The iris's distinctive arrangement of furrows, freckles, and crypts produces a specific pattern that remains consistent throughout a person's life. Iris recognition systems collect these patterns using high-resolution cameras, giving a quick and accurate form of identification verification. This technique is used in border control, airports, and security institutions where precision is critical.
- **Voice recognition**: Behavioral biometrics, such as speech recognition, are concerned with distinguishing patterns in an individual's behavior. To build a unique voiceprint, voice recognition algorithms analyze vocal properties such as pitch, tone, and cadence. This type of biometric authentication is commonly used in call centers, mobile banking, and voice-controlled gadgets, providing a smooth and safe way of verifying identification.
- **Behavioral biometrics**: In addition to physiological features, behavioral biometrics include a variety of attributes relating to how people engage with technology. Typing dynamics, mouse motions, and even how people traverse touchscreen devices may all be used as behavioral biometric identifiers. These characteristics give an added degree of protection to identity verification, especially in cases involving ongoing authentication.

  Biometric identity verification has invaded many aspects of our everyday lives, becoming essential in a variety of applications. Biometric uses range from improving physical security to optimizing digital interactions and are both broad and transformational.

- **Access control**: Biometrics have revolutionized access control systems in the field of physical security. Fingerprint scanners, iris recognition devices, and facial recognition technologies are used to guarantee that only authorized persons obtain access to protected facilities, business offices, and even residences. The speed and precision of biometric access control systems enhance security and convenience.
- **Mobile devices**: The use of biometric identification in mobile devices has completely transformed how people interact with their smartphones and tablets. Fingerprint sensors, facial recognition, and voice recognition technologies have largely supplanted conventional PINs and passwords, making device unlocking and transaction authorization more secure and user-friendly.

- **Financial services**: Biometric identification verification is revolutionizing how people access their accounts and make transactions in the financial services industry. In mobile banking apps, fingerprint and face recognition technologies are routinely used, allowing users to check in securely and authorize transactions with a simple biometric scan. This not only improves security but also streamlines the user experience in the financial services industry.
- **Healthcare**: Biometrics are used in the healthcare business to improve patient identification and safeguard access to medical information. Biometric solutions, such as fingerprint or palm vein recognition, guarantee that patient data is kept private, and that sensitive information is only accessed by authorized individuals. This improves the efficiency and security of healthcare operations.
- **Border control and travel**: Biometric identification verification has emerged as a critical component of border control and travel management. Many nations have installed biometric systems at immigration checkpoints, which use facial recognition or fingerprint scans to verify travelers' identities. This not only improves border security but also speeds up the immigration process by cutting lines and wait times.

Having had an overview of the various biometric options and permissible use options, let's take a deeper look specifically at the most widely used biometric recognition method, which is facial recognition.

Facial recognition extensive usage

Already used to catch killers and let students log in to their student portals, businesses and governments are endlessly finding new ways to use this technology, which uses urbane algorithms and scanning to recognize your facial biometrics.

Figure 14.2 – Biometric identity card

While there is a fiery debate on regulation, the pure simplicity and worth of facial recognition technology imply that you'll be seeing a lot more of it in the future.

The first thing nobody ever told you about digital identity verification is that contemporary digital identity verification is simply not safe without the application of government-grade facial recognition.

Among the several kinds of biometric identification, facial recognition technology stands out as a cutting-edge invention that is transforming the identity verification scene. In the digital arena, the human face, with its distinct features and expressions, becomes a potent identifier. Facial recognition technology analyzes facial traits and creates a unique biometric template, enabling a plethora of applications in security, convenience, and user experience.

Facial recognition is a leading biometric technology that has transformed the landscape of identity verification and access management. This section digs into the concepts that underpin facial recognition, beginning with the complex process of face enrollment and database matching. We will look at the dynamics of real-time recognition and its critical role in user authentication, as well as the many different applications of facial recognition in many industries. From its applications in law enforcement to its integration into cutting-edge payment systems and access control mechanisms, this section reveals the intricate tapestry of facial recognition technology and its pervasive impact on how we authenticate, secure, and control access in our rapidly evolving digital world:

- **Principles of facial recognition**: Facial recognition technology is based on pattern recognition and artificial intelligence concepts. Advanced algorithms analyze important facial traits, including eye distance, nose shape, and jawline, to create a unique template or *faceprint*. This faceprint is a digital depiction of a person's facial traits, allowing for precise and quick identification.
- **Enrollment and database matching**: The enrolling step is where an individual's facial traits are recorded and transformed into a faceprint. This template is then safely saved in a database. The individual's face is collected and analyzed throughout the verification process, and the resulting faceprint is checked against the database's stored templates. The identification is validated if a match is found.
- **Real-time recognition**: Real-time uses of facial recognition technologies allow for rapid identification. Surveillance cameras with facial recognition capabilities may identify and follow people in real time in security systems, providing a proactive approach to security monitoring. This is especially useful in public areas, airports, and high-security institutions.
- **User authentication**: Facial recognition has become an essential component of digital device user identification. Many smartphones, tablets, and laptops include front-facing cameras and facial recognition technology, allowing users to unlock their devices with a glance. This improves security while also streamlining the user experience by eliminating the need for passwords or PINs.

  The widespread use of face recognition technology covers a wide range of applications, including security, consumer convenience, and even artistic expression. Because of its adaptability, facial recognition has emerged as a transformational force in a variety of sectors.
- **Security and surveillance**: Facial recognition technology has evolved into a powerful weapon in the field of security and surveillance. Surveillance cameras fitted with facial recognition capabilities can identify and follow people anywhere, from airports to shopping malls, adding an extra degree of protection. This proactive strategy enables authorities to respond to possible threats quickly.

- **Law enforcement**: Facial recognition technology is used by law enforcement to help in criminal investigations. Authorities can identify and catch criminals by comparing face pictures acquired from surveillance films or public databases with existing records. However, this use raises ethical concerns about privacy and potential exploitation.
- **User authentication in devices**: Facial recognition integration in gadgets has become associated with user authentication. Smartphones, in particular, have adopted facial recognition as a safe and simple way to unlock devices and authorize transactions. This departure from typical passwords improves security, as well as the user experience.
- **Payment authorization**: Payment authorization using facial recognition technology is becoming more common, particularly in retail and banking. Transactions may be authorized with a simple facial scan by connecting a user's facial biometric data to their payment accounts. This not only enhances security but also speeds up the checkout process.
- **Human–computer interaction**: Facial recognition has cleared the path for novel human–computer interactions. Facial recognition, for example, may detect and analyze facial emotions in gaming and virtual reality, allowing users to control characters or modify virtual settings through their expressions. This innovative application broadens the scope of human-machine interaction.
- **Attendance and access control**: Facial recognition is used in educational institutions and corporate settings for attendance monitoring and access management. Facial recognition systems can identify people quickly and precisely, allowing for smooth admission into secure areas or tracking attendance in huge meetings. This improves security, as well as operational efficiency.

While facial recognition technology has several advantages, its widespread use has not been without controversy. Ethical problems and privacy concerns have emerged as major grounds of dispute, sparking debates regarding the appropriate and transparent deployment of facial recognition in a variety of fields.

Some of the major elements to consider while implementing facial recognition methodologies for the identification of authenticity of individuals are as follows:

- **Privacy concerns**: The major worry with facial recognition technology is one of privacy. The growing usage of surveillance cameras with facial recognition technology raises concerns about people being continually watched without their knowledge or agreement. The possibility of widespread monitoring violates the right to privacy, and the lack of clear restrictions exacerbates these fears.
- **Biases and accuracy**: Facial recognition algorithms have been criticized for biases, particularly toward certain gender and ethnic groupings. According to studies, these algorithms can be less accurate when recognizing people from specific demographic backgrounds. Biased face recognition results in misidentification, reinforcement of prejudices, and potentially discriminatory outcomes, particularly in law enforcement applications.

- **Consent and data security**: Clear permissions and adequate data security procedures are required for the collection and storage of facial biometric data. Users should be informed about how their face data will be used, and organizations should put in place strict security processes to protect this sensitive information. Breach of facial recognition databases might have serious consequences for individuals, resulting in identity theft or abuse of personal information.
- **Legislation and regulation**: The changing environment of facial recognition technology needs thorough laws and regulations to ensure ethical usage and to defend the rights of users. Many countries are wrestling with the necessity to set clear criteria controlling the deployment of face recognition, balancing innovation and privacy protection.

Facial recognition technology is a pinnacle in the biometric identity verification scene, providing a multidimensional approach to security, convenience, and the user experience. Its many applications range from enhancing security and law enforcement to streamlining user authentication in common gadgets. However, the ethical issues and privacy problems related to facial recognition necessitate caution.

As technology advances, the appropriate use of facial recognition in identity verification necessitates the harmonious integration of innovation, ethics, and legal frameworks. To navigate a future where the advantages of this technology can be utilized without sacrificing privacy or perpetuating prejudices, it is critical to strike a balance between the revolutionary potential of facial recognition and the preservation of individual rights. The voyage into the future of facial recognition in identity verification unfolds like a complicated tapestry, with technological innovation and ethical issues weaving together to define the shapes of a safe and egalitarian digital world.

Blockchain and identity verification

The introduction of blockchain technology has heralded a new era in identity verification, dramatically altering how people maintain and safeguard their digital identities. This technical investigation digs into blockchain identity verification functions, revealing the ideas, procedures, and significant influence on privacy and security. The use of blockchains results in a decentralized, tamper-resistant structure that not only strengthens identity verification but also gives individuals more control over their personal information.

At its foundation, blockchains are decentralized and distributed ledgers that securely and transparently record transactions over a network of computers. Their three defining characteristics—immutability, decentralization, and cryptographic security—lay the groundwork for a paradigm change in identity verification.

Blockchains rely heavily on the idea of immutability. Once data is recorded in a block, it is nearly hard to change it in the future. Each block includes a cryptographic hash of the preceding block, forming a chain of blocks in which any change in a single block would necessitate changing all following blocks, which is a computationally infeasible endeavor. This immutability guarantees the integrity of data recorded on the blockchain.

Blockchains, unlike traditional centralized databases, run on a peer-to-peer network. Every network participant has a copy of the complete blockchain, removing the need for a central authority. This decentralization not only improves resilience against single points of failure but also encourages transparency and trust among players.

To safeguard data and regulate access, blockchains use cryptographic algorithms. Each network member has a pair of cryptographic keys: a public key that others may see and a private key that only the owner knows. Consensus methods validate and add transactions to the blockchain, guaranteeing that only individuals with the necessary private keys may access and edit the data.

Traditional identity verification systems frequently rely on centralized databases, raising issues regarding data breaches, identity theft, and personal information abuse. By offering a decentralized, trustless environment for identity verification, blockchains introduce a paradigm change. The evolution occurs in three distinct stages:

1. **Centralized identity management**: Identity information is traditionally held and handled by central authorities such as government agencies, financial organizations, or service providers. Individuals are exposed to the hazards of large-scale data breaches since a single breach can damage a significant reservoir of sensitive information.
2. **Federated identity management**: The federated approach entails numerous entities working together to share identification information while retaining separate databases. While it enhances interoperability and eliminates the need for recurrent identification checks, it still relies on trust among participating companies, and the danger of data breaches remains.
3. **Decentralized identity management**: Blockchains present a decentralized identity management concept in which people retain ownership over their personal data. Users' identifying traits are stored on the blockchain as self-sovereign identities. These identities are cryptographically protected, and users may selectively provide pertinent information without relying on a centralized authority.

In summary, blockchains revolutionize identity verification by providing a decentralized, tamper-resistant ledger for securely maintaining digital identities. They create a visible and immutable record of identification transactions, lowering the risk of fraud and increasing confidence in the verification process. Blockchains improve the efficiency, security, and privacy of identity verification in a variety of industries by combining smart contracts and cryptography approaches.

Privacy and security enhancements

In an increasingly linked digital world, ID verification is a vital part of our online security. While you might not use it too much today, in the coming years, it will become as predominant as using a (now comparatively insecure) username and password.

Identity verification encompasses some of the most thrilling new tech developments.

The decentralization of identification information is the foundation of blockchain-based identity verification. Individuals take on the role of data stewards, limiting the danger of unauthorized access and allowing them granular control over who may access certain aspects of their identity. By reducing reliance on centralized institutions for identity verification, this user-centric method improves privacy.

Blockchains facilitate the establishment of self-sovereign identities, a paradigm in which people have complete control over the aspects of their identity. SSI enables users to selectively control and distribute their identity information, removing the requirement for a central authority to vouch for their identity. This improves user privacy while simultaneously lowering the danger of identity theft and fraud.

A blockchain's immutability protects the integrity of identification data. Identity information stored on the blockchain cannot be changed without network consensus. This function protects against tampering and fraudulent operations, laying the groundwork for identity verification.

For safe authentication, blockchain uses cryptographic algorithms. Users have private keys, which allow them to access their identification information and sign transactions. The use of cryptographic signatures guarantees that only the correct owner of the private key may authenticate and authorize identity-related operations.

Data sharing becomes a consent-based procedure with blockchain-based identity verification. Users can offer entities particular rights to access their identity traits, which are enforced using smart contracts. This level of data-sharing control alleviates privacy issues connected with indiscriminate access to personal information.

Blockchain-based identity verification

Digital signatures are critical components of blockchain-based identity verification. When users establish a transaction or request access to their identity attributes, they use their private key to sign the request. The public key, which is available on the blockchain, acts as an identification and lets others verify the signature's legitimacy.

Smart contracts are self-executing agreements that are encoded in the blockchain with established rules. Smart contracts enforce data sharing and access permissions regulations in the context of identity verification. A smart contract, for example, can specify that a user's identification information can only be accessed by a specified service provider with the user's explicit approval.

DIDs are an essential part of blockchain-based identification systems. On the blockchain, they serve as a unique identification for people and companies. DIDs are accompanied by DID documents, which provide public keys, authentication mechanisms, and service endpoints. The combination of DIDs and DID papers guarantees that identity verification is standardized and interoperable.

Verifiable credentials are an essential component of blockchain-based identification systems. They represent a series of assertions made by a trustworthy entity regarding a subject's identification. These credentials are cryptographically signed, and the receiver can use them to prove specified qualities without releasing the raw data. Verifiable credentials ensure a safe and private identity verification procedure.

Scalability becomes an issue when blockchain networks expand in size. It is critical for widespread adoption to be able to manage a large number of transactions and identity verifications per second. To solve this difficulty, several scaling options, such as layer-2 protocols and consensus algorithm upgrades, are being investigated.

Creating a fluid and inclusive identity verification environment requires interoperability across multiple blockchain networks and identification systems. Interoperability efforts are aided by standards such as the DIF's Universal Resolver and Verifiable Credentials Data Model.

The growing environment of blockchain-based identity verification must be compatible with both present and upcoming regulatory regimes. Balancing privacy, security, and compliance with regional and international standards is a continuing problem that necessitates coordination among blockchain developers, regulators, and legal experts.

To transition from traditional identity verification approaches to blockchain-based solutions, user education and mass acceptance are required. Users must comprehend the advantages of self-sovereign identities, cryptographic security principles, and the control they obtain over their data. Adoption is driven by user-friendly interfaces and instructional programs.

Real-world applications

Blockchain-based identity verification is being used in a variety of sectors, including the following, demonstrating the revolutionary potential of decentralized identification solutions:

- **Financial services**: Blockchain improves **Know Your Customer** (**KYC**) operations in financial services by enabling safe and private identification verification. Users may exchange verified credentials without providing sensitive information, such as proof of address or income.
- **Healthcare**: Blockchain-based identity verification facilitates the secure sharing of patient information among authorized healthcare providers in the healthcare industry. Patients retain ownership over their medical records, with access granted depending on certain criteria.
- **Education**: A blockchain enables the safe and tamper-proof authentication of academic qualifications. Individuals may save their educational achievements on the blockchain as verified credentials, allowing businesses to streamline the employment process.
- **Government services**: Governments may use blockchains to handle identities in a safe and transparent manner. Blockchain-based identification systems improve the accuracy of citizen records, decrease identity fraud, and increase the efficiency of government services.
- **Cross-border verification**: The decentralized nature of blockchains and their cryptographic security make it a suitable alternative for cross-border identity verification. Individuals can use verified credentials given by one country to access services in another, expediting processes such as visa applications.

The usage of blockchains in identity verification indicates a paradigm shift toward user-controlled, privacy-centric solutions. Blockchain-based identification solutions hold the promise of building a more secure, transparent, and inclusive digital identity environment as the technology matures, solving obstacles such as scalability and interoperability.

Individuals' self-sovereign identities, together with blockchains' cryptographic security and decentralized nature, signal a new era in which privacy is not only a right but a design concept. The continuous advancements and activities in the blockchain identity verification arena demonstrate a shared commitment to negotiating the complexity of digital identification while adhering to the ideals of security, privacy, and user empowerment. In this scenario, the future is a landscape in which individuals have unprecedented power over their digital selves, and the immutable ledger of blockchain serves as the foundation of confidence in the developing field of identity verification.

Summary

This chapter presented a thorough review of identity verification, including its historical development, issues with old approaches, and technological advancements. It explored the historical roots of identity verification, from personal recognition in ancient civilizations to the usage of formal identification papers in the digital era. The limitations of traditional identity verification systems, such as reliance on static data and human verification processes, were underlined, as were the possible security and privacy problems associated with centralized data storage. The chapter also explored the revolutionary power of technical developments such as machine learning, artificial intelligence, digital signatures, and blockchain technology in reshaping the identity verification environment. It emphasized the need for ethical considerations, regulatory guidance, and user education in managing ethical challenges and guaranteeing responsibility. In the next chapter, we take a look at biometric security in the distributed identity management landscape.

Part 4 - Digital Identity Era: A Probabilistic Future

This part explores the role of biometrics in distributed identity management. *Chapter 15* delves into biometric security algorithms, secure storage and transmission, user acceptance and education, and the scalability and flexibility of biometric systems.

This part has the following chapter:

- *Chapter 15, Biometrics Security in Distributed Identity Management*

15

Biometrics Security in Distributed Identity Management

Biometric security is an increasingly popular method of authentication in distributed **identity management** (**IdM**) systems. Fingerprints, face recognition, iris scans, and voice recognition are all examples of biometrics. In this chapter, we will cover the following topics:

- Principles of biometric security in **distributed identity and access management** (**DIAM**)
- Securing biometrics with blockchain
- Mechanisms for biometric authentication
- Real-world applications of biometrics in DIAM
- Emerging technologies and trends in biometrics

Principles of biometric security in DIAM

In a DIAM system, biometric security can be used to provide a high level of security and convenience. Instead of relying on traditional passwords or tokens, users can authenticate their identity using their unique biometric characteristics. This can help prevent identity theft and fraud, as biometric traits are difficult to forge or replicate.

The introduction of biometrics has ushered in a new era in IdM, revolutionizing the way people verify themselves in digital contexts. Biometric technologies employ unique physiological or behavioral traits to verify and authenticate users' identities. Biometrics plays a critical role in guaranteeing security, privacy, and a smooth user experience in the field of distributed IdM, where decentralized and user-centric methods are gaining traction. This investigation dives into biometric concepts, with an emphasis on technologies such as fingerprint, iris, and voice recognition, while also exposing the critical function of encryption approaches in ensuring privacy in identification data:

- **Fingerprint recognition**: This is one of the oldest and most extensively used biometric authentication technologies and is based on an individual's fingerprint's distinctive ridges and valleys. With its various ridges and valleys, the human fingerprint generates a unique and highly identifiable pattern that acts as a digital signature for identification verification. Fingerprint recognition is a stable and reliable method of authentication in dispersed IdM systems. As users engage with various decentralized services, their fingerprint data, which is frequently saved in the form of cryptographic templates, is used to open secure access points. This solution not only improves security but also streamlines user interactions in decentralized contexts, providing a smooth and user-friendly experience. Because each person's fingerprint is unique, it is an excellent option for safe identification verification. Fingerprints are recorded and translated into digital templates in biometric systems, which are typically scrambled for increased protection. When a user requests authentication, their fingerprint is matched to a saved template, providing a safe and quick verification procedure. Fingerprint recognition is not only dependable but also easy, providing consumers with a consistent experience across a wide range of applications, from unlocking devices to accessing secure facilities.
- **Iris identification**: This makes use of distinctive patterns seen in the colored region of the eye, providing a degree of precision and accuracy that sets it apart in the biometric environment. Iris recognition contributes to safe authentication in distributed IdM by collecting and analyzing the detailed details of an individual's iris. The generated iris templates, which are often encoded as cryptographic hashes, serve as the foundation for identity verification. Iris recognition thrives in settings requiring a high degree of precision, making it a vital tool in decentralized systems requiring user faith in authentication. Iris identification systems digitize the iris's distinguishing characteristics, such as furrows and freckles. The provided iris is compared to the saved template during authentication, allowing for exact identity verification. Iris recognition shines in situations requiring great precision, such as secure access control in key infrastructure or border control.
- **Voice recognition**: A behavioral biometric that focuses on specific patterns in a person's voice. The specific mix of pitch, tone, cadence, and other vocal characteristics results in a voiceprint that acts as a unique identity. Voice recognition provides an additional degree of protection in the context of dispersed IdM. Users can communicate with decentralized services by authenticating with their voice. This not only offers a **multi-factor authentication** (**MFA**) component but also a straightforward and natural means for users to authenticate their identities in a variety of decentralized settings. Individuals' distinctive vocal features, such as pitch, tone, and cadence, are used in voice recognition. Each individual has a unique voiceprint, and advances in voice

recognition technology allow for advanced analysis for reliable identification. The technology records and analyzes sound patterns during voice identification, providing a digital representation that acts as a unique identity. Voice recognition is used in secure telephone banking, voice-activated gadgets, and even distant customer service identification verification.

- **Face recognition**: This uses an individual's facial traits, such as eye distance, nose shape, and jawline, to generate a unique facial template. This technology has grown in prominence in recent years, particularly for smartphone unlocking and airport security systems. **Artificial intelligence** (**AI**) advancements have improved facial recognition accuracy, allowing it to adapt to fluctuations in facial expressions, haircuts, and even aging. While face recognition is a handy and non-intrusive identification mechanism, privacy issues and potential biases in particular implementations highlight the importance of responsible and ethical deployment.

In conclusion, biometrics play a crucial role in DIAM systems by enhancing the privacy and authenticity of the data contained and presented by these systems. In the next few sections, we will cover topics on how cryptography plays a critical parallel role in enhancing the security of such frameworks and platforms.

Cryptography as a guardian of privacy

While biometric principles improve the security of identity verification, the value of privacy cannot be overemphasized. By definition, distributed IdM solutions prioritize user-centric methods in which people retain control over their personal information. Encryption techniques are crucial in striking this delicate balance between security and privacy.

In distributed IdM, cryptography, the study of secure communication, acts as a privacy defender. Cryptographic techniques are used in biometrics to secure the integrity and secrecy of identification data. When people engage with decentralized systems, their biometric information is frequently converted into cryptographic representations known as hashes or templates. These cryptographic representations act as digital locks, assuring the security of sensitive biometric data even in the case of unauthorized access.

Two of the most efficient cryptographic algorithms that add value to DIAM systems are the following:

- **Homomorphic encryption**: This is a cutting-edge cryptographic technology that allows calculations on encrypted data to be done without the requirement for decryption. Homomorphic encryption appears as a strong technique in safeguarding privacy during identity verification procedures in the field of distributed IdM. Biometric data may be safely processed while remaining secured using homomorphic encryption. This not only safeguards sensitive information's secrecy but also generates a higher degree of trust among users who maintain control over their encrypted identification traits. Homomorphic encryption enables computations on encrypted data to be done without the requirement for decryption. This implies that sensitive identification information may be handled while still being encrypted in the context of secure identity. Homomorphic encryption protects data confidentiality, preserving user privacy even in a decentralized context.

- **Zero-knowledge proofs (ZKPs)**: By allowing one party (the prover) to demonstrate knowledge of a specific piece of information without disclosing the information itself to another party (the verifier), ZKPs provide a privacy-preserving approach to identity verification. In the context of biometrics, ZKPs allow users to confirm their identities without revealing their raw biometric data. This technique is consistent with the ideas of user-centric distributed IdM, in which individuals may establish their identity without jeopardizing the privacy of their biometric data. ZKPs allow a statement to be verified without disclosing the underlying information. They also enable users to verify themselves without revealing their basic identification attributes in the field of secure identity. For example, a person can prove their age without giving their precise birthday. This privacy-focused strategy is consistent with the self-sovereign identification ethos in dispersed systems.

Cryptography serves as a stalwart guardian of privacy, employing advanced mathematical techniques to secure sensitive information. Through the use of encryption algorithms, it transforms data into unreadable formats, ensuring that only authorized entities with the appropriate keys can decipher the information. As a key component in privacy protection, cryptography plays a pivotal role in safeguarding digital communications and sensitive data from unauthorized access.

Balancing security and privacy

The concepts of biometrics become critical in striking the delicate balance between security and privacy in the dispersed landscape when a centralized authority is substituted by user-centric models. With their distinct capabilities, fingerprint, iris, and voice recognition technologies contribute to a diverse and secure identity verification ecosystem. The cryptographic foundations of these biometric technologies safeguard the secrecy and integrity of their identifying traits while users interact with decentralized services.

Furthermore, encryption approaches such as homomorphic encryption and ZKPs emerge as critical instruments for protecting user privacy. These cryptographic algorithms give people authority over their identification data, which is consistent with the ideas of **self-sovereign identities** (**SSIs**) in decentralized systems. The combination of biometric principles and encryption methodologies in distributed IdM heralds a new era in which users can navigate digital spaces with confidence, knowing that their identities are secure, their privacy is respected, and the decentralized landscape is strengthened by biometric principles.

Securing biometrics with blockchain

Blockchain emerges as the backbone of security in the ever-changing field of digital identity, encouraging trust, transparency, and user empowerment. Blockchain is, at its core, a decentralized, tamper-resistant ledger that provides a secure foundation for managing digital transactions. This investigation delves into blockchain's position as the foundation of safe identity in distributed systems, with an emphasis on the transformative potential of smart contracts and identity rules in ensuring security while keeping the decentralized ethos.

The following are three basic elements that underpin blockchain's expertise in safe IdM:

- **Immutability**: The immutability of blockchain is proof of the durability of recorded information. Once data is added to a block, it becomes unchangeable. This protects the integrity of identifying information by making it tamper-proof and preventing unauthorized alterations or manipulations.
- **Decentralization**: The dependence on a central authority in conventional IdM entails inherent dangers. The decentralized structure of blockchain eliminates the need for a central governing body, instead dispersing control over a network of nodes. This not only improves system resilience but also reduces the chance of a **single point of failure** (**SPOF**), making it an excellent choice for dispersed situations.
- **Cryptographic security**: To safeguard data and regulate access, blockchain uses cryptographic algorithms. Every network member has a pair of cryptographic keys: a public key that others may see and a private key that only the key holder knows. Cryptographic signatures validate and authorize transactions, guaranteeing that only individuals with the necessary private keys may access and alter the data.

Blockchain serves as the backbone of secure identity by providing a decentralized and tamper-resistant ledger for storing and managing digital identities. Its immutable nature ensures the integrity and transparency of identity records, enhancing security in the digital realm.

Smart contracts – self-executing agreements enforcing security

The notion of smart contracts is at the forefront of blockchain's disruptive influence on secure identity. Smart contracts are self-executing agreements that are encoded in the blockchain with established rules. These contracts are the foundation of distributed IdM, automating procedures, enforcing regulations, and boosting security. Let's take a closer look at this:

- **Automating processes**: Manual interventions and third-party verifications are common in traditional identification verification. Smart contracts automate these operations by following predetermined rules and eliminating the need for middlemen. A smart contract may validate identity attributes when a user asks for access to a decentralized service, guaranteeing a seamless and speedy procedure.
- **Enforcing identity rules**: Smart contracts are critical for enforcing identity rules in dispersed systems. These rules can govern the circumstances in which identity data is accessed, shared, or updated. A smart contract, for example, may specify that some identity attributes can only be accessed with explicit user authorization, ensuring privacy and conforming with user-centric identification principles.
- **Decentralized governance**: The decentralized network's consensus governs the execution of smart contracts. This decentralized governance paradigm guarantees that smart contract rules are visible, verifiable, and independent of a single authority. It promotes a trustless environment in which participants can rely on predetermined rules rather than a central entity.

In summary, smart contracts are self-executing contracts with the terms directly written into code. They automatically execute and enforce contractual agreements when predefined conditions are met, eliminating the need for intermediaries in various digital transactions.

Building blocks of secure identity

Decentralized identifiers (**DIDs**) and verified credentials are core aspects of blockchain-based secure identity. Let's take a look at why they are so important:

- **DiDs**: On the blockchain, DIDs function as unique identifiers for persons or companies. They serve as the foundation for connecting a user's identity to cryptographic keys, ensuring a safe and standardized method of identity representation. DIDs, when paired with public and private key pairs, allow people to exercise control over their identity without relying on a centralized authority.
- **Verifiable credentials**: They are an essential component of secure IdM. These credentials, which are granted by trustworthy bodies, constitute a collection of assertions regarding a subject's identification. They are signed cryptographically, assuring their legitimacy. Verifiable credentials, when offered, serve as verification of specified qualities without disclosing the underlying raw data. This privacy-preserving feature is consistent with user-centric identification concepts and is useful in ensuring safe interactions in dispersed situations.

DIDs serve as the foundation of secure identity by providing a unique and self-owned identifier for individuals in the digital realm. When coupled with verifiable credentials, which cryptographically attest to the authenticity of information, they form a robust framework for individuals to assert and control their identities, enhancing security, privacy, and interoperability across digital interactions.

Scalability, interoperability, and regulatory compliance

While blockchain provides unparalleled security for distributed IdM, it is not without its drawbacks.

A blockchain network's capacity to process a large number of transactions per second is critical for mainstream adoption. As networks expand, scalability issues occur, needing new solutions such as layer 2 protocols and consensus algorithm modifications to maintain efficient operation.

Interoperability across multiple blockchain networks and identification systems is critical for building a unified identity verification ecosystem. Standards such as the **Decentralized Identity Foundation's** (**DIF's**) Universal Resolver and Verifiable Credentials Data Model help to facilitate interoperability.

Identity systems based on blockchain must comply with existing and upcoming regulatory frameworks. Balancing innovation, security, and compliance with regional and international legislation is a continuous problem that necessitates collaboration among blockchain developers, legislators, and legal experts.

The influence of blockchain on secure identity spans across several industries, demonstrating its disruptive potential:

- **Financial services**: Blockchain improves **Know Your Customer** (**KYC**) operations in the banking sector, ensuring safe and privacy-preserving identification verification. Users may share valid credentials without disclosing sensitive information, making onboarding and compliance procedures easier.
- **Healthcare**: The use of blockchain-based IdM guarantees the safe and transparent exchange of patient information across authorized healthcare providers. Patients retain control over their medical information, providing access depending on particular criteria while protecting their health data's privacy.
- **Education**: Blockchain enables safe authentication of academic qualifications. Individuals may save their educational accomplishments as verified credentials, making the employment process easier for businesses and lowering the danger of credential fraud.
- **Government services**: Governments use blockchain to handle identities in a safe and transparent manner. Blockchain improves the integrity of citizen records, decreases identity fraud, and increases the efficiency of government services.

With its immutable and decentralized character, blockchain provides the foundation of secure identity in distributed systems. Smart contracts and identity rules, enabled by cryptographic security, automate procedures, enforce rules, and raise decentralized IdM security standards. Decentralized identities and verified credentials enable a standardized and safe representation of identity when users interact with multiple services.

Privacy-preserving technologies such as homomorphic encryption and ZKPs contribute to secure identity's user-centric nature. Scalability, interoperability, and regulatory compliance challenges pave the way for continued development and cooperation.

Blockchain disrupts businesses in the real world by offering a safe framework for identity verification. The decentralized and safe nature of blockchain reshapes how individuals engage with digital services, encouraging a future in which users have control over their identities in a transparent, secure, and decentralized world.

Mechanisms for biometric authentication

Biometric authentication, which uses unique physical or behavioral traits to identify users, has emerged as a critical component in improving identity security. Biometrics provides a diverse and safe solution to authentication in an environment driven by digital interactions. This investigation dives into various biometric authentication systems, emphasizing the need for digital signatures and public-key cryptography to provide safe authentication.

Digital signatures and public-key cryptography are foundational technologies for ensuring the security of identity verification processes. Public-key cryptography employs a pair of cryptographic keys: a public key that may be freely disseminated and a private key that is kept private. These key pairs are used to produce digital signatures, which assure the validity and integrity of digital messages or documents.

In the context of biometric authentication, public-key cryptography is critical for safeguarding biometric data transmission and storage. When a user's biometric data is collected, it can be encrypted with their public key. Only the user's matching private key can decode and expose the original biometric data. This procedure assures that even if the encrypted data is intercepted during transmission or stored on a server, it remains unreadable in the absence of the user's private key.

Digital signatures give an extra degree of security by allowing the origin and integrity of a message or document to be verified. A digital signature can be applied to biometric data or the authentication request in the context of biometric authentication. This signature is created with the user's private key, and its validity can be verified by anybody with access to the user's public key. This process verifies that the biometric authentication request is genuine and has not been tampered with.

The pillars of identity verification

Smart contracts and DIDs emerge as basic factors in changing identity verification paradigms as the digital world decentralizes. Smart contracts, which are self-executing agreements with specified rules stored on the blockchain, and DIDs, which are unique identifiers rooted in decentralized networks, collaborate to create a safe, user-centric, and transparent solution to identity verification.

The four pillars of identity verification are the following:

- **Smart contracts**: Smart contracts, which are implemented on blockchain networks, automate and enforce identity verification processes. These agreements contain particular rules and conditions that regulate how identification data is accessed, shared, and used. Smart contracts in the field of biometric authentication can automate the verification process based on established rules. A smart contract, for example, might specify that a biometric authentication request is only granted if specific requirements are satisfied, such as the production of a valid digital signature or the observance of privacy-preserving procedures. Smart contracts improve the efficiency and security of identity verification in a decentralized setting by automating these operations.
- **DIDs**: DIDs provide a standardized and decentralized method of representing identity. DIDs serve as unique identifiers anchored in decentralized networks such as blockchain in a world where users frequently interact with numerous services and platforms. DIDs, as opposed to traditional identifiers attached to a central authority, provide people sovereignty over their identification. DIDs, when used in conjunction with biometric authentication, guarantee that the relationship between a user's biometric data and their identification stays safe and user-centric. DIDs, which are frequently linked with valid credentials, allow users to selectively release information without jeopardizing their entire identity.

- **User-centric control**: The combination of smart contracts and DIDs substantially moves control of identity verification away from centralized agencies and toward users. Users may utilize smart contracts to specify rules that govern the usage of their biometric data, ensuring that their privacy choices and consent requirements are satisfied. DIDs, which serve as the anchor for user identities, provide a unified and compatible method for identity verification across numerous decentralized systems.
- **Privacy-preserving biometric authentication**: The combination of smart contracts with DIDs encourages privacy-centric practices in the context of biometric authentication. Smart contracts can include rules that govern how biometric data is handled, ensuring that only authorized organizations have access to data. DIDs, which operate as unique identifiers, allow users to keep control over the relationship between their biometric features and identity, preventing critical information from being exposed inadvertently.

When combined with digital signatures, public-key cryptography, smart contracts, and DIDs, biometric authentication technologies create the groundwork for a safe, user-centric, and decentralized identity environment. Fingerprint recognition, iris recognition, voice recognition, and face recognition are all valid biometric identification methods. Digital signatures and public-key cryptography strengthen biometric data transmission and storage, assuring its secrecy and authenticity.

As the foundation of identity verification, smart contracts and DIDs provide users control over their identities and the rules that govern the usage of their biometric data. Identity verification in the digital era is being reshaped by a user-centric approach combined with privacy-preserving methods.

Ethical issues, biometric spoofing, regulatory compliance, and interoperability are all challenges that highlight the importance of continued cooperation, innovation, and responsible implementation of biometric authentication technology. As the digital environment advances, the integration of biometric identification with cryptographic and decentralized technologies offers a path for individuals to navigate safe, user-centric, and transparent identity ecosystems.

Challenges and considerations

While combining biometric authentication with digital signatures, public-key cryptography, smart contracts, and DIDs improves security, it is critical to handle issues and concerns that come with it.

However, implementing biometric security in a distributed IdM system also poses some challenges. For example, biometric data is highly sensitive and must be protected from unauthorized access or disclosure. In addition, different biometric technologies may have different levels of accuracy and reliability, and some individuals may not be able to use certain biometric methods due to physical or medical limitations.

To address these challenges, distributed IdM systems that incorporate biometric security should implement strong security and privacy measures, such as encryption, secure data storage, and user consent mechanisms. They should also provide multiple biometric options to accommodate users with different physical and medical needs.

Overall, biometric security has the potential to provide a highly secure and convenient authentication method in distributed IdM systems, but careful consideration must be given to security and privacy concerns and the practical limitations of different biometric technologies.

Biometrics has proven to be one of the most significant technologies nowadays for digital IdM. However, with its cumulative adoption in big-scale projects such as national security and banking, there's a rising concern over the susceptibility of the system that information can be escaped or used by a third party without the consciousness of the owners.

The following diagram depicts the components that are built upon a trusted standardized infrastructure stack:

Figure 15.1 – Trusted infrastructure components

In the age of the Fourth Industrial Revolution, blockchain is regarded as a crucial technology for digital transformation and the development of a future information technology platform. According to a lot of industry experts, blockchain technology is mostly intended to maintain distributed decentralized databases, which are made up of an increasing volume of data records that reserve the integrity, singularity, and cogency of kept information without involving any third party for verification purposes. The information on the blockchain cannot be modified or hacked, and it can only be added with the agreement of all the system's nodes.

Let's take a look at a few critical aspects of the use of biometrics in DIAM systems:

- **Ethical use of biometrics**: The ethical use of biometric data is still an important topic. As biometric authentication becomes increasingly common, it is critical to ensure that the collection, storage, and use of biometric data adhere to ethical guidelines. Transparent regulations, user consent methods, and effective security measures are critical for establishing confidence and addressing ethical concerns.

- **Biometric spoofing and accuracy**: The possibility of biometric spoofing, in which attackers attempt to imitate or trick biometric systems, emphasizes the significance of putting in place strong anti-spoofing mechanisms. Biometric technology developments, such as liveness detection and multimodal biometrics, aim to improve accuracy and counter-spoofing attempts.
- **Regulatory compliance**: When using biometric authentication systems, it is critical to navigate the regulatory landscape. Compliance with regional and international data protection standards, such as the **General Data Protection Regulation** (**GDPR**) or the **California Consumer Privacy Act** (**CCPA**), necessitates a thorough understanding of how biometric data is collected, kept, and disseminated.
- **Interoperability and standards**: Interoperability across different biometric systems, as well as adherence to standards, becomes critical in developing a smooth and globally recognized approach to biometric authentication. By establishing industry-wide standards, biometric data may be safely exchanged and validated across several platforms, contributing to a unified and interoperable identity verification ecosystem.

By uniting biometrics and blockchain, identity information will be kept in a distributed ledger system, which implies that the authorities don't have any access to users' personal information anymore, and they now have control over their information to select which information and authorities to share with. In addition, the mixture eliminates the requirement of having a biometric entrenched physical proof of identity such as a passport or ID card, which is disposed to fraud and information exposure since all you need is to show up yourself. Hence, it aids in securing the citizen's identity and rationalizes the IdM process as well, at the same time. The following diagram depicts the critical components of a biometric system leveraged in a DIAM system:

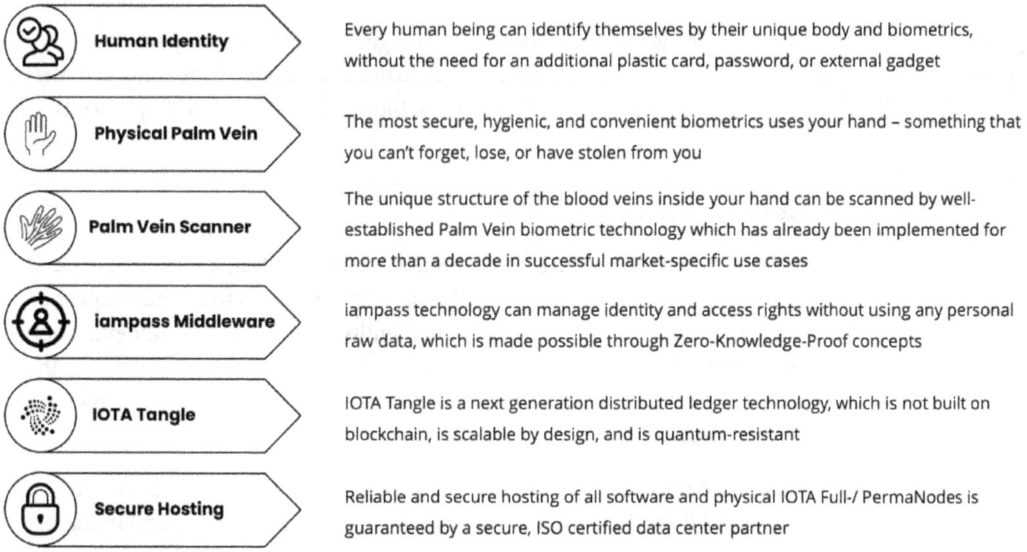
Human Identity
Every human being can identify themselves by their unique body and biometrics, without the need for an additional plastic card, password, or external gadget
Physical Palm Vein
The most secure, hygienic, and convenient biometrics uses your hand - something that you can't forget, lose, or have stolen from you
Palm Vein Scanner
The unique structure of the blood veins inside your hand can be scanned by well-established Palm Vein biometric technology which has already been implemented for more than a decade in successful market-specific use cases
iampass Middleware
iampass technology can manage identity and access rights without using any personal raw data, which is made possible through Zero-Knowledge-Proof concepts
IOTA Tangle
IOTA Tangle is a next generation distributed ledger technology, which is not built on blockchain, is scalable by design, and is quantum-resistant
Secure Hosting
Reliable and secure hosting of all software and physical IOTA Full-/ PermaNodes is guaranteed by a secure, ISO certified data center partner

Figure 15.2 – Components of biometrics in DIAM

In terms of banking, installing biometrics along with blockchain technology can be a great novelty in today's financial industry. Although biometric identification makes it quite impossible for criminals and fraudsters to impersonate, the present data storage offerings leave the transaction logs susceptible to exploitation. Thanks to the unchallengeable and transparent nature of the distributed ledger, blockchain-based biometric authentication makes it perfect for upholding the integrity of transactions and activity logs, which is the topmost priority in the financial sector. Transaction records having information such as a user ID, transaction ID, date, time, amount, accounts, beneficiary, and so on cannot be changed or hacked and can be tracked down easily.

Real-world applications of biometrics in DIAM

The incorporation of biometrics and the advancement of KYC processes indicate a dramatic paradigm change in the ever-changing world of financial services. Biometrics, which uses unique physical or behavioral qualities to identify individuals, in conjunction with improved KYC procedures, is revolutionizing how people access and interact with financial institutions. This investigation digs at biometrics' varied influence on financial services, the dynamic growth of KYC, and the symbiotic interaction between these two pillars in constructing a safe, efficient, and user-centric financial environment.

Transforming financial services with KYC

The KYC procedure, a critical component of client due diligence in the banking sector, has evolved dramatically. KYC has traditionally entailed time-consuming documentation, human verifications, and significant time delays. The emergence of digital technology, fueled by biometrics, has revolutionized KYC, making it more efficient, safe, and conducive to the quick speed of digital banking.

Digital onboarding, aided by biometric authentication, has emerged as a critical component of KYC progression. New consumers can create financial accounts, seek loans, and engage in investing activities without physically being there. During the onboarding process, biometric data such as fingerprints or facial characteristics serve as a safe and unique identity. This not only speeds up the client acquisition process but also improves the overall customer experience by removing the need for time-consuming documentation and in-person verifications.

Biometrics plays an important role in bolstering KYC standards by providing a more robust and dependable method of validating consumer identities. Traditional KYC processes frequently depended on document-based verification, which was prone to fraud and identity theft. Biometric authentication offers an additional degree of protection by guaranteeing that the person presenting the papers is the rightful owner. Biometrics improve the accuracy and integrity of the KYC process, whether they are captured during a video chat or used to validate identification papers.

The use of biometrics in KYC not only speeds up client onboarding but also functions as a strong deterrent against identity fraud and financial crimes. With its resistance to counterfeiting and imitation, biometric identification technologies considerably lower the danger of fraudulent account openings or unauthorized transactions. Financial organizations may use biometrics to build a more secure and trustworthy environment, inspiring faith in both consumers and regulators.

KYC's changing landscape is inextricably related to the frameworks that regulate financial services. As digital onboarding and biometric authentication grow more common in KYC, financial institutions must traverse a complicated web of rules, including **anti-money laundering** (AML) and **counter-terrorist financing** (CTF) obligations. It is critical to strike a balance between innovation and compliance, and sophisticated biometric technologies help fulfill demanding regulatory standards while encouraging a smooth consumer experience.

Biometrics and KYC work together to provide a dynamic and responsive financial ecosystem that prioritizes security, compliance, and user experience. Integrating biometrics into KYC processes provides several benefits to financial companies. These include increased efficiency and cost savings, improved client contacts, and a solid foundation of trust and security in financial transactions, depicted as follows:

- **Efficiency and cost reduction**: Biometrics integration in KYC not only improves security but also simplifies operational efficiency for financial institutions. The manual labor associated with traditional KYC processes is reduced via automated biometric verification, allowing financial institutions to spend resources more effectively. Cost savings from increased efficiency contribute to a more sustainable and adaptable financial infrastructure.
- **Frictionless interactions**: Biometric-based KYC improves the consumer experience by reducing friction in contracts with financial institutions. Customers may use a touch, a look, or a spoken command to access their accounts, make transactions, or apply for financial products. Biometrics' user-centric nature coincides with changing customer expectations, driving loyalty and engagement.
- **Building trust and security**: In financial services, trust is vital, and the combination of biometrics and KYC serves as a bedrock of client confidence. Financial organizations increase client trust in their services by providing safe and simple methods of identification verification. The apparent commitment to security and compliance via sophisticated biometric solutions builds a trusting connection, which is critical for client retention and brand reputation.

While the use of biometrics in KYC heralds a new age in financial services, there are certain problems and considerations:

- **Privacy concerns**: As biometric data becomes more prevalent in KYC, it is critical to provide strong privacy safeguards. To protect biometric information from unauthorized access or breaches, financial institutions must establish strong security processes. To address privacy issues and create consumer confidence, transparent privacy policies, informed consent processes, and compliance with data protection rules are required.
- **Navigating ethical dilemmas**: The ethical application of biometrics in financial services needs careful consideration of potential pitfalls. Institutions must find a balance between the advantages of increased security and ethical concerns about permission, data storage, and potential biases in particular biometric technology. Adopting ethical frameworks and industry best practices is critical for appropriately handling these ethical issues.

- **Technological advancements**: While biometrics provide a strong defense against identity fraud, financial institutions must remain cautious in the face of emerging technical threats. Biometric spoofing techniques and deepfake technologies are constantly evolving, posing problems that necessitate continuing innovation in biometric authentication approaches. To remain ahead of fraudsters, financial institutions must invest in adaptable and advanced biometric technologies.

A few examples of biometric spoofing are as follows:

- **Presentation attacks**: Using forged or manipulated biometric samples (such as fingerprints, iris scans, or face photos) to trick biometric systems
- **Replay attacks**: Recording and replaying previously collected biometric data to mislead the system into authenticating the incorrect user
- **Synthetic biometric generation**: Synthetic biometric data is created by mimicking authentic biometric features, sometimes utilizing advanced picture or audio manipulation techniques
- **Spoofing with prosthetics**: Using prosthetic devices or artificial reproductions of body parts to simulate biometric attributes such as fingerprints or facial features
- **Gummy fingers**: Creating synthetic fingerprints with materials such as gelatin or silicone to avoid fingerprint recognition systems
- **Biometric data interception**: Biometric data is intercepted during transmission and manipulated before it reaches the authentication system
- **3D mask attacks**: Creating very detailed 3D masks or reproductions of a person's face in order to fool facial recognition systems
- **Voice synthesis**: Voice synthesis technology is used to produce speech patterns that resemble an authorized user's voice for voice recognition systems
- **Eye scan spoofing**: Using high-resolution photos or contact lenses with printed patterns to deceive iris recognition systems
- **Behavioral biometric spoofing**: To fool behavioral biometric systems, mimic an authorized user's unique behavioral attributes (such as typing habits or gait)

Biometrics and KYC are altering the future of financial services, ushering in an era of safe, efficient, and user-centric interactions. Biometrics integration fortifies financial institutions against identity fraud while upgrading the client experience, from quicker digital onboarding to strengthened security standards. As financial services expand, the dynamic interplay between biometrics and KYC presents itself as an innovation accelerator, laying the groundwork for a trustworthy and technologically sophisticated financial environment.

Blockchain solutions for patient identity

Patient identification verification is a vital cornerstone in healthcare, guaranteeing accurate and secure access to medical information and treatments. The use of blockchain technology provides new solutions to patient IdM difficulties while conforming to the severe criteria of the **Health Insurance Portability and Accountability Act** (**HIPAA**). This investigation digs at real-world blockchain applications aimed at improving patient identity verification and HIPAA compliance, ushering in a new age of safe, interoperable, and patient-centric healthcare.

The traditional healthcare system has patient IdM difficulties, such as fragmented medical records, data silos, and the danger of identity theft or medical fraud. Incorrect patient identification not only impedes timely and effective treatment delivery but also endangers patient safety. Furthermore, HIPAA requirements mandate the safeguarding of patients' sensitive health information, adding another degree of complexity to healthcare IdM.

Blockchain technology, known for its decentralized and tamper-proof nature, brings disruptive solutions to patient identification verification by providing a secure and interoperable platform.

DIDs on the blockchain offer a unique and interoperable mechanism for verifying patient identity. Each patient is issued a DID that is recorded on a blockchain, ensuring an unchangeable record of their identity. DIDs give patients a portable and secure digital identity that they may use across many healthcare providers and systems. This removes the need for duplicate identification verification processes at each healthcare interaction, speeding up patient onboarding and enhancing the entire healthcare experience. Smart contracts, which are self-executing agreements with established rules that are inscribed on the blockchain, are critical in patient identification verification. These contracts can automate the verification process by defining the circumstances in which patient identification can be accessed or disclosed. A smart contract, for example, may ensure that only authorized healthcare practitioners with the patient's explicit consent have access to certain medical details. This automated and rule-based solution improves security, transparency, and patient identification control.

Permissioned blockchains can be used by healthcare consortiums to build a network of trustworthy institutions such as healthcare providers, insurers, and regulatory authorities. These organizations create a consortium in which access to patient-identifying information is governed by consensus processes. This consortium-based method ensures that the blockchain network is only used by authorized businesses, reducing the danger of unauthorized access or data breaches. It also conforms with HIPAA rules' requirement for safe and compliant data sharing.

The capacity of blockchain to allow interoperability is critical in the context of patient identification verification. Interoperability issues in traditional healthcare systems frequently result in fragmented patient data, impeding the easy interchange of information between various healthcare providers. The decentralized and standardized method of blockchain allows for the construction of a single patient identification that can be accessed and modified by many authorized parties, encouraging a comprehensive perspective of the patient's health history.

HIPAA compliance in blockchain solutions

Given its emphasis on protecting patients' **protected health information** (**PHI**), HIPAA compliance is critical in healthcare. By implementing privacy-preserving techniques and assuring safe data management, blockchain systems for patient identification verification comply with HIPAA rules:

- **Privacy-preserving measures**: To secure patients' PHI, blockchain systems employ privacy-preserving features. ZKPs and homomorphic encryption, for example, enable safe data transfer without disclosing the underlying information. A ZKP, for example, might establish the legitimacy of a patient's identification without exposing personal information, guaranteeing HIPAA compliance.
- **Immutable audit trails**: The immutability of blockchain guarantees a visible and tamper-proof audit record of access to patient-identifying information. Every contact with patient data is recorded on the blockchain, including identity verification requests and updates. This audit trail improves accountability by providing a detailed record of who accessed the data, when was it accessed, and for what purpose—essential components for confirming HIPAA compliance.
- **Data minimization and patient consent**: Through the deployment of consent management on the blockchain, blockchain technologies provide patients with more control over their data. Smart contracts enforce patients' permission choices for the use and sharing of their identifying information. This is consistent with HIPAA's emphasis on data minimization and the requirement for people to have a voice in how their health information is used.
- **Secure data sharing**: The safe and encrypted data exchange protocols built into the blockchain ensure that patient-identifying information is communicated in a secure and compliant way. Blockchain technologies permit authorized data sharing while retaining the integrity and security of patient-identifying information by providing a safe and interoperable network.

HIPAA compliance is critical in blockchains to ensure the safe and confidential management of sensitive health data. Healthcare organizations may build confidence and integrity in the handling of patient information by complying with HIPAA regulations inside blockchain frameworks, protecting against unauthorized access, and fostering transparency. This compliance not only meets regulatory standards but also establishes a foundation of security and privacy in the ever-changing field of healthcare technology.

Real-world examples

Several real-world examples demonstrate the efficacy of blockchain technology for patient identification verification and HIPAA compliance. Here are a few examples:

- MedRec: A permissioned blockchain is used by MedRec, a blockchain-based **electronic health record** (**EHR**) system, to improve patient identity verification and data exchange. Patients have access to their encrypted health records, and healthcare practitioners can access the information with patient permission. This technology protects and safeguards patient-identifying information while promoting interoperability among healthcare organizations.

- **Medicalchain:** This is a secure and decentralized health data platform built with blockchain technology. Patients control their health information and have the ability to authorize access to healthcare professionals via smart contracts. This patient-centered approach is in accordance with HIPAA standards, ensuring that patient-identifying information is handled safely and in accordance with privacy rules.
- **ProCredEx:** This is a blockchain-based credentialing and identity verification service for healthcare professionals. The technology simplifies the verification process and ensures that healthcare providers comply with regulatory obligations. ProCredEx improves the efficiency and security of identity verification in the healthcare workforce by employing a consortium-based blockchain.

Real-world blockchain solutions for patient identification verification not only solve the shortcomings of existing healthcare systems but also open the way for a more secure, interoperable, and patient-centric healthcare environment. These systems match HIPAA principles by using DIDs, smart contracts, permissioned blockchains, and privacy-preserving mechanisms, assuring compliance while improving patient IdM efficiency. The promise for a seamless, secure, and patient-driven healthcare ecosystem becomes more evident as the healthcare industry embraces blockchain innovation.

Emerging technologies and trends in biometrics

Biometric security is at the forefront of the ever-changing environment of cybersecurity, delivering a paradigm shift in how people are validated, and identities are confirmed. The landscape of biometric security evolves with technology, with emerging technologies and trends determining the future of this crucial subject. This investigation goes into cutting-edge biometric and blockchain security, revealing the most recent technologies and trends that promise to revolutionize authentication, improve security, and transform the digital identity environment:

- **Behavioral biometrics**: The emergence of behavioral biometrics is one of the most exciting topics in biometric security. This method entails analyzing distinct patterns in human behavior, such as typing rhythm, mouse motions, or even an individual's walk. In contrast to physiological biometrics, which relies on bodily characteristics, behavioral biometrics offers another degree of complexity to identification, making it more difficult for bad actors to imitate or reproduce.
- **Vascular biometrics**: This is gaining popularity as a safe and non-contact authentication approach. This method creates a unique identifier by analyzing vein patterns under the skin's surface, generally in the palm or finger. The benefit of vascular biometrics is that vein patterns are difficult to recreate, making it resistant to spoofing efforts. Vascular biometrics appears as a powerful answer as organizations seek more strong anti-spoofing methods.
- **DNA biometrics**: The investigation of DNA as a form of identification is at the bleeding edge of biometric research. While still in its infancy, DNA biometrics has the potential to provide an unsurpassed level of uniqueness in identification. The DNA sequence of an individual acts as a biological identification, giving an immutable and extremely secure method of authentication. However, ethical and privacy problems accompany the development of DNA biometrics, prompting a thorough examination of its ramifications.

- **Multimodal biometrics**: The integration of various modalities, known as multimodal biometrics, is the future of biometric security. This method integrates biometric factors such as fingerprint scanning, facial recognition, and iris scanning to build a more robust and accurate identification system. Organizations can improve security while reducing the risk associated with depending on a single biometric element by merging modalities. As worries about biometric spoofing and presentation assaults grow, liveness detection becomes an increasingly important component of biometric systems. In IdM, liveness detection is the process of determining if a biometric sample, such as a face picture or fingerprint, is being supplied by a real person rather than a static or faked representation. Its goal is to avoid fraudulent efforts to fool biometric systems by assuring that biometric data is acquired from a living, present human in real time. To identify authentic data from fraudulent biometric data, liveness detection algorithms frequently entail measuring physiological responses or dynamic features, such as face movements or pulse detection. The goal of liveness detection is to verify that the biometric data supplied is from a live and present human rather than a static picture or duplicate. Facial movement analysis, heartbeat detection, and voice-based liveness checks all help to design more secure and robust biometric systems.

Biometric security is quickly evolving, resulting in a shift in identity authentication and validation. Emerging technologies such as behavioral, vascular, and DNA biometrics, as well as multimodal techniques, are improving security by increasing complexity and resistance to spoofing. Furthermore, liveness detection is becoming increasingly important in IdM because it ensures that biometric data is gathered from live persons rather than static or phony representations, hence improving biometric system security.

AI and machine learning in biometrics

Biometric security is evolving in response to advances in AI and **machine learning** (**ML**). Adaptive biometric systems, **deep learning** (**DL**), and the usage of **generative adversarial networks** (**GANs**) are making biometric authentication stronger and more resistant to spoofing assaults. A few important methodologies to consider for building a strong AI-based biometric model are the following:

- **Adaptive biometric systems**: By allowing adaptive systems, the combination of AI and ML is revolutionizing biometric security. These systems learn and adapt to changes in biometric data over time, allowing for variances due to aging, injury, or other circumstances. Adaptive biometric systems improve accuracy and reliability by updating the biometric model in real time depending on an individual's changing attributes.
- **DL and feature extraction**: DL techniques, notably **convolutional neural networks** (**CNNs**), are changing how characteristics in biometric recognition are retrieved. Traditional biometric systems rely on human-designed feature extraction techniques, whereas DL enables the system to discover discriminative features autonomously from raw biometric data. This enhances accuracy while also making biometric systems more resistant to variances in input data.

- **GANs for anti-spoofing**: The fight against biometric spoofing is being aided by GANs. GANs are made up of a generator and a discriminator that are trained together. GANs may be used to produce realistic synthetic biometric data for training anti-spoofing models in the context of biometric security. This adversarial training improves biometric systems' capacity to discriminate between genuine and fraudulent efforts.

As AI continues to transform biometric security, the combination of adaptive systems, DL methods, and GANs presents intriguing opportunities to improve biometric authentication systems' accuracy, dependability, and anti-spoofing capabilities. These improvements are paving the way for more secure and trustworthy identity verification methods across several sectors.

Secure biometric template protection

To address privacy problems connected with maintaining biometric templates, homomorphic encryption allows calculations on encrypted data to be done without decryption. This means that in the world of biometrics, matching may take place without disclosing the actual biometric template, boosting the privacy and security of stored biometric information.

Secure multi-party computation (**SMPC**) allows many participants to compute a function concurrently while keeping their inputs confidential. SMPC in biometric security allows several institutions to work together to authenticate an individual's identification without releasing the actual biometric data. This joint method improves security while protecting biometric data privacy.

Concerns regarding centralized storage and potential data breaches are addressed by the use of blockchain in biometric security. Biometric templates may be securely saved and retrieved by exploiting the blockchain's decentralized and irreversible nature. Each user keeps ownership of their biometric data, with cryptographic principles protecting the integrity and validity of saved templates.

The incorporation of fingerprint sensors directly into mobile device displays is a key development in biometric security. Without the need for a separate sensor, in-display fingerprint sensors provide a smooth and safe identification experience. This technology, which is frequently based on ultrasonic or optical principles, improves user ease while ensuring excellent security.

As a main authentication mechanism in mobile devices, facial recognition is evolving. Advanced facial recognition systems build detailed and precise facial maps using 3D sensing technologies such as structured light or **time-of-flight** (**ToF**) cameras. Face identification on mobile platforms is secure and efficient because of the convergence of hardware and software improvements.

Biometric authentication is quickly becoming the preferred way to protect mobile payment transactions. Fingerprint, face, and even behavioral biometrics are critical in guaranteeing the user's identification during financial transactions. As mobile payment systems develop, the incorporation of strong biometric security becomes increasingly important.

The use of biometric technology raises ethical concerns about privacy, consent, and potential abuse. To strike a balance between the advantages of increased security and the preservation of individual rights, continual ethical inspection and clear regulatory frameworks are required.

Biases in training data can result in biased outputs in biometric systems, affecting particular demographic groups disproportionately. To mitigate prejudice, varied and representative datasets must be carefully curated, and fairness concerns must be incorporated into the design and assessment of biometric algorithms.

As biometric systems grow increasingly common, they become attractive targets for nefarious actors. To protect biometric data from security threats such as spoofing attempts, data breaches, and system vulnerabilities, effective countermeasures and proactive security measures are required.

As technology breakthroughs continue, the future of biometric security presents intriguing potential. The integration of developing modalities and the fusion of biometric elements, as well as the usage of AI and safe template protection, all lead to a more complex, secure, and user-friendly authentication landscape. As organizations embrace these improvements, it is critical to negotiate ethical and security concerns in order to guarantee that biometric technologies evolve responsibly and inclusively, leading to a safer and more secure digital future.

The evolution of identity verification has been an enthralling journey, spanning biometrics and concluding in the paradigm-shifting landscape of distributed IdM driven by blockchain technology. This epic journey represents not just technological progress but also a fundamental shift in how we view, safeguard, and manage digital identities.

The journey began with the introduction of biometrics, a forerunner in identity verification. Biometrics provided a dramatic break from old authentication approaches by harnessing unique physiological or behavioral features. Fingerprint scanning, iris recognition, and face biometrics emerged as stalwarts, giving hitherto unheard levels of accuracy and security. These biometric identifiers, which functioned similarly to a person's digital fingerprint, offered not just safe access but also a more smooth and user-friendly experience.

However, as the digital world grew, so did the difficulties. Concerns about the security of biometric databases, possible spoofing, and privacy problems prompted a rethinking of identity verification techniques. The voyage took an introspective turn, encouraging the pursuit of more resilient, decentralized, and privacy-protecting solutions.

Enter blockchain, a game-changing force that will change the course of identity verification. The fundamental notion of DIDs arose as a pillar in this transition. DIDs, or unique identifiers based on blockchain technology, provided a decentralized and tamper-proof method of managing digital identities. The immutability of the blockchain protected the integrity of identification information, and its decentralized nature decreased the possibility of a SPOF or unauthorized access. The combination of blockchain and DIDs paved the way for a distributed IdM system, ushering in a shift away from a centralized authority and placing individuals in charge of their digital selves. The transition from biometrics to blockchain-powered identity verification represented a paradigm shift toward self-sovereign identification.

In the current constantly evolving landscape, identity has emerged as a crucial notion, with SSI giving individuals authority over their personal information. At its foundation, SSI allows individuals to own their credentials and decide how and when to distribute them rather than depending on many centralized databases and authorities. This not only improves privacy but also accords with the digital autonomy ideal. The function of blockchain in SSI is critical. Because blockchain is dispersed and decentralized, no one entity has access to an individual's identity. Smart contracts, which are implemented on the blockchain, allow for safe and automated verification procedures, promoting confidence between parties without the need for middlemen. This advancement in IdM transforms the narrative away from a system of licenses issued by central authorities.

The path from biometrics to distributed IdM has not been easy. Obstacles include technical complexity, interoperability problems, and the requirement for mass adoption. Because blockchain technology is still in its early stages, it must overcome skepticism and develop collaboration among a wide range of stakeholders, including governments, businesses, and individuals. Furthermore, as the voyage unfolds, ethical issues become more prominent. Questions on permission, data ownership, and inclusion highlight the importance of developing and deploying ethical frameworks to guide the development and deployment of blockchain-based identification systems. Learning from the lessons of biometrics, the road toward distributed IdM emphasizes the significance of developing systems that are not just safe but also ethical.

The focus switches to the future possibilities of blockchain-based identity verification as views on this revolutionary journey unfold. The future route envisions a society in which individuals traverse digital interactions with ease, armed with safe, privacy-preserving identities and SSI. Interoperability becomes a primary focus, imagining a future in which blockchain networks interact effortlessly, guaranteeing that identity credentials are universally recognized and accepted. Efforts to standardize gain popularity, establishing the basis for a unified and interoperable identification infrastructure.

The incorporation of new technology enhances the environment even further. AI-powered identity verification, SMPC, and advances in biometric modalities combine with blockchain to form a multidimensional approach to IdM. Blockchain-based identity solutions have the potential to go beyond individual interactions in the future. They have uses in healthcare, banking, education, and a variety of other fields where safe and verified identities are critical. The possibility of a worldwide, decentralized identification infrastructure has the potential to transform how societies function in the digital age.

Reflecting on the path from biometrics to distributed IdM reveals a mosaic of identities woven by technical progress, ethical considerations, and the unwavering quest for safe and user-centric solutions. The narrative shifts from dependence on biometric markers to a distributed, decentralized, and self-sovereign approach that puts people at the center of their digital identity. As the trip progresses, it encourages teamwork, contemplation, and adherence to ethical practices. The future holds the promise of a digital environment in which identity is more than simply a set of credentials, but a testament to individual autonomy, privacy, and the limitless potential of blockchain-driven innovation. The journey's reflections serve as a compass, directing the path toward a future in which identities are accepted, safeguarded, and cherished in the enormous expanse of the digital sphere.

Summary

This chapter gave an overview of biometric security integration in DIAM systems. It underlined the importance of biometric technology, such as fingerprint recognition, iris identification, and face recognition, in delivering high levels of security and convenience in DIAM systems. It also emphasized the necessity of ethical issues, technical improvements, and the integration of biometrics and blockchain technology to establish a safe, user-centric, and transparent identity verification ecosystem. The chapter also discussed challenges and considerations in biometric security implementation, such as ethical issues, regulatory compliance, and interoperability, with a focus on biometrics and blockchain's potential to improve privacy, authenticity, and security in distributed IdM systems.

And finally, to every reader who joined me on this trip and added immeasurable value to the experience, I hope you enjoyed reading this book as much as I enjoyed authoring it.

Index

Symbols

A

E

F

J

K

L

M

N

O

P

Q

R

T

U

V

W

Z

Other Books You May Enjoy

If you enjoyed this book, you may be interested in these other books by Packt:

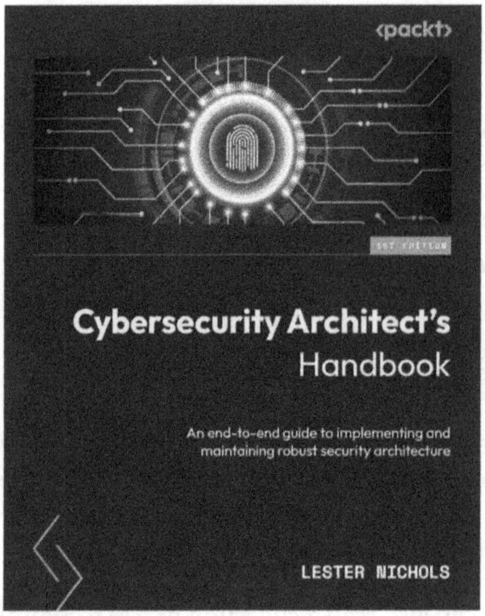

Cybersecurity Architect's Handbook

Lester Nichols

ISBN: 978-1-80323-584-4

- Get to grips with the foundational concepts and basics of cybersecurity
- Understand cybersecurity architecture principles through scenario-based examples
- Navigate the certification landscape and understand key considerations for getting certified
- Implement zero-trust authentication with practical examples and best practices
- Find out how to choose commercial and open source tools
- Address architecture challenges, focusing on mitigating threats and organizational governance

Keycloak - Identity and Access Management for Modern Applications

Stian Thorgersen, Pedro Igor Silva

ISBN: 978-1-80461-644-4

- Understand how to install, configure, and manage the latest version of Keycloak
- Discover how to obtain access tokens through OAuth 2.0
- Utilize a reverse proxy to secure an application implemented in any programming language or framework
- Safely manage Keycloak in a production environment
- Secure different types of applications, including web, mobile, and native applications
- Discover the frameworks and third-party libraries that can expand Keycloak

Packt is searching for authors like you

If you're interested in becoming an author for Packt, please visit `authors.packtpub.com` and apply today. We have worked with thousands of developers and tech professionals, just like you, to help them share their insight with the global tech community. You can make a general application, apply for a specific hot topic that we are recruiting an author for, or submit your own idea.

Share Your Thoughts

Now you've finished *Decentralized Identity Explained*, we'd love to hear your thoughts! Scan the QR code below to go straight to the Amazon review page for this book and share your feedback or leave a review on the site that you purchased it from.

`https://packt.link/r/1804617636`

Your review is important to us and the tech community and will help us make sure we're delivering excellent quality content.

Download a free PDF copy of this book

Thanks for purchasing this book!

Do you like to read on the go but are unable to carry your print books everywhere?

Is your eBook purchase not compatible with the device of your choice?

Don't worry, now with every Packt book you get a DRM-free PDF version of that book at no cost.

Read anywhere, any place, on any device. Search, copy, and paste code from your favorite technical books directly into your application.

The perks don't stop there, you can get exclusive access to discounts, newsletters, and great free content in your inbox daily

Follow these simple steps to get the benefits:

1. Scan the QR code or visit the link below

   `https://packt.link/free-ebook/9781804617632`

2. Submit your proof of purchase
3. That's it! We'll send your free PDF and other benefits to your email directly

www.ingramcontent.com/pod-product-compliance
Lightning Source LLC
Chambersburg PA
CBHW080609060326
40690CB00021B/4635